COMPREHENSIVE

CHEMICAL KINETICS

EDITED BY

R.G. COMPTON
M.A., D. Phil. (Oxon.)
Oxford University
The Physical and Theoretical Chemistry Laboratory
Oxford, England

G. HANCOCK
Oxford University
The Physical and Theoretical Chemistry Laboratory
Oxford, England

VOLUME 38

KINETICS OF
HOMOGENEOUS MULTISTEP REACTIONS

F.G. HELFFERICH
Department of Chemical Engineering
The Pennsylvania State University
University Park, PA 16802-4400
USA

ELSEVIER
AMSTERDAM – BOSTON – LONDON – NEW YORK – OXFORD – PARIS
SAN DIEGO – SAN FRANCISCO – SINGAPORE – SYDNEY – TOKYO

ELSEVIER SCIENCE B.V.
Sara Burgerhartstraat 25
P.O. Box 211, 1000 AE Amsterdam, The Netherlands

First edition 2001
Second impression 2003

Library of Congress Cataloging in Publication Data
A catalog record from the Library of Congress has been applied for.

ISBN: 0-444-41631-5 (Series)
          0-444-82606-8 (Vol. 38)
ISSN: 0069-8040

COMPREHENSIVE CHEMICAL KINETICS

ADVISORY BOARD

Volumes in the Series

Preface

> *Grey is all theory,*
> *but green is life's golden tree.*
> Johann Wolfgang v. Goethe, *Faust*

> *For, his reason's razor slant*
> *rules: what must not happen, can't.*
> Christian Morgenstern, *Galgenlieder**

May the reader who studies this book, or goes as far as trying to work with it, keep in mind what these two wise poets had to say: one reminding us that nature weaves an infinitely finer, more intricate, more colorful tapestry than the best of all theories can project, the other whimsically warning against doctrinaire conclusions from what we have come to perceive as right. In no other field of science and engineering are their words more to the point than in reaction kinetics.

Even so, I have written a book full of theory of reaction kinetics. I have done so in the firm belief that sound theory can at least serve as a sturdy framework, ready to be fleshed out with all the vagaries we encounter; that it can help us to acquire insight, a "feel" for what is apt to happen and why, a subconscious knowledge and perspective that springs from familiarity. If we throw a stone into a quiet pond, we don't even have to look: In our mind we will see the picture of expanding rings of waves the stone's impact has set in motion. Ideally, the kineticist will have learned to "read" a network and see in his mind how the reaction will evolve, how it will respond to changes in conditions, much like a conductor can read the score of a symphony and in his mind hear the orchestra play it. I believe such subconscious comprehension of complex reaction kinetics is within our grasp, and hope my book will help to bring it closer.

The origins of this book date all the way back to the 1960s and 70s, when I worked in, and for a time directed, grass-roots development of large-scale processes in chemical industry. I spent untold hours, days, and weeks struggling to unravel mechanisms, derive rate equations, understand cause and effect, finally telling myself there *had* to be a better way. Ever since, I have worked on and off trying to find better, shorter, easier ways in practical reaction kinetics, and this book by an old man is the culmination of my efforts. It is the book I dearly wish I had had at hand when still in the front lines of development. Yet, I see it as only a step toward true mastery of its subject, and am hoping others will carry on where I left off.

---

* *Songs from the gallows*, translation by Walter Arndt, Yale University Press, 1993.

This book tries to go beyond collecting and compiling accepted wisdom. In every instance it attempts to evaluate prior art critically, and in large parts it presents new methodology that has yet to withstand the acid test of extensive practical application. All this has required judgment calls. With much own experience and good advice to draw on, I am confident the big picture is correct. If there are errors in detail, I must accept sole responsibility, even where other sources are quoted.

To expedite publication and reduce cost, this book has been reproduced photographically from the manuscript. I beg the reader's forbearance if the visual appearance of its pages does not in all places meet exacting standards.

I am deeply indebted and grateful for gracious help, expert advice, and invaluable suggestions, first and foremost to Phillip E. Savage (University of Michigan), who has closely and patiently worked with me throughout most of the years this project has taken, and furthermore to Shao-Tan Hsieh (Mitsubishi Chemical), Yng-Long Hwang (Union Carbide), Jia-Ming Chern (Tatung Technical University), Robert L. Albright (Albright Consulting), Joe E. Hightower (Rice University), and many others, too many to name them all. Special thanks are also due to George Selembo for preparation of the illustrations and formulas, to my friends at Elsevier for their unwavering patience, to the editors for their encouragement and constructive criticism, and last but not least to the Pennsylvania State University and its Department of Chemical Engineering for lending their resources.

This book has been almost ten years in the making. It will be my last such endeavor. If I ever take up the pen again, it will be to write science fiction, a field for which I now feel well equipped.

Friedrich G. Helfferich
State College, Pennsylvania
September 2000

Contents

Contents

power-law or polynomial equations that best fit the data. Ask a physical organic chemist and he is apt to conjure up Woodward-Hoffmann exclusion rules or electrons that pair in different ways. Ask a plant engineer and he will think of how yield and purity in his reactor's effluent respond to changes in control settings. This book is devoted to still a different facet of kinetics, to what is sometimes called "fundamental kinetics," that is, the study of reactions as composites of elementary molecular steps and the mathematics reflecting the latter.

The pioneer work in fundamental reaction kinetics—by Bodenstein, Michaelis, Lindemann, Hinshelwood, Rice, Christiansen, and Semenov, to name only the most prominent—was done in the first six decades of the twentieth century. Since then, surprisingly few advances have been made in the state of the art of fundamental kinetics, with notable exceptions mostly in heterogeneous catalysis, polymerization, and on esoteric topics such as periodic and chaotic reactions. Perhaps this can be attributed to our preoccupation with thermodynamics. In any other field of science and engineering, the excitement is in dynamics, and statics is left to the more pedestrian minds. Only in chemical engineering and physical chemistry have we let our technical thinking and education be dominated by thermodynamics, which is not dynamics by any stretch of the imagination and should rightly be called thermostatics. Just because this wonderful and enormously successful tool exists, we have even tried to use it for dynamic phenomena, an application for which it was not designed and is not too well suited. In a way, we are now paying the price for the genius of Gibbs, Clausius, and company, who created for us this admirable edifice that has placed dynamics in its shadow for a century.

Today, however, we see a resurgence of interest in reaction kinetics. Chemical industry has matured and its competitive pressures keep increasing. A chemical plant must produce to pay for its construction, its operation and maintenance, the raw materials it consumes, the disposal of the by-products and wastes it generates, the development of its process, the attempted developments of maybe half a dozen other processes that came to naught, the salaries of the company's managers and business staff, taxes, and dividends for the stockholders. The plant generates income only while it is in operation. Unlike a car, a plant does not die from old age or corrosion, it is shut down because a better or cheaper process has been invented, the need for the product has disappeared, a raw material has become too expensive, or some other event has made its operation unprofitable. That point in time is quite independent of when the plant was taken on stream. The only way to prolong the plant's productive life is to move its start-up date forward by shortening the time span between conception of the process and start of production. Accordingly, there is a great incentive to cut process development time by replacing traditional scale-up through several intermediate stages—demonstration units and pilot plants of increasing sizes—by a direct scale-up from the laboratory bench to the eventual, full-sized commercial plant. To be safe, any scale-up by a very large factor cannot rely on empiricism, it must be based on mathematics that correctly reflect the individual molecular phenomena, among them the elementary steps of which the chemical reactions consist. This requires a sovereign command of fundamental reaction kinetics. Even routine operation of a plant is safer if the fundamental kinetics of its chemistry is fully understood.

To be sure, the fundamental approach to process development cannot obviate demonstration units and pilot plants. They are still needed as final proof of operability and to ascertain long-term effects such as catalyst life or build-up of minor impurities, effects that cannot be measured in short-duration bench-scale experiments. They also serve well for producing representative samples ahead of time for potential future customers. Moreover, they are invaluable for fine-tuning and provide excellent opportunities for corrosion tests and piloting envisaged process control. However, fundamental kinetics can free them of the obligation to scan wide ranges of potential operating conditions for optimization and design.

This is not to say that the fundamental approach to reaction kinetics is automatically the best in every situation. At least today, if the scale is small, the process likely to be short-lived, the chemistry complicated, and timing more important than cost, the work to elucidate the mechanisms may not be warranted or entail unacceptable delay. An empirical scale-up then is preferable. In industry, the fundamental approach is at its best and fundamental kinetics in greatest demand if the scale is large and construction of successive plants over years to come is envisaged. This book has been written chiefly with such applications in mind.

Almost every chemical reaction of practical interest consists of a network of elementary steps, each with its own contribution to kinetics. Single-step reactions are most often found in textbooks. Until fairly recently, computers were not efficient enough to permit reactor design and optimization to be based on rate equations reflecting individual steps, except in quite simple cases. Today, computation has become so fast and cheap that capacity and execution time are no longer limiting. The problem is not how to program and solve the simultaneous equations for the reaction steps, mass transfer, heat transfer, etc., but to verify the presumed reaction network and obtain numerical values for all its rate coefficients and their activation energies. To make fundamental reaction kinetics a *practical* tool, it must be streamlined without unacceptable sacrifices in accuracy. Chemical engineers are known for loving to construct complicated theories of simple phenomena. Here, the opposite is needed: a simple theory of a complicated phenomenon. In the words of Ian Stewart: "Science is *not* about devising hugely complex descriptions of the world. It is about devising descriptions that illuminate the world and make it comprehensible" [2]. Progress is the progression from the primitive to the complicated to the simple, to a clarity that arises from true comprehension of the subject. Nowhere is this more true than in practical reaction kinetics, where much of that last step is still to be taken. May this book help to speed us along.

All emphasis on simplification notwithstanding, the development of reliable reaction mathematics for design is an exacting job. Because scale-up by large factors may be involved, mechanisms that are merely plausible working hypotheses will not do; the basis must be established beyond reasonable doubt. Incorrect kinetics is worse than none. In process development, kinetics is not a game for amateurs. Even in research, misinterpretation of kinetic observations may result in futile efforts and missed opportunities. The presentation here is geared toward the demand for exactitude.

This book is intended as an aide and guide for the hands-on chemist and engineer in development. While stressing accuracy, it is kept as simple as possible. It addresses methodology rather than science and glosses over many of the finer points of kinetic theory. Its goals are those of reaction kinetics in practice and can be summarized as follows:

- establishment of reliable mathematics of reactions by means of short-duration bench-scale experiments, and

- construction of simple but sufficiently accurate mathematical models of reaction kinetics for design, scale-up, optimization, on-line control, and trouble shooting.

The book is on kinetics, not reaction engineering: It focuses on reactor-independent behavior, that is, on reaction rates under given momentary and local conditions (concentrations, temperature, pressure). Reactor-dependent, global behavior is included only to the extent necessary for evaluation of kinetic experiments, which, of course, require reactors, and in a few instances in which vagaries of multistep kinetics produce uncommon behavior or impact reactor choice.

The coverage is also essentially restricted to *homogeneous* reactions and so does not include one very important topic: heterogeneous catalysis. Not that the principles, concepts, and methodology developed in this book are not fully applicable to reactions on surfaces of solid catalysts. They are, and the practitioner of heterogeneous catalysis can benefit from them. However, the additional complications in that field are so massive and so important that a balanced and manageable treatment must drastically simplify the mechanistic aspects emphasized here. Excellent such treatments can be found in recent reaction engineering texts and therefore are not replicated. On the other hand, the book includes gas-liquid reactions such as hydrogenation, hydroformylation, hydrocyanation, air oxidation, etc., in which the reaction occurs in the liquid phase although a reactant must be resupplied from the gas phase.

The book addresses mostly the concerns of the industrial chemist and engineer. It does not include an in-depth coverage of very fast reactions of biochemistry or methods for their study.

The book is structured to supplement modern texts on kinetics and reaction engineering, not to present an alternative to them. It intentionally concentrates on what is not easily available from other sources. Facets and procedures well covered in standard texts—statistical basis, rates of single-step reactions, experimental reactors, determination of reaction orders, auxiliary experimental techniques (isotopic labeling, spectra, etc.)—are sketched only for ease of reference and to place them in context. Emphasis is on a comprehensive presentation of strategies and streamlined mathematics for network elucidation and modeling suited for industrial practice.

While concentrating on methods, the book uses a number of reactions of industrial importance for illustration. However, no comprehensive review of multistep homogeneous reactions is attempted, simply because there are far too many reactions and reaction mechanisms to present them all. Instead, the book aims at providing the tools with which the practical engineer or chemist can handle his specific reaction-kinetic problems in an efficient manner, and examples of how problems unique to a specific reaction at hand can be overcome. Some examples drawn from my own laboratory experience have been construed or details have been left out, in order to protect former employers' or clients' proprietary interests. In particular, the omission of information on exact structure and composition of catalysts is intentional.

Each chapter concludes with a summary. Before he delves into the main text, the user may want to check it to see whether what he seeks is indeed covered.

I expect a rapid evolution of fundamental reaction kinetics in the years to come and a growing awareness of its enormous practical value. I hope and trust that this book will contribute its share. Although the publisher does not agree, I wish it will help to stimulate advances in practical kinetics so swift that it will soon become obsolete.

References

1.   S. M. Walas, *Reaction kinetics*, Chapter 7 in *Perry's chemical engineers' handbook*, 7th ed., D. W. Green, and J. O. Maloney, eds., McGraw-Hill, New York, 1997, ISBN 0070498415, pp. 3-15.

2.   I. Stewart, *Life's other secret: The new mathematics of the living world*, Wiley, New York, 1998, ISBN 0471158453, p. 9.

Chapter 1

Concepts, Definitions, Conventions, and Notation

For ease of reference this chapter outlines and explains the essential concepts and the formalism of presentation.

1.1. Classification of reactions

This book is about *homogeneous reactions*, that is, all kinds of reactions that occur within a single fluid phase. The term excludes reactions at interfaces, among them reactions of solids with fluids, heterogeneous catalysis, and phase-transfer catalysis. It does not exclude reactions in which a dissolved reactant is resupplied from another phase, as is the case, for example, in homogeneous hydrogenation or air oxidation reactions in the liquid phase in contact with a gas phase.

The title of the book refers to *multistep reactions*, defined as all kinds of reactions that involve more than a single molecular event such as rearrangement or break-up of a molecule or transformation resulting from a collision of molecules. Some standard texts speak instead of *complex reactions* and *multiple reactions*, depending on whether or not the mechanism involves trace-level intermediates. The term multistep reactions comprises both these categories.

On the other hand, a distinction exists between multistep and *simultaneous reactions*. The latter are independent reactions that occur side by side in the same reactor. For example, a reaction in which one and the same reactant A can undergo two different reactions leading to different products

$$A \begin{array}{c} \nearrow P \\ \searrow Q \end{array}$$

qualifies as multistep because the same reactant is involved in two different molecular events. In contrast, two reactions

$$A \longrightarrow P$$
$$B \longrightarrow Q$$

occurring side by side in the same reactor are simultaneous (but each may be multistep, namely, if it involves one or more intermediates).

References

1.   S. W. Churchill, *The interpretation and use of rate data: The rate concept*, Scripta, Washington, DC, ISBN 0070108455, 1974, p. 8.
2.   F. G. Helfferich and P. E. Savage, *Reaction kinetics for the practical engineer*, Course #195, AIChE Educational Services, New York, 7th ed., 1999, Section 2.1.
3.   S. Arrhenius, *Z. physik. Chem.*, **4** (1889) 226; see also any text on physical chemistry or reaction engineering.
4.   N. A. Bhore, M. T. Klein, and K. B. Bischoff, *I&EC Research*, **29** (1990) 313.

Chapter 2

Fundamentals

The picture we have formed of our world is based on reproducible observation. This is how we have discovered and established our laws of nature. In fact, it has been said that human intelligence could not have developed in a world without such observable, reproducible regularities. In this respect, reaction kinetics poses a problem: It has no universal laws. Rather, reactions differ widely in their kinetics and so resist the attempt to formulate equations that are generally valid. Indeed, even one and the same reaction may obey different rate equations in different media. At the root of this difficulty are differences in mechanism, that is, in the combination and interplay of reaction steps. For an orderly approach to kinetics we are therefore best served by one of the engineer's favorite methods of problem solving: If a problem is difficult, to forget it and, instead, solve a simpler one of the same kind and then try to generalize. In the case of reaction kinetics, a simpler problem is that of elementary reaction steps, that is, conversion as a result of a single rearrangement or break-up of a molecule or of a single collision of two or more molecules. The kinetics of single elementary reaction steps nicely follow the laws of statistics, provided a sufficiently large population is observed. In this chapter, the kinetics of single steps, extensively covered in standard texts on reaction kinetics and reaction engineering [G1-G8], is briefly reviewed and then used to formulate in general terms the mathematics of multistep reactions.

2.1. Statistical basis: molecularities and reaction orders

The relationships between stoichiometry, molecularity, and reaction orders of elementary steps arise from statistics and are illustrated in Table 2.1.

In a *unimolecular* elementary step, a reactant molecule undergoes re-arrangement or break-up. The reaction is a matter of inherent probability. At a given temperature, each reactant molecule has the same probability to react (however, see a qualifying comment at the end of this section). Accordingly, the reaction rate—defined as the number of molecules converted per unit volume and unit time—is proportional to the number of reactant molecules per unit volume at the respective time, that is, to the local and momentary reactant concentration. The proportionality factor reflecting the probability of the event is the *rate coefficient*, denoted k.

Table 2.1. Molecularities, rate equations, and reaction orders of elementary steps.

molecularity	step	rate equation*	reaction order
unimolecular	A $\longrightarrow$ product(s)	$-r_A = kC_A$	1st
bimolecular	A + B $\longrightarrow$ product(s)	$-r_A = kC_AC_B$	1st in A 1st in B 2nd overall
	2A $\longrightarrow$ product(s)	$-r_A = 2kC_A^2$	2nd
trimolecular	A + B + C $\longrightarrow$ product(s)	$-r_A = kC_AC_BC_C$	1st in A 1st in B 1st in C 3nd overall
	2A + B $\longrightarrow$ product(s)	$-r_A = 2kC_A^2 C_B$	2nd in A 1st in B 3rd overall
	3A $\longrightarrow$ product(s)	$-r_A = 3kC_A^3$	3rd

* Numerical factor (stoichiometric coefficient) indicates relative amount of A consumed; see Section 2.4.

      In a *bimolecular* step, two molecules of the same or different reactants collide and form one or several products. For any one reactant molecule, the probability of collision with another is proportional to how many of the other are around, that is, to the concentration of the other reactant. The overall probability of the event occurring, summed over all molecules of the first reactant, is in addition proportional to the concentration of that reactant. If the step involves two molecules of the same reactant, the overall probability is proportional to the square of the reactant concentration. Not every collision of molecules capable of reacting is successful. Both the number of collisions and the probability of their success are reflected in the rate coefficient.

      A *trimolecular* step requires three molecules of the same or different reactants to collide simultaneously to form one or several products. If molecules were ideal billiard balls, the time of contact of two colliding partners would be infinitesimal, and so would be the probability of a third partner making contact at exactly the same time. However, molecules are somewhat soft and deformable. Moreover, bonds and bond angles are distorted in the force field of a near-by other molecule. In a collision, this happens before actual contact is made and lasts for a time after contact has been broken. Two colliding molecules thus are in states of contact and distortion for a finite time, during which a third may join in. However, the probability is low. This makes trimolecular steps rare because, more often than not, faster uni- or bimolecular steps consume the reactant molecules before three of them collide simultaneously. As in bimolecular steps and for the same reasons, the rate is proportional to the respective reactant concentrations or their powers. Again, not every qualifying ternary collision is successful, and the rate coefficient reflects both the number of such collisions and the probability of their success.

A *quadrimolecular* step would require four molecules to collide with one another. Such an event is even much more improbable than a collision of three. Indeed, no quadrimolecular elementary step has ever been identified.

Provided the sample is large enough for statistics to be reliable, the rate equations in Table 2.1 are valid under all circumstances, regardless of the presence or absence of other molecules not involved in the respective step, and regardless of whether other reactions occur simultaneously in the same volume element. However, these "ideal" rate equations in terms of concentrations and with concentration-independent coefficients are only approximations. Deviations must be expected, but are relatively minor in most cases of practical interest and will be disregarded in this book, except for a brief discussion of nonideality in the next section.

As seen in Table 2.1, the overall order of an elementary step and the order or orders with respect to its reactant or reactants are given by the molecularity and stoichiometry and are always integers and constant. For a multistep reaction, in contrast, the reaction order as the exponent of a concentration, or the sum of the exponents of all concentrations, in an empirical power-law rate equation may well be fractional and vary with composition. Such apparent reaction orders are useful for characterization of reactions and as a first step in the search for a mechanism (see Chapter 7). However, no mechanism produces as its rate equation a power law with fractional exponents (except orders of one half or integer multiples of one half in some specific instances, see Sections 5.6, 9.3, 10.3, and 10.4). Within a limited range of conditions in which it was fitted to available experimental results, an empirical rate equation with fractional exponents may provide a good approximation to actual kinetics, but it cannot be relied upon for any extrapolation or in scale-up. In essence, fractional reaction orders are an admission of ignorance.

Except for reactions known to be single-step, molecularities or mechanisms cannot be deduced from observed reaction orders, nor can orders be predicted from stoichiometries.

The statement that each reactant molecule in a unimolecular reaction has the same probability to react is an acceptable and convenient simplification. In reality, molecules of the same species can exist at different energy levels, and only those "activated" to occupy a high energy level are capable of reacting. Energy is exchanged between molecules when they collide. If collisions are frequent, as in dense fluids, the energy distribution is practically in equilibrium. The simplified statement then applies to the average molecule, that is, it includes the probability of the molecule being activated. In contrast, in gases at very low pressures, collisions

may be so rare that activated molecules react before having had a chance to lose their
high energies in subsequent collisions. If so, activation by collision (a bimolecular
event) becomes rate-controlling [1-3]. The simplified statement then applies only to
the activated molecules, and the reaction should be viewed as a two-step event.

2.2. Nonideality

Rate equations in terms of concentrations and with presumably concentration-
independent rate coefficients, as used in this book, are idealizations. In the real
world, matters are more complex. For example, a change in polarity of the medium
with progressing conversion may cause a variation of rate coefficients. Such effects
are hard to predict and, as a rule, not overly serious. For practical purposes they
can often be disregarded. Where this is not so, an experimental determination of
the composition dependence of the coefficients is usually the best way to proceed.

At first glance one might think that deviations from ideality could be
accounted for by substitution of thermodynamic activities for concentrations in the
rate equations. However, this is not so. In fact, it can even make matters worse.

According to the theory of absolute reaction rates [4-9], the rate is the
product of a universal frequency factor and the concentration of the *activated
complex* or *transition state*, $M^{\ddagger}$ (the system in the transient state of highest potential
energy), crossing the energy barrier in the direction toward the products. The
activated complex, in turn, is postulated to be in equilibrium with the reactants.
Say, for a single-step reaction $A + B \longrightarrow P$:

$$\frac{\gamma_{M^{\ddagger}} C_{M^{\ddagger}}}{\gamma_A C_A \gamma_B C_B} = K^{\ddagger}$$

where $K^{\ddagger}$ is the thermodynamic equilibrium constant of the process $A + B \longleftrightarrow$
$M^{\ddagger}$, and the γ_i are the activity coefficients. With the rate proportional to $C_{M^{\ddagger}}$, it
follows that

$$r_P = k_{AP} \left[\frac{\gamma_A \gamma_B}{\gamma_{M^{\ddagger}}} \right] C_A C_B \tag{2.1}$$

where the coefficient k_{AP} is the product of the frequency factor and the equilibrium
constant $K^{\ddagger}$.

A comparison of eqn 2.1 with the ideal rate equation $r_P = k_{AP} C_A C_B$ shows
the deviation from ideality to appear as the factor $\gamma_A \gamma_B / \gamma_{M^{\ddagger}}$. A replacement of the
reactant concentrations C_A and C_B by thermodynamic activities $\gamma_A C_A$ and $\gamma_B C_B$ thus
does not necessarily constitute an improvement. Rather, if the activity coefficient
ratio $\gamma_A \gamma_B / \gamma_{M^{\ddagger}}$ is close to unity but the coefficients themselves are not, the ideal rate

law is more accurate than that with thermodynamic activities. Only if the activity coefficient of the activated complex, $\gamma_{M\ddagger}$, is closer to unity than are the activity coefficients of the reactants does the substitution of reactant activities for concentrations lead to a better approximation. This may or may not be the case.

Nonidealities are important for the study of kinetics as a science, However, activity coefficients of activated complexes cannot be measured directly and are difficult to predict with any degree of certainty. Therefore, in practice, rate equations are almost always stated in terms of concentrations, as is done in this book. This may be frowned upon by thermodynamic purists, but we could counter that, here, the pot calls the kettle black, that their science deals with equilibrium, a state that in our world is only approached, nowhere and never exactly attained.

2.3. Temperature dependence

Reactions differ from human beings. We humans tend to slow down when it becomes hot, reactions speed up. At least this is true in general for elementary steps. With increase in temperature, bond vibrations in molecules become stronger and collisions occur more often and with greater vigor, increasing the probability of molecules to react. As a rule, the rates of single-step reactions therefore increase with increasing temperature. In the great majority of cases this is also true for rates of multistep reactions, but not without exceptions: The overall rate may decrease with increasing temperature if the rates of reverse steps increase more sharply with temperature than those of forward steps. Such anomalies will be discussed in detail in a later chapter (see Section 12.1).

An approximate formula for the temperature dependence of the reaction rate coefficients, k, is given by the Arrhenius equation

$$k \;=\; A \, \exp(-E_a/RT) \qquad (1.3)$$

or, in differential form

$$\frac{\mathrm{d}\ln k}{\mathrm{d}(1/T)} \;=\; -\frac{E_a}{R} \qquad (2.2)$$

These equations apply regardless of the reaction order. If the activation energy E_a is taken to be constant, integration of eqn 2.2 yields

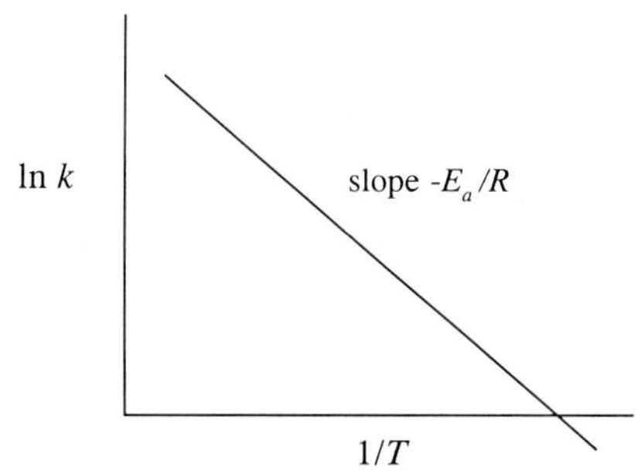

Figure 2.1. Arrhenius plot.

$$k(T) \;=\; k(T^\circ)\,\exp\left[\frac{E_a}{R}\left(\frac{1}{T^\circ} - \frac{1}{T}\right)\right] \qquad (2.3)$$

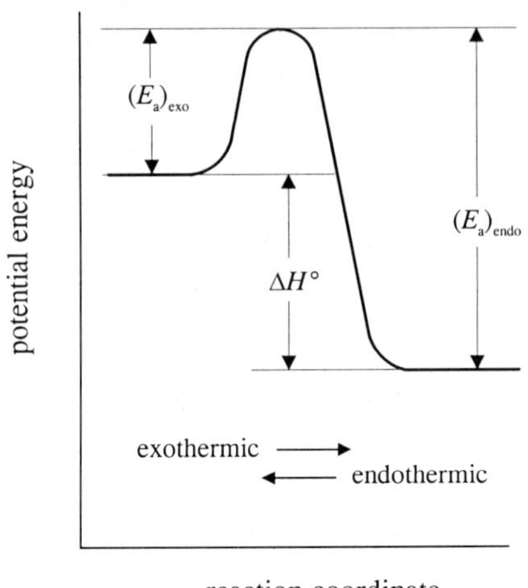

Figure 2.2. Potential-energy profile of a single-step reaction.

where $T°$ is a reference temperature. If the activation energy is constant, a plot of $\ln k$ versus $1/T$ gives a straight line with slope $-E_a/R$, as shown in Figure 2.1.

The activation energy can be viewed as an energy barrier which the reaction must overcome. This is illustrated in Figure 2.2, in which the potential energy of the reaction system is plotted versus a "reaction coordinate," an imaginary quantity characteristic of how far the reaction has progressed. If the reaction is exothermic, the system moves from a higher initial to a lower final potential-energy level (left to right in Figure 2.2); if the reaction is endothermic, the system moves in the opposite direction. As the illustration shows, the activation energy of an endothermic reaction must be higher than the standard-enthalpy change $\Delta H°$ of the reaction; in a reversible reaction, the activation energy must be higher in the endothermic than in the exothermic direction; and the $\Delta H°$ value must equal the difference between the activation energies of the forward and reverse reactions. (Figure 2.2 is for a single elementary step; the plot for a multistep reaction would show a local potential-energy maximum for each step.)

In contrast to the formally analogous van't Hoff equation [10] for the temperature dependence of equilibrium constants, the Arrhenius equation 1.3 is empirical and not exact: The pre-exponential factor A is not entirely independent of temperature. Slight deviations from straight-line behavior must therefore be expected. In terms of collision theory, the exponential factor stems from Boltzmann's law and reflects the fact that a collision will only be successful if the energy of the molecules exceeds a critical value. In addition, however, the frequency of collisions, reflected by the pre-exponential factor A, increases in proportion to the square root of temperature (at least in gases). This relatively small contribution to the temperature dependence is not correctly accounted for in eqns 2.2 and 2.3. [For more detail, see general references at end of chapter.]

For an elementary step, the straight-line Arrhenius plot with negative slope is a reasonably good approximation, corresponding to a positive and approximately constant apparent activation energy (E_a in eqn 2.2) and an increase in rate with increase in temperature. A multistep reaction, however, may show a different behavior. Its Arrhenius plot may have a positive slope, corresponding to a negative apparent activation energy and a decrease of the reaction rate with increasing temperature. In other cases, the apparent activation energy may "jump" within a relatively narrow temperature span from one value to another, giving rise to an Arrhenius plot with two connected, almost straight-line portions of different slopes. All these deviations from normal behavior are relatively rare. Their nature and mechanistic causes will be discussed in Section 12.1.

Because deviations can occur, extrapolations with the Arrhenius equation or plot to temperature ranges not covered by experiments may not be reliable.

2.4. Compilation of rate equations of multistep reactions

Mathematically, a multistep reaction corresponds to a set of simultaneous rate equations r_i , one for each participant (reactant, intermediate, product, catalyst, silent partner). Each of these rate equations describes the respective *net* formation or consumption, to which all steps involving that species contribute. The present section explains how the set of rate equations for a given network can be compiled.

The steps of a reaction are statistically independent, coupled only through their mutual dependence on the concentrations of the participants they have in common. Accordingly:

• *The algebraic form of the rate equation of a step and the value of its rate coefficient are not affected by any other steps; they are the same as they would be if the step were the only reaction occurring.*

Each rate equation may consist of contributions from several steps:

• *The rate equation r_i of a participant i is obtained by summation over the contributions from all steps involving that participant.*

For the purpose of this rule, the forward and reverse directions of a reversible step are counted as two steps. Each arrowhead in the network, except if pointing at a co-product or co-reactant, thus corresponds to one step. Steps in which a species i does not participate as reactant or product do not contribute to the respective r_i. For example, in the network

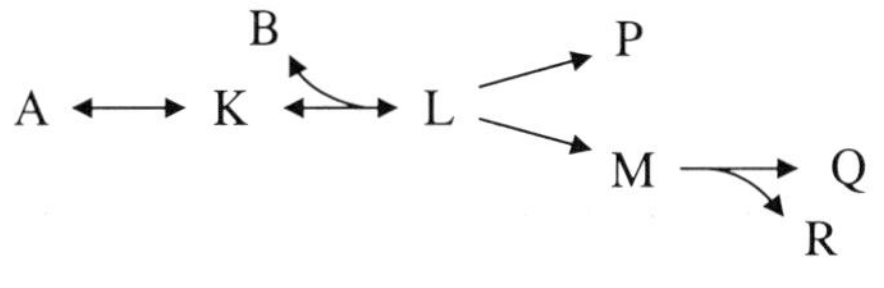

$$A \longleftrightarrow K \underset{B}{\overset{B}{\longleftrightarrow}} L \overset{P}{\underset{M \longrightarrow Q + R}{\diagdown}}$$

the rate equation r_L for the intermediate L is obtained by summation of the contributions from the steps K + B $\longrightarrow$ L, L $\longrightarrow$ K + B, L $\longrightarrow$ P, and L $\longrightarrow$ M; the steps A $\longrightarrow$ K, K $\longrightarrow$ A, and M $\longrightarrow$ Q + R do not contribute.

where $\{r\}$ and $\{C\}$ are vectors with the r_i and C_i, respectively, as elements (i = A, K, L, P, Q).

While useful for computer programming, the matrix notation does not contribute to better understanding.

2.5. Consistency criteria

Rate equations and their coefficients in networks are not entirely independent. They are subject to two constraints: thermodynamic consistency and so-called microscopic reversibility. For reversible reactions, the algebraic form of the rate equation of the forward reaction imposes a constraint on that of the rate equation of the reverse reaction. In addition, the requirements of thermodynamic consistency and microscopic reversibility can be used to verify that the postulated values of the coefficients constitute a self-consistent set, or to obtain a still missing coefficient value from those of the others.

2.5.1. Thermodynamic consistency

At equilibrium, there is no net formation or consumption of reactants and products, that is, the forward and reverse reaction rates must be equal. This is true no matter how many steps the reaction involves. Therefore:

Equating forward and reverse rates must lead to an expression
that is compatible with the mass-action law of equilibrium.

This fact can be used as a self-consistency check of postulated equations for the forward and reverse rates and their coefficients; or as a help in deriving the reverse rate equation from the forward one; or to calculate the reverse rate coefficient from the forward one and the equilibrium constant (or the forward rate coefficient from the reverse one and the equilibrium constant) [11,12].

Example 2.2. Thermodynamic consistency of an association reaction. To see how the thermodynamic consistency criterion can help in the search for a reverse rate equation compatible with a given empirical forward rate equation, consider the following reaction:

$$\text{stoichiometry:} \qquad 2A \longrightarrow P \qquad\qquad (2.4)$$

$$\text{equilibrium requirement:} \qquad C_P/C_A^2 \;=\; \text{const.} \;=\; K_{AP} \qquad (2.5)$$

$$\text{empirical forward rate:} \qquad \overrightarrow{r}_P \;=\; k_a\, C_A^{1.35} \qquad\qquad (2.6)$$

A likely reverse rate equation is

$$\text{reverse rate:} \qquad -\overleftarrow{r}_P \;=\; k_b\, C_P/C_A^{0.65}$$

It is compatible because equating the forward and reverse rates gives

$$k_a/k_b \;=\; \text{const.} \;=\; C_P/C_A^{1.35}\,C_A^{0.65} \;=\; C_P/C_A^2 \;=\; K_{AP}$$

meeting the equilibrium requirement 2.5.

However, this reverse rate equation is not unique. Rather, any equation of the form

$$\text{reverse rate:} \qquad -\overleftarrow{r}_P \;=\; k_b\, C_P^n/C_A^{2n-1.35}$$

with constant n chosen at will meets the equilibrium requirement 2.5 because it gives

$$k_a/k_b \;=\; \text{const.} \;=\; C_P^n/C_A^{2n} \;=\; (C_P/C_A^2)^n \;=\; K_{AP}^n \qquad (2.7)$$

(if K_{AP}^n is constant, so is K_{AP}).

If the forward rate equation contains an expression with additive terms, the reverse rate equation must contain that same expression. For example, if the forward rate of the reaction 2.4 is

$$\text{forward rate} \qquad \overrightarrow{r}_P \;=\; \frac{k_a\,C_A^2}{1 + k_c\,C_A}$$

only a reverse rate equation of the form

$$\text{reverse rate} \qquad -\overleftarrow{r}_P \;=\; \frac{k_b\,C_P^n\,C_A^{2-2n}}{1 + k_c\,C_A}$$

(same denominator with additive terms) can meet the equilibrium requirement.

If the equilibrium constant and the algebraic forms of both the forward and reverse rate equations are known, the reverse rate coefficient k_b can be calculated from the forward coefficient k_a or vice versa:

$$k_b \;=\; k_a/K_{AP}^n \qquad (2.8)$$

with n in accordance with eqn 2.7.

Equation 2.8 can also be applied to forward and reverse rate equations with denominators containing additive terms; this is so because the denominators cancel when the ratio is formed. Moreover, of course, eqn 2.8 is equally valid for single-step reversible reactions.

2.5.2. *Microscopic reversibility*

"Microscopic reversibility" as used in chemical kinetics is a classic misnomer. The name stems from a complicated derivation based on Onsager's axiom of reversibility at the molecular level [13-17]: At that level there is no preferential direction of time and, therefore, all events are in principle reversible. What is called microscopic reversibility in chemical kinetics is the statement that there can be no net circular reaction in a loop at equilibrium. For example, at equilibrium there can be no net circulation A $\longrightarrow$ B $\longrightarrow$ C $\longrightarrow$ D $\longrightarrow$ A in the loop 2.9:

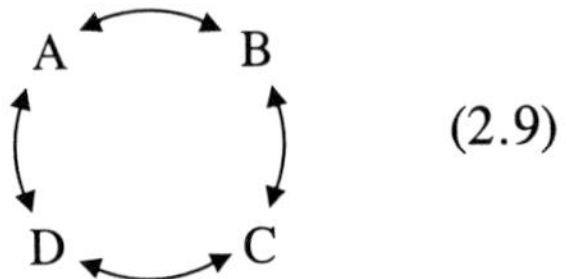

$$(2.9)$$

Figure 2.3. Waterfall by M. C. Escher, © 2000 Cordon Art B.V., Baarn, Holland. All rights reserved. Reproduced with permission.

Without invoking reasoning at the molecular level, that fact can easily be proved with the following argument. For a net forward reaction to occur (i.e., for the forward rate to exceed the reverse rate), the Gibbs free energy of the product or products must be less than that of the reactant or reactants. This makes a circular reaction in a loop impossible because the free energy would have to decrease from step to step as on a spiral staircase, yet reach its starting level again with the step that closes the loop, a feat that can be accomplished only in the world of M. C. Escher (see Figure 2.3).

The principle of microscopic reversibility can be used to check a set of postulated rate coefficients for self-consistency or to calculate the still unknown value of one rate coefficient from those of all others. To this end, most texts on kinetics prescribe a procedure called detailed balancing. However, a much simpler rule will do:

> The product of the "clockwise" rate coefficients in a closed loop must equal the product of the "counter-clockwise" rate coefficients.

Derivation. The equilibrium constants of the reversible steps of the loop are related to the standard Gibbs free energies ΔG_f° of formation of the members. For the four-membered loop 2.9:

$$-RT \ln K_{AB} = (\Delta G_f^\circ)_B - (\Delta G_f^\circ)_A$$

$$-RT \ln K_{BC} = (\Delta G_f^\circ)_C - (\Delta G_f^\circ)_B$$

$$-RT \ln K_{CD} = (\Delta G_f^\circ)_D - (\Delta G_f^\circ)_C$$

$$-RT \ln K_{DA} = (\Delta G_f^\circ)_A - (\Delta G_f^\circ)_D$$

Adding these equations to one another one obtains

$$-RT \,(\ln K_{AB} + \ln K_{BC} + \ln K_{CD} + \ln K_{DA}) = 0$$

so that

$$K_{AB} K_{BC} K_{CD} K_{DA} = 1$$

Each equilibrium constant equals the ratio of the forward to the reverse rate coefficient of the respective reaction step:

$$K_{ij} = k_{ij}/k_{ji} \tag{2.10}$$

Accordingly:

$$k_{AB} k_{BC} k_{CD} k_{DA} = k_{BA} k_{CB} k_{DC} k_{AD} \tag{2.11}$$

as stated in the rule above.

The argument leading to the conclusion that there can be no net circular reaction in a closed loop is based on the free energies of the members. As thermodynamic quantities, these are independent of whether or not the species involved also undergo other reactions. Accordingly, the rule is valid also for loops that are parts of larger networks, say, for ABC in

$$\ldots \longleftrightarrow D \longleftrightarrow A \quad B \longleftrightarrow E \longleftrightarrow \ldots \quad C \longleftrightarrow F \tag{2.12}$$

Likewise, the rule that the product of the clockwise coefficients must equal that of the counter-clockwise ones, based on the standard free energies, also holds even in such cases. The only restriction is that a full-circle reaction, if it were to occur, may not entail any net conversion of co-reactants to co-products.

In contrast to the traditional derivation based on microscopic reversibility, that given here does not invoke equilibrium. It thus shows that no net circular reaction is possible even under non-equilibrium conditions. However, there is one very important qualification: A net circular reaction *does* occur if it entails the conversion of co-reactants to co-products of lower Gibbs free energy. For example, the cycle 2.13 (next page) converts reactants A and B to product P while undergoing a net circular reaction K $\longrightarrow$ L $\longrightarrow$ M $\longrightarrow$ N $\longrightarrow$ K. This is a typical catalytic cycle. Such a cycle is *not* a loop, a term to be reserved for circular pathways in which any

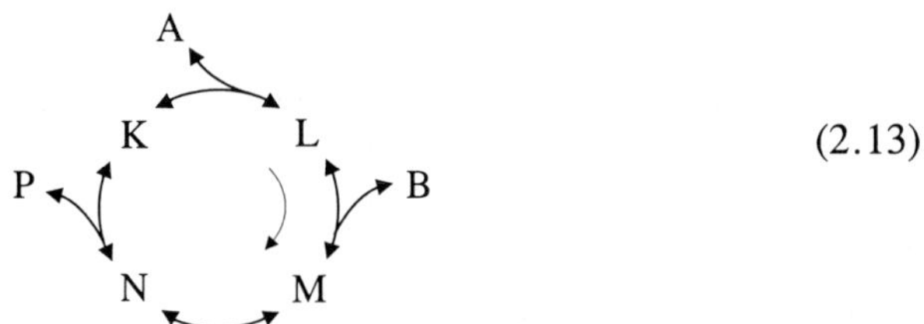

$$(2.13)$$

co-reactants are restored (a loop of two parallel pathways converting same reactants to same products meets this condition). Net circulation in the catalytic cycle occurs only if it entails a drop in free energy, that is, if the reactant or reactants (A and B in the cycle 2.13) are not in equilibrium with the product or products (P in 2.13). If so, the cycle is also not at equilibrium. At equilibrium, there is no net conversion of reactants to products to cause circulation. The principle of no net circular reaction at equilibrium thus applies to catalytic cycles as well as loops. However, the rule describing the constraint on the rate coefficients must be modified:

> The product of the forward rate coefficients in a catalytic cycle must equal the product of the reverse rate coefficients multiplied with the equilibrium constant of the catalyzed reaction.

For example, for the catalytic cycle 2.13:

$$k_{KL}k_{LM}k_{MN}k_{NK} \;=\; k_{LK}k_{ML}k_{NM}k_{KN}K_{AP} \tag{2.14}$$

where K_{AP} is the equilibrium constant of the reaction $A + B \longleftrightarrow P$.

Derivation. At equilibrium of the four reversible steps in the cycle 2.13:

$$\frac{C_L}{C_K C_A} = K_{KL}, \qquad \frac{C_M}{C_L C_B} = K_{LM}, \qquad \frac{C_N}{C_M} = K_{MN}, \qquad \frac{C_K C_P}{C_N} = K_{NK}$$

so that

$$K_{KL}K_{LM}K_{MN}K_{NK} \;=\; C_P/C_A C_B$$

Replacing $C_P/C_A C_B$ by the equilibrium constant K_{AP} of the reaction $A + B \longrightarrow P$ and the other equilibrium constants by the respective ratios of the forward and reverse rate coefficients (eqn 2.10) and then multiplying both sides by the product of the reverse rate coefficients one obtains eqn 2.14.

It is often loosely said that a catalyst "drives" a reaction. As the consideration here demonstrates, a picture more true to nature shows the *reaction* driving the *catalyst*. With its free-energy decrease, the reaction puts the catalyst system through its paces, much like water drives a water wheel. Rather than "driving" the reaction, the catalyst makes it possible by providing a pathway.

Summary

Although reactions differ widely in their kinetics in a manner that is unpredictable unless their mechanisms are known, rate equations of elementary steps obey simple laws of statistics: Rate equations of steps are power laws with integer reaction orders that can be directly deduced from molecularities.

Rate equations are conventionally written in terms of concentrations. This is an idealization, but substitution of thermodynamic activities for concentrations is not a proper way to account for nonidealities and may even make matters worse.

The temperature dependence of rate coefficients of elementary steps generally follows the Arrhenius equation in good approximation. That of apparent rate coefficients of empirical rate equations of multistep reactions, however, may deviate in several ways: The activation energy may be negative (slower rate at higher temperature) or have different values in different temperature regions.

The kinetics of a multistep reaction is described by a set of simultaneous rate equations, one for each participant. The equations are independent of one another in their algebraic forms and values of their coefficients. Each equation is the summation of the contributions from all steps in which the respective species participates. Each such contribution is the product of the stoichiometric coefficient of the species, the rate coefficient of the step, and the concentration (or concentrations) of the reactant (or reactants) of the step, raised to the power corresponding to the molecularity.

Self-consistency of postulated forward and reverse rate equations can be tested with the principles of thermodynamic consistency and so-called microscopic reversibility. The former invokes the fact that forward and reverse rates must be equal at equilibrium; the latter is for loops in networks and can be stated as requiring that the products of the clockwise and counter-clockwise rate coefficients of the loop must be equal, or, for catalytic cycles, that the product of the forward coefficients must equal that of the reverse coefficients multiplied with the equilibrium constant of the catalyzed reaction.

References

General references

G1.   M. Boudart, *Kinetics of chemical processes*, Prentice-Hall, Englewood Cliffs, 1968.

G2.   K. G. Denbigh, *The principles of chemical equilibrium: with applications in chemistry and chemical engineering*, Cambridge University Press, 4th ed., 1981, ISBN 0521236827.

G3.   H. S. Fogler, *Elements of chemical reaction engineering*, Prentice-Hall, Englewood Cliffs, 3nd ed., 1999, ISBN 0135317088.

G4.   G. F. Froment and K. B. Bischoff, *Chemical reactor analysis and design*, Wiley, New York, 2nd ed., 1990, ISBN 0471510440.

G5.   C. G. Hill, Jr., *An introduction to chemical engineering kinetics & reactor design*, Wiley, New York, 1977, ISBN 0471396095.

G6.   K. J. Laidler, *Chemical kinetics*, McGraw-Hill, New York, 3rd ed., 1987, ISBN 0060438622.

G7.   O. Levenspiel, *Chemical reaction engineering*, Wiley, New York, 3nd ed., 1999, ISBN 047125424X.

G8.   J. W. Moore and R. G. Pearson, *Kinetics and mechanism. A study of homogeneous chemical reactions*, Wiley, New York, 3rd ed., 1981, ISBN 0471035580.

Specific references

1.   J. A. Christiansen, *Reaktionskineticske studier*, thesis, University of Copenhagen, 1921, p. 58.

2.   F. A. Lindemann, *Trans. Faraday Soc.*, **17** (1922) 598.

3.   C. N. Hinshelwood, *Proc. Roy. Soc.*, **A 113** (1927) 230.

4.   H. Eyring, *J. Chem. Phys.*, **3** (1935) 107.

5.   W. F. K Wynne-Jones and H. Eyring, *J. Chem. Phys.*, **3** (1935) 492.

6.   S. Glasstone, K. J. Laidler, and H. Eyring, *The theory of rate processes*, McGraw-Hill, New York, 1941, Chapter VIII.

7.   H. Eyring, S. H. Lin, and S. M. Lin, *Basic chemical kinetics*, Wiley, New York, 1980, ISBN 0471054968, Chapter 4.

8.   Moore and Pearson (ref. G8), Chapter 5.

9.   Froment and Bischoff (ref. G4), Section 1.6.

10.   J. H. van't Hoff, *Études de dynamique chimique*, Muller, Amsterdam, 1884; see also any text on physical chemistry or chemical engineering.

11.   Denbigh (ref. G2), Section 15.5.

12.   Hill (ref. G5), Section 5.1.3.

13.   R. C. Tolman, *Phys. Rev.*, **23** (1924) 693.

14.   R. C. Tolman, *The principles of statistical mechanics*, Clarendon Press, Oxford, 1938 (reprint Dover, New York, 1979, ISBN 0486638960), p. 163.

15.   K. G. Denbigh, *The thermodynamics of the steady state*, Methuen, London, 1951, p. 31.

16.   Denbigh (ref. G2), p. 448.

17.   Hill (ref. G5), Section 4.1.5.4.

Chapter 3

Determination of Rates, Orders, and Rate Coefficients

The first order of business in the study of a new reaction in the context of process research and development is to measure reaction rates, establish approximate reaction orders for empirical power-law rate equations, and obtain values of their apparent rate coefficients. This chapter presents a brief overview of laboratory equipment, design of kinetic experiments, and evaluation of their results. It is intended as a tour guide for the practical chemist or engineer. More complete and detailed descriptions can be found in standard texts on reaction engineering and kinetics [G1-G7].

3.1. Research reactors

The four principal types of reactors used for bench-scale kinetic studies are batch, continuous stirred-tank (CSTR), tubular, and differential reactors. Which of these to choose is essentially a matter of the reaction conditions, available equipment, and the chemist's or engineer's predilections. The discussion here will focus on facets that pertain specifically to quantitative kinetic studies of homogeneous reactions.

Whatever reactor is chosen, the reaction should be conducted at constant temperature, if at all possible. A variation of temperature in the course of the reaction makes the evaluation much more difficult and less reliable.

3.1.1. *Batch reactors*

Typical batch reactors for kinetic studies of liquid-phase reactions at ambient and elevated pressures are shown in Figures 3.1 and 3.2, respectively. They are equipped with stirrer, heating or cooling arrangement, temperature sensor, reflux condenser, and sampling ports.

Batch kinetic results are easiest to evaluate if the reactor is operated at constant volume. A variation of reaction volume with conversion contributes to the variation of concentrations with time (even the concentrations of inerts will vary!).

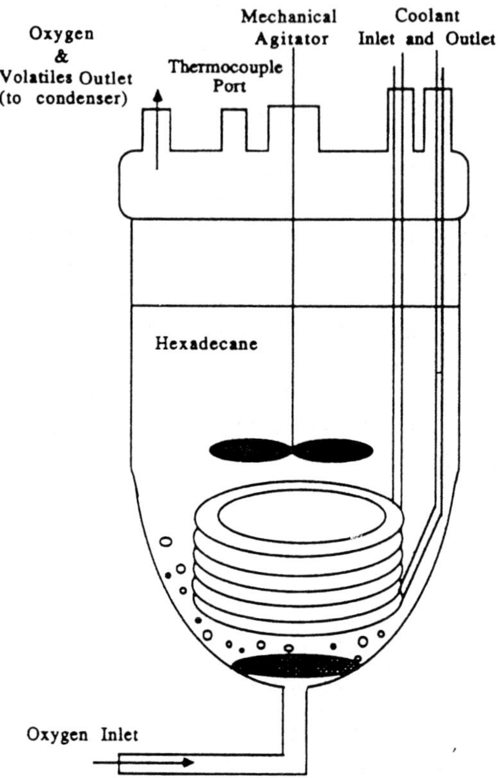

Figure 3.1.   Batch reactor for kinetic studies of oxidation at ambient pressure (from Blaine and Savage [1]).

This not only complicates the evaluation, but also introduces a source of error (see also Section 3.3.4).

   Volume variations with conversion are large for constant-pressure gas-phase reactions with change in mole number. Here, as a rule, operation at constant volume poses no difficulties. Liquid-phase reactions may also entail volume contraction or expansion. However, these are not related to changes in mole number and can be predicted only if information on partial molar volumes is at hand. Because liquids are essentially incompressible, even at elevated temperature, it is unsafe to conduct liquid-phase reactions without a gas cap in a closed reactor. Some variation of liquid-phase volume with conversion therefore is apt to occur. Fortunately, the variation at constant temperature is usually so small that it can be neglected in the evaluation or accounted for by a minor correction.

An advantage of batch over continuous stirred-tank and tubular reactors is that a single experiment with samples taken at frequent intervals in effect scans the entire conversion range. A disadvantage is that samples must be taken from the reactor itself.

For elucidation of mechanisms, rate data at very low conversions may be highly desirable. They can be obtained more easily from a batch reactor than from a CSTR or plug-flow tubular reactor. A standard CSTR would have to be operated at very high flow rates apt to cause fluid-dynamic and control problems. The same is true for a standard tubular reactor unless equipped with a sampling port near its inlet, a mechanical complication apt to perturb the flow pattern. If the problem of confining the reaction to a very small flow reactor can be solved—as is possible, for example, for radiation-induced reactions—a differential reactor operated once-through or with recycle may be the best choice.

The two main problems in the design of batch reactors for quantitative kinetic studies are effective removal or supply of heat to keep the temperature constant, and establishment of a sharp zero time.

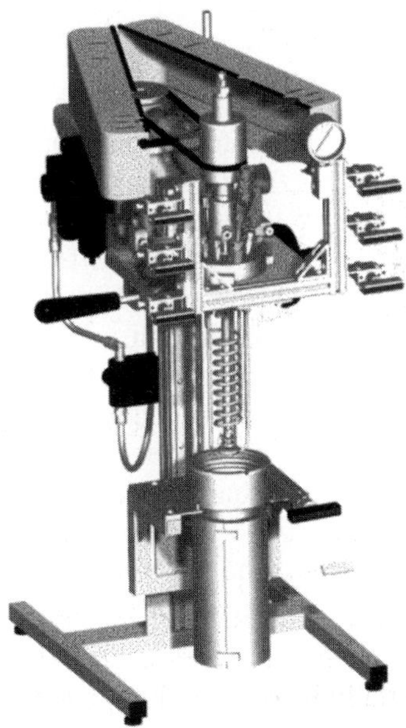

Figure 3.2. Commercial batch reactor for kinetic studies at elevated pressures (courtesy Autoclave Engineers, Erie, PA, U.S.A., reproduced with permission).

Heat transfer is more of a problem in batch reactors than in other types of equipment because of their small surface-to-volume ratio and because rates are high initially. Effective stirring is essential. If the reaction is highly exothermic or endothermic, cooling or heating coils are usually needed.

A sharp zero time calls for fast attainment of reaction conditions. No significant conversion can be tolerated before the reactants are completely mixed and the mixture is at reaction temperature and pressure. If reaction conditions are reached too slowly, rate coefficients can no longer be calculated accurately. Moreover, while the reactant ratios, temperature, and pressure are not yet on target, reactions different from those under the eventual reaction conditions may occur at least to some extent and falsify the results. Especially for reactions at elevated temperatures and pressures and for studies using unstable intermediates as starting materials, fast attainment of reaction conditions usually requires special set-ups. For example, the starting materials may be divided into two unreactive portions, which are pressured and heated separately in different vessels and then mixed as quickly as possible to initiate the reaction. An intermediate that is unstable under reaction conditions may be held in a cooled and pressured vessel, to be injected quickly into a much larger amount of a medium already at reaction temperature and pressure or at a temperature such that mixing of the cold injected fluid with the hot reactor contents produces the desired reaction temperature (see Figure 3.3).

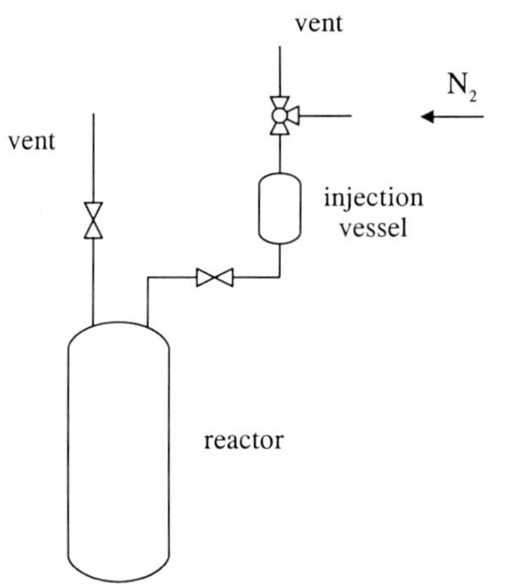

Figure 3.3. Arrangement for injection at zero time (schematic).

Batch reactors can also be used for studies of gas-liquid reactions. A common procedure, sometimes called "semi-batch," is to conduct the reaction as batch with respect to the liquid and bubble the gas through at constant composition and pressure. Effective gas-liquid contacting is essential in order to avoid mass-transfer limitation with respect to the reactant supplied by the gas phase. Good ways of introducing the gas are through a fine-pore sieve plate or through the hollow shaft and arms of a stirrer.

In a constant-volume batch reaction, the concentration variations with time are caused exclusively by chemical reaction, so that

$$r_i = dC_i/dt$$

(process rate equals rate of change, see Section 1.3). However, the rate cannot be measured directly. What can be measured are concentrations at different times, and a finite-difference approximation is needed to obtain the rate:

$$r_i \;\simeq\; \Delta C_i / \Delta t \qquad\qquad (3.1)$$

This has important implications for the evaluation of results (see Section 3.3).

If a gas-phase reaction with change in mole number is conducted at constant pressure rather than constant volume, the governing equation is

$$r_i \;=\; (1/V)\,dN_i/dt$$

or, in finite-difference form:

$$r_i \;\simeq\; (1/V)\,\Delta N_i/\Delta t \qquad\qquad (3.2)$$

where V is the (varying) reaction volume.

3.1.2. Continuous stirred-tank reactors

Continuous stirred-tank reactors (CSTRs, see Figure 3.4) have the advantage over batch reactors that they are easier to keep at constant temperature. This is because they do not see the high initial reaction rate of batch reactors, and because continuous flow into and out of the reactor helps to consume or supply heat. For example, if the reaction is highly exothermic, the entering fluid may be introduced at a temperature below that of the reactor, so that it consumes heat as it is heated up. Another advantage is that samples can be taken from the effluent rather than the reactor itself. A disadvantage is that any single experiment in a CSTR at steady state provides information only at one conversion level, whereas a batch experiment in effect scans an entire conversion range.

Reactors like those shown in Figures 3.1 and 3.2 can be used as CSTRs for kinetic studies if equipped with entry and exit ports.

The main problems in the design of CSTRs for quantitative kinetic studies are to provide effective mixing and excellent control of the flow rate. As a rule, the evaluation presumes that the entering fluid is instantaneously mixed with the reactor contents, so that the latter is uniform. Incomplete mixing can falsify results. Corrections are complex and require detailed knowledge

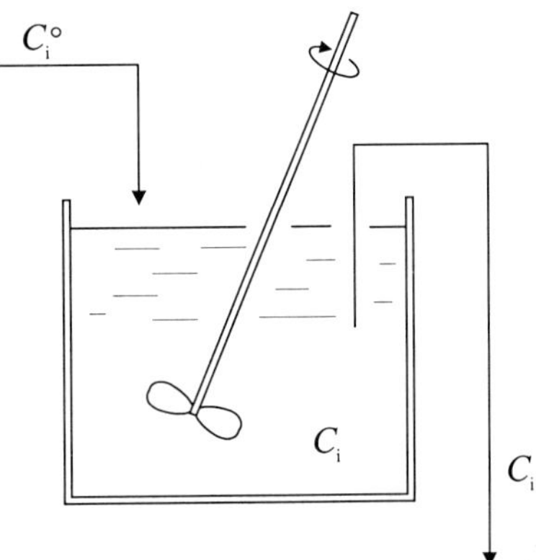

Figure 3.4. Continuous stirred-tank reactor (schematic).

$$\ln(-r_A) \;=\; \ln k_{app} + n \ln C_A \qquad (3.11)$$

so that a plot of $\ln(-r_A)$ versus $\ln C_A$ gives a straight line with slope n and intercept $\ln k_{app}$. Because of the logarithmic scales, this method is much less reliable.

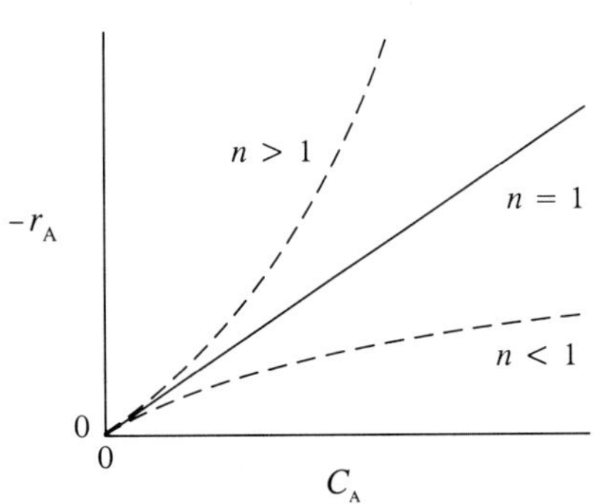

Figure 3.9. Straight-line plot for first-order reaction in CSTR, also showing curves given by reactions of higher and lower orders.

Note that there also is a concentration method for evaluation of CSTR results (see farther below).

Rate methods are much less suited for evaluation of results from batch, tubular, and differential recycle reactors because for these the rate must be obtained by a finite-difference approximation (see eqns 3.1, 3.5, and 3.8). In particular the method based on eqn 3.11 should not be used for such a purpose because of its high sensitivity to even minor experimental errors.

For ease of reference, plots are summarized in Table 3.1.

Concentration methods. For results from batch, tubular, or differential recycle reactors at constant fluid density, the most common procedures are based on integrated forms of the rate equation.

For a first-order, constant-volume batch reaction A $\longrightarrow$ P:

$$\ln(C_A/C_A^o) \;=\; -k_{app}\, t \qquad (3.12)$$

or

$$C_A \;=\; C_A^o \exp(-k_{app}t) \qquad (3.13)$$

so that

$$-r_A \;=\; r_P \;=\; k_{app}\, C_A^o \exp(-k_{app}t) \qquad (3.14)$$

and for the product:

$$\ln(1 - C_P/C_A^o) \;=\; -k_{app}\, t \qquad (3.15)$$

$$C_P \;=\; C_A^o \left(1 - \exp(-k_{app}t)\right) \qquad (3.16)$$

For constant-volume batch reactions with orders other than first, one has

$$(1/C_A)^{n-1} - (1/C_A^o)^{n-1} \;=\; (n-1)k_{app}\, t \qquad (3.17)$$

(not applicable to first order, error-sensitive for orders close to first).

For tubular reactors and reactions with no fluid-density variation, the reactor space time, $\tau = V/\dot{V}$, takes the place of the actual time, t.

Table 3.1. Common straight-line plots for determination of reaction orders and apparent rate coefficients.

RATE PLOTS (applicable to all reactor types)

reaction rate $A \longrightarrow$ products $-r_A = k_{app} C_A^n$

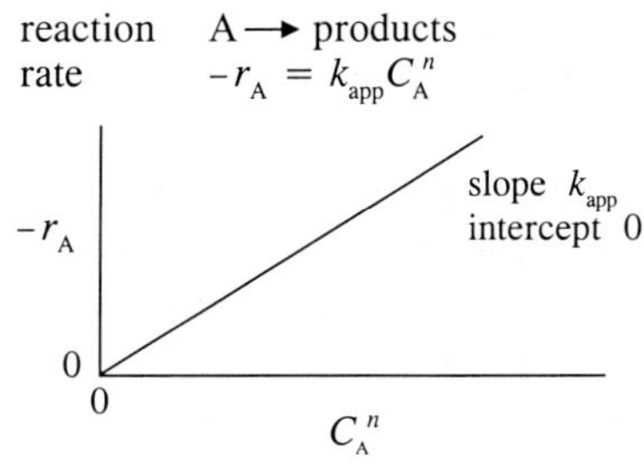

reaction rate $n_A A + n_B B + \ldots \longrightarrow$ products $-r_A/n_A = k_{app} C_A^m C_B^n \ldots$

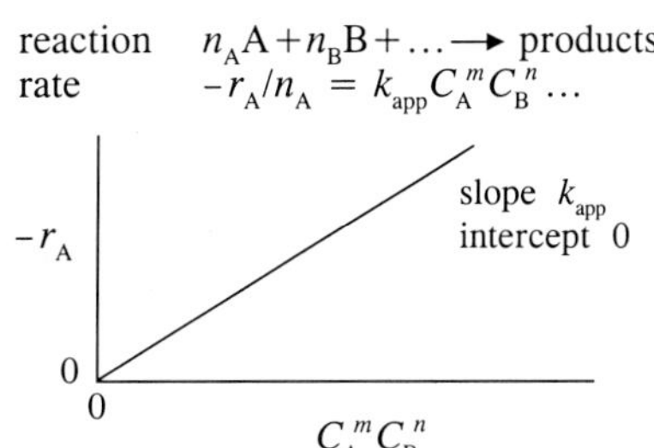

CONCENTRATIONS PLOTS

constant-volume batch

reaction rate $A \longrightarrow$ products $-r_A = k_{app} C_A$

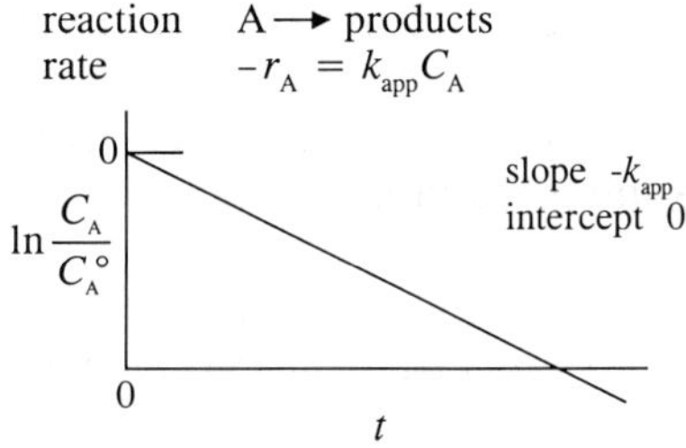

reaction rate $A \longrightarrow$ products $-r_A = k_{app} C_A^n$

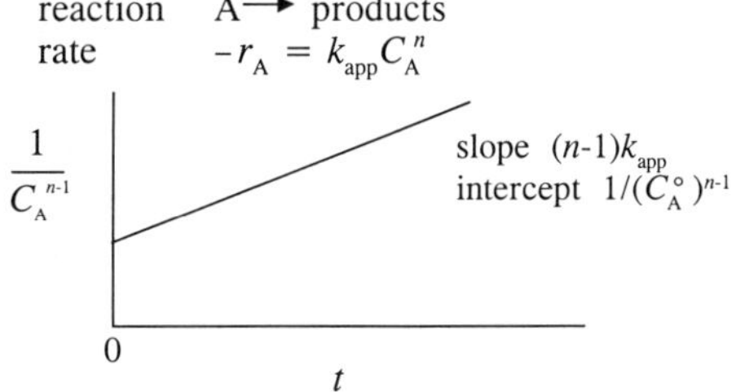

reaction rate $A + B + \ldots \longrightarrow$ products $-r_A = k_{app} C_A C_B$

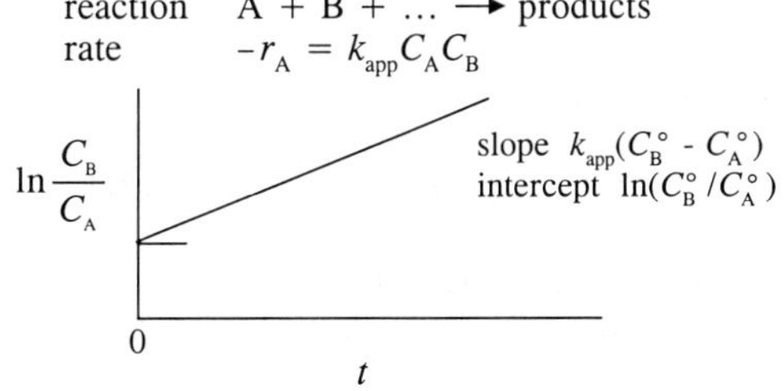

continuous stirred tank

reaction rate $A \longrightarrow$ products $-r_A = k_{app} C_A$

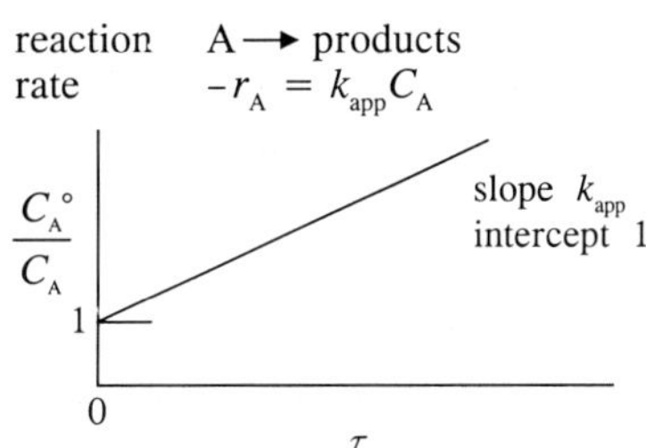

reaction rate $A \longrightarrow$ products $-r_A = k_{app} C_A^n$

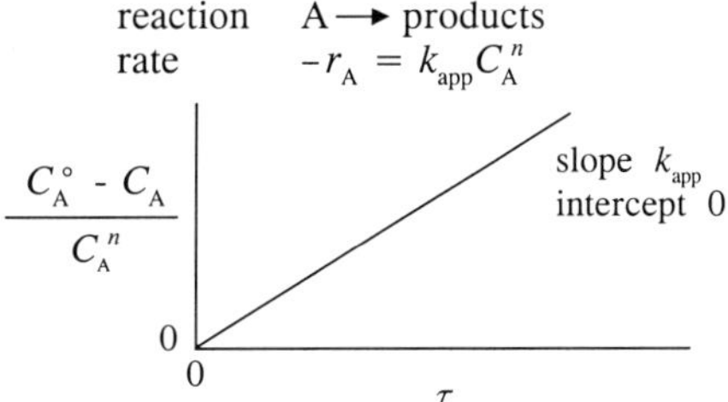

plug-flow tube

If fluid density does not vary with conversion, use same plots as for constant-volume batch, with reactor space time τ substituted for time t.

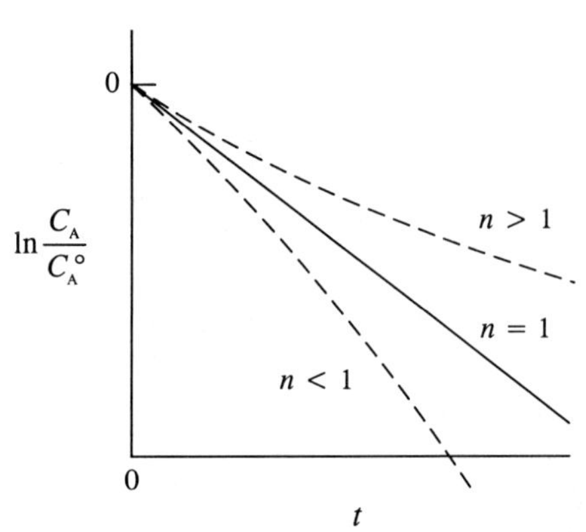

Figure 3.10. First-order concentration plot for constant-volume batch reactor, also showing curvatures given by reactions of higher and lower orders.

Equations 3.15 and 3.16 presume that no product is present initially; otherwise, C_P must be replaced by $C_P - C_P^o$. Also, for stoichiometries $n_A A \longrightarrow n_P P$, multiply C_A^o in eqns 3.15 and 3.16 with n_P / n_A).

Straight-line plots are summarized in Table 3.1. Reactions of higher or lower order than assumed give concave-up or convex-up curves, respectively, as shown for a first-order plot in Figure 3.10.

In eqn 3.12 and Figure 3.10, $\ln(C_A / C_A^o)$ can be replaced by $\ln C_A$. The intercept then is at $\ln C_A^o$, but the slope of the curve remains the same.

A more general form of eqn 3.12 for first-order reactions in batch or differential recycle reactors is

$$\ln(1 - f_A) = -k_{app} t \tag{3.18}$$

where $f_A \equiv 1 - N_A / N_A^o$ is the fractional conversion (see definition 1.4). In this form the equation is valid regardless of any fluid-density variation.

For a continuous stirred-tank reactor at steady state and a reaction with no fluid-density variation, the evaluation can be based on a plot of concentrations instead of rates. Replacement of $-r_A$ by $k_{app} C_A^n$ in eqn 3.3 for A yields

$$(C_A^o - C_A)/C_A^n = k_{app} \tau \tag{3.19}$$

For a first-order reaction A $\longrightarrow$ P, eqn 3.19 reduces to

$$C_A^o/C_A = 1 + k_{app} \tau \tag{3.20}$$

The respective straight-line plots are included in Table 3.1.

From eqn 3.20 and $C_P = C_A^o - C_A$ one can obtain

$$C_A = \frac{C_A^o}{1 + k_{app} \tau}, \qquad C_P = \frac{C_A^o \, k_{app} \tau}{1 + k_{app} \tau} \tag{3.21}$$

so that

$$-r_A = r_P = \frac{k_{app} C_A^o}{1 + k_{app} \tau} \tag{3.22}$$

Incorrect guesses of the reaction order lead to curved plots. As in Figures 3.9 and 3.10, orders higher or lower than assumed give concave-up or convex-up curves, respectively.

Fractional-life methods. If a reaction is known to be first order and at constant fluid density, its apparent rate coefficient can be found very quickly. For batch and differential recycle reactors, the relationship between the rate coefficient and the time t_y required for all but a fraction y of the reactant to be consumed is

$$k_{app} = -(\ln y)/t_y \qquad (3.23)$$

Specifically, for the half-time $t_{1/2}$ and the time $t_{1/e}$ ($e = 2.71828$):

$$k_{app} = (\ln 2)/t_{1/2} \qquad \text{and} \qquad k_{app} = 1/t_{1/e} \qquad (3.24)$$

For tubular reactors and reactions with no fluid-density change, the same equations apply with reactor space time substituted for real time. For CSTRs at steady state:

$$k_{app} = \frac{1-y}{\tau_y y} \qquad \text{and} \qquad k_{app} = 1/\tau_{1/2} \qquad (3.25)$$

as follows from the material balance $-r_A = k_{app}C_A = (C_A^\circ - C_A)/\tau$ with $C_A = yC_A^\circ$.

Some texts describe fractional-life methods for reactions other than first order and with more than one reactant. However, the effort their application requires is not in proportion to the limited objectives of the evaluation in the present context.

3.3.2. Irreversible reactions with two or more reactants

reaction: $\qquad\qquad n_A C_A + n_B C_B + \ldots \longrightarrow$ product(s) $\qquad (3.26)$

power-law rate equation: $\qquad -r_A/n_A = k_{app}\, C_A^m\, C_B^n \ldots \qquad (3.27)$

Rate methods. As for reactions with a single reactant, the most primitive and convenient rate method is to guess and then adjust the reaction orders until an approximately straight-line plot of $-r_A$ versus the rate versus $C_A^m C_B^n \ldots$ is obtained (see Table 3.1). For example, if a reaction is expected to be second order in A and first order in B, the rate would be plotted versus $C_A^2 C_B$. For first guesses of reaction orders, the available data can be subdivided so that the concentration of all but one participant are the same within each group. The dependence of the rate on the concentration of that participant then is an indication of the respective order. Often, the evaluation need not be carried beyond this stage.

If the development chemist or engineer has the opportunity to design the kinetic experiments, the evaluation can often be simplified by establishment of pseudo-orders, as shown farther below.

As for reactions with a single reactant, rate methods are not well suited for evaluation of results from batch, tubular, and differential recycle reactors. For complex mechanisms, however, no convenient concentration methods might exist, so that a rate method could be the lesser of two evils.

Concentration methods. Results from constant volume-batch or differential recycle reactors for a reaction $n_A A + n_B B \longrightarrow$ product(s) give a linear plot of $\ln(C_B/C_A)$ versus t with slope $k_{app}(n_A C_B^\circ - n_B C_A^\circ)$ and intercept $\ln(C_B^\circ/C_A^\circ)$ if the reaction is first order in both A and B [41]. However, this procedure is inapplicable to experiments with stoichiometric amounts of A and B (i.e., $n_B C_A^\circ = n_A C_B^\circ$), and is highly error-sensitive for amounts that are close to or extremely different from stoichiometric.

For tubular reactors, the same procedure can be used with reactor space time τ substituted for time t.

Pseudo- and overall reaction orders. Kinetic textbooks describe other, more complicated methods applicable to other forms of proposed rate equations, mostly for evaluation of results from batch reactors. However, if the development chemist or engineer can commission experiments—as opposed to having to evaluate existing data—he can often save himself much effort by determination of pseudo- and overall reaction orders. For example, for a reaction $A + B \longrightarrow$ product(s) and power-law rate equation $-r_A = k_{app} C_A^m C_B^n$, three series of experiments suggest themselves:

(1)   with a very large excess of A over B,

(2)   with a very large excess of B over A, and

(3)   with stoichiometric amounts of A and B.

In series (1), the concentration of A decreases only insignificantly from its starting value while B becomes depleted, so that the rate equation 3.27 can be reformulated with a new, approximately constant coefficient k_b:

$$\text{if } C_B \ll C_A, \qquad -r_A \cong k_b C_B^n \qquad \left(k_b \equiv k_{app} (C_A^\circ)^m \cong \text{const.} \right)$$

Under these condition, the kinetic behavior is analogous to that of n'th order with single reactant B, and n can be determined as for such a reaction. The reaction is said to be pseudo-n'th order in B.

Similarly, under the conditions of series (2),

$$\text{if } C_A \ll C_B, \qquad -r_A \cong k_a C_A^m \qquad \left(k_a \equiv k_{app} (C_B^\circ)^n \cong \text{const.} \right)$$

The reaction is pseudo-m'th order in A, and m can be determined as for a reaction with A as the sole reactant.

Lastly, in series (3), the concentrations of A and B remain equal because both reactants were charged in equal amounts and are consumed at the same rate. With $C_A = C_B$, the rate equation 3.27 becomes

$$-r_{\mathrm{A}} \;=\; k_{\mathrm{app}}\, C_{\mathrm{A}}^{\,n+m} \;=\; k_{\mathrm{app}}\, C_{\mathrm{B}}^{\,n+m}$$

The reaction behavior now is analogous to that of a reaction of order $m+n$ with A (or B) as the sole reactant, and the overall order $m+n$ can be determined as for such a reaction.

Thus, two of the three series of experiments suffice to determine all reaction orders and the apparent rate coefficient.

Pseudo-orders can also be established by semi-batch experiments in which one or several reactants are replenished to keep their concentration or concentrations constant. In this way, the reaction order or orders with respect to the other reactant or reactants can be established. The order with respect to a replenished reactant can then be found by comparison of the rates obtained with different concentrations of that reactant. This method is particularly convenient for gas-liquid reactions such as homogeneous oxidation, halogenation, hydrogenation, hydroformylation, and hydrocyanation. Here, the gaseous reactant or reactants can be admitted to the reactor on demand as they are consumed, or by bubbling an excess of the gas at constant pressure through the reactor. Some later examples of network elucidation are of cases in which this method was used.

3.3.3. Reversible reactions

reaction:
$$n_{\mathrm{A}}C_{\mathrm{A}} + n_{\mathrm{B}}C_{\mathrm{B}} + \ldots \;\longleftrightarrow\; n_{\mathrm{P}}C_{\mathrm{P}} + n_{\mathrm{Q}}C_{\mathrm{Q}} + \ldots \qquad (3.28)$$

power-law rate equation:
$$-r_{\mathrm{A}}/n_{\mathrm{A}} \;=\; k_{\mathrm{a}}\, C_{\mathrm{A}}^{a}\, C_{\mathrm{B}}^{b}\ldots \;-\; k_{\mathrm{b}}\, C_{\mathrm{P}}^{p}\, C_{\mathrm{Q}}^{q}\ldots \qquad (3.29)$$

Reversible reactions are best considered as consisting of forward and reverse steps. Such step combinations will be examined in Section 5.1. Simple plots suited for evaluation exist only for first order-first order reactions A $\longleftrightarrow$ P (see Figure 5.2). In more complex situations, the best approach often is to isolate the forward from the reverse reaction. This can be done to some extent with experiments that measure initial rates.

Initial rates. As long as conversion has still remained insignificant, the reverse reaction contributes little to the net rate: For a short initial time span the reaction behaves as though it were irreversible. This allows the previously described methods for irreversible reactions to be used to establish the reactions orders and rate coefficient of the forward reaction. Often, initial reverse rates can be measured in the same fashion by using the reaction products as starting materials. Otherwise, a likely reverse rate equation and its coefficient can be established with the principle of thermodynamic consistency (see Section 2.5.1).

The procedure based on measurement of initial rates is not entirely reliable. It can give misleading results if fast early reaction steps cause one or several intermediates to build up to substantial concentrations during the time span in which product concentrations are measured for calculation of initial rates.

3.3.4. *Gas-phase reactions at constant pressure*

The chemist or engineer designing his experiments to establish quantitative kinetics of gas-phase reactions will do his best to look for constant-volume equipment. However, occasionally he may have to work with data obtained at constant pressure. The complication here is that a change in mole number affects the reaction volume and, thereby, the concentrations of the participants, distorting their histories from which reaction orders and rate coefficients are deduced: Volume variation disguises kinetics and must be corrected for.

Reaction time and rate in a batch reactor are related by [42,43]

$$ t \;=\; p_A^o \int_{f_A=0}^{f_A} \frac{\mathrm{d}f_A}{(-r_A)\,(V/V^\circ)} \tag{3.30} $$

where f_A is the fractional conversion of the limiting reactant, V is the (variable) reaction volume, and V° is that volume at start. A strictly analogous equation with reactor space time τ instead of reaction time t applies to an ideal plug-flow reactor. Standard practice is to postulate that the volume varies linearly with conversion; this is correct if the ideal gas laws are obeyed, any intermediates remain at very low concentrations, and the reaction does not switch pathways. If so:

$$ V/V^\circ \;=\; 1 + \delta_A f_A \tag{3.31} $$

where

$$ \delta_A \;=\; (N_{f_A=1} - N^\circ)/N^\circ \tag{3.32} $$

is the relative mole-number (or volume) variation resulting from complete conversion of the limiting reactant A [44,45]. For example, if the mole number N increases from 2 to 3, causing a 50% increase in volume, δ_A is $+1/2$; or if the mole number decreases from 3 to 2, δ_A is $-1/3$.

The partial pressures of the participants, corrected for volume variation, are

$$ p_A = p_A^o\,\frac{1-f_A}{1+\delta_A f_A}\,, \qquad\qquad p_i = \frac{p_i^o - (\nu_i/\nu_A)p_A^o f_A}{1+\delta_A f_A} $$

Equation 3.30 may have to be integrated numerically, but that is no problem at a time when even pocket calculators can integrate. Since reaction orders hardly ever need to be determined exactly, the following general rule may come in handy:

> Volume expansion, if not corrected for, lets the reaction order appear farther from first than it is; volume contraction has the opposite effect.

For example, if the mole number doubles, the concentration histories up to about 50% conversion look much the same for a reaction of order 1.5 at constant pressure as for one of order 1.8 at constant volume; or for order 0.5 at constant pressure as for order 0.2 at constant volume. If the reaction is first order, a volume variation does not distort the concentration history.

Unfortunately, the conditions that guarantee the volume to vary linearly with conversion are often not met in reactions with complex mechanisms. This makes the evaluation harder and less reliable and provides an even stronger incentive to conduct the work at constant volume.

3.4. Numerical work-up, error recognition, and reliability

Today, easy-to-use computer software for evaluation of experimental kinetic data to obtain reaction orders and apparent rate coefficients by regression is commercially available [46-50].

Short of resorting to computerized evaluation, the chemist or engineer may want to stay with more pedestrian long-hand numerical methods that are perfectly adequate for the immediate purpose of obtaining approximate orders and coefficients in power-law equations that serve only as temporary expedients. In essence, these methods are the numerical equivalents of the plots that were shown earlier in Table 3.1. Two detailed examples will illustrate typical work-up, including error recognition and relative reliability of experimental data.

Example 3.1. Decomposition of hexaphenylethane in constant-volume batch. The decomposition of hexaphenylethane into two triphenylmethyl radicals

$$(C_6H_5)_3C - C(C_6H_5)_3 \longrightarrow 2\ (C_6H_5)_3C\cdot$$

in liquid chloroform at 0° C has been studied in a batch reactor [51]. The volume variation is negligible. The results are shown in the first two columns of Table 3.2 (next page).

A quick first inspection of the experimental data suggests that the times required to reduce the fractional reactant concentration by one half from 1.0 to 0.5, from 0.5 to 0.25, and from 0.25 to 0.125 are about the same. This indicates that, at least in good approximation, the reaction is first order, as its chemistry makes one expect. Application of eqn 3.23 to a data point in the mid-conversion range, say, at 209 seconds, gives

$$k_{app} \cong -(\ln 0.471)/209 = 3.6 * 10^{-3}\ s^{-1}$$

Since the reaction may indeed occur in a single step and so be accurately first order, a more thorough evaluation with eqn 3.12 may be desired. This is done in columns 3 and 4 of Table 3.2. In each row, the logarithm of the fractional concen-

Table 3.2. Decomposition of hexaphenylethane in constant-volume batch reactor; global evaluation.

time s	concentration C_A/C_A°	$-\ln(C_A/C_A^\circ)$	$k_{app} =$ $-\ln(C_A/C_A^\circ)/t$ s^{-1}
17.4	0.941	0.0608	$3.49*10^{-3}$
35.4	.883	.1244	3.51
54.0	.824	.1936	3.59
174	.530	.6349	3.65
209	.471	.7529	3.60
313	.324	1.1270	3.60
367	.265	.3280	3.62
434	.206	.5799	3.64
584	.118	2.1371	3.66
759	.059	.8302	3.73

tration is calculated and divided by the time to give the apparent first-order rate coefficient. The coefficient is seen to be constant within reason, confirming that the reaction is first order. This "global" evaluation procedure is the numerical equivalent of plotting $\ln(C_A/C_A^\circ)$ versus t and obtaining a k_{app} value for each data point as the negative slope of the straight line from the origin to that point. The values so found are characteristic of the time intervals from start of the experiment to the moments the respective samples were taken. Confirming the quick first estimate, the average value is $3.61*10^{-3}$ s^{-1}.

Alternatively, the original data can be evaluated point by point by calculating the differences between successive values of $\ln(C_A/C_A^\circ)$ and dividing them by the differences between the respective times, as is done in Table 3.3. This procedure is the numerical equivalent of connecting successive points in the plot of $\ln(C_A/C_A^\circ)$ versus t by straight lines and obtaining k_{app} values for the respective time intervals from the slopes of the connecting lines. The average value found by this procedure is a little higher, $3.65*10^{-3}$ s^{-1}.

For comparison, the interested reader might want to evaluate the experimental data with the method based on eqn 3.11. He would find a reaction order that varies between 0.24 and 1.24 in point-by-point evaluation, and between 0.52 and 1.06 in global evaluation. This demonstrates the high sensitivity of this method to even minor experimental errors.

It is instructive to compare the two evaluation methods used in Tables 3.2 and 3.3. The global method gives a more nearly constant value of the rate coefficient because each value is, in effect, averaged over the entire time span from start to sampling. In contrast, the point-by-point method is particularly suited for

Table 3.3. Decomposition of hexaphenylethane in constant-volume batch reactor; point-by-point evaluation.

time s	concentration C_A/C_A°	$\ln(C_A/C_A^\circ)$	$\Delta\ln(C_A/C_A^\circ)$	Δt s	$k_{app} =$ $\dfrac{-\Delta\ln(C_A/C_A^\circ)}{\Delta t}$ s^{-1}
0	1.000	0			
			-0.0608	17.4	$3.49*10^{-3}$
17.4	0.941	-0.0608			
			.0636	18.0	3.53
35.4	.883	.1244			
			.0692	18.6	3.72
54.0	.824	.1936			
			.4413	120	3.68
174	.530	.6349			
			.1180	35	3.37
209	.471	.7529			
			.3741	104	3.60
313	.324	-1.1270			
			.2010	54	3.72
367	.265	.3280			
			.2519	67	3.76
434	.206	.5799			
			.5572	150	3.71
584	.118	-2.1371			
			.6931	175	3.96
759	.059	.8302			

spotting suspect data points. For instance, the coefficient value for the time interval from 174 to 209 seconds is out of line, making one suspect an error in one of the two times or concentrations used for calculating it.

Apart from that value, the greatest deviations from the average are at the lowest and highest conversions. In general, these results are the least reliable and might be given less weight. Batch data at very short reaction times suffer most from a lack of ideal sharpness of the zero time. In fact, if all listed sampling times were decreased by half a second, the coefficient values from the global method would vary less and be more in line with those from the point-by-point method (in the latter, only the first value would change, to $3.60*10^{-3}\ s^{-1}$). A half-second delay might well have resulted from the finite time needed to bring the reactant from liquid-nitrogen temperature to $0°$ C.

At the other end of the scale, results at very high conversion suffer most from analytical inaccuracies. This is because the relative error of analytical methods is largest at low concentrations. (If the product concentration had been determined, the situation would be no better because the evaluation at high conversion then would depend on small differences between large concentration values.)

Even if the time scale were adjusted by half a second and suspect values omitted, a very slight tendency of the rate coefficient to increase with conversion remains. A reaction order slightly less than one would give a better fit. A deviation in the opposite direction could be explained with the assumption that the reaction is not entirely irreversible (see also Section 5.1.1), but the trend observed

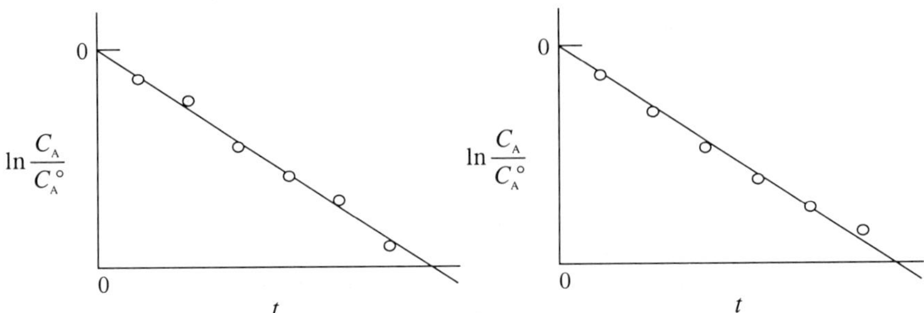

Figure 3.11. First-order concentrations plots of data with same quality of statistical correlation: left with random scatter, right with systematic bias.

the best straight line. In the plot at left the deviation is random and can confidently be attributed to experimental error. In the plot at right the deviation is systematic and leads one to suspect accurate data but a reaction order slightly higher than first. As Connors puts it, "the human eye, in combination with chemical knowledge, is a more subtle qualitative judge of data than is regression analysis" [53].

This section has concentrated on relatively simple cases. More detail can be found in texts on kinetics and reaction engineering (see general references). Establishment of empirical rate equations and coefficients for multistep reactions will be discussed in Chapter 7.

Summary

Bench-scale kinetic experiments can be conducted in batch, continuous stirred-tank, tubular plug-flow, or differential reactors. The last of these can be operated with once-through flow or recycle. The advantages and disadvantages of the various types are discussed in Section 3.1.

Fast reactions (in the millisecond to second range) require special reactors with efficient mixing chambers. Faster reactions (down to the microsecond range and below) call for special techniques; most of these are based on relaxation after an equilibrium state has been disturbed by an instantaneous pulse or step variation of conditions. With laser and photon-echo techniques the range has been extended down to femtoseconds.

Analytical support is essential. A wise program shifts its work load as much as possible from kinetic experiments, which required highly skilled personnel, to analytical routine work that can be performed by technicians. Gas chromatography and spectra are the most useful tools.

Reactions orders and rate coefficients can be established with methods that use either rate or concentration data. Batch, tubular plug-flow, and differential recycle reactors yield concentrations as directly measured quantities, and calculation of rates requires finite-difference approximations. To avoid these, concentration methods should be used. In contrast, continuous stirred-tank reactors allow rates to be calculated without approximation from material balances. Here, evaluation based on rates is equally suited.

In reactions with two or more reactants, reaction orders are best established by experiments with stoichiometric initial concentrations (to give the overall reaction order) and with all but one of the reactants in large excess (to give the order with respect to that minority reactant). For reversible reactions, measurement of the initial rate allows the forward reaction to be studied largely unencumbered by the reverse one.

Rate equations of multistep reactions often are not power laws. Reaction orders therefore may vary with concentrations, and attempts at accurate determination would be futile. Unless reaction orders are integers, or integer multiples of one half in special cases, only their ranges (such as between zero and plus one) are of interest.

Long-hand numerical work-up, error recognition, and reliability is illustrated with two examples: thermal decomposition of hexaphenylethane and an accelerated decomposition test of an herbicide.

References

General references

G1.   S. W. Churchill, *The interpretation and use of rate data: the rate concept*, revised printing, Hemisphere, New York, 1979, ISBN 0891162348.

G2.   K. A. Connors, *Chemical kinetics: the study of reaction rates in solution*, VCH Publishers, New York, 1990, ISBN 3527218223, Chapters 2 and 3.

G3.   J. H. Espenson, *Chemical kinetics and reaction mechanisms*, McGraw-Hill, New York, 2nd ed., 1995, ISBN 0070202605, Section 2-8.

G4.   H. S. Fogler, *Elements of chemical reaction engineering*, Prentice-Hall, Englewood Cliffs, 3nd ed., 1999, ISBN 0135317088, Chapter 5.

G5.   C. G. Hill, Jr., *An introduction to chemical engineering kinetics and reactor design*, Wiley, New York, 1977, ISBN 0471396095, Chapter 3.

G6.   K. J. Laidler, *Chemical kinetics*, Harper & Row, New York, 3rd ed., 1987, ISBN 0060438622, Chapter 2.

G7.   O. Levenspiel, *Chemical reaction engineering*, Wiley, New York, 3nd ed., 1999, ISBN 047125424X, Chapters 2 and 3.

Specific references

1.   S. Blaine and P. E. Savage, *I&EC Research*, **30** (1991) 2185.

2.   E. F. Caldin and F. W. Trowse, *Disc. Faraday Soc.*, **17** (1954) 133.

3.   H. Hartridge and F. J. W. Roughton, *Proc. Roy. Soc.*, **A 104** (1923) 376.

4.   B. Chance, *J. Franklin Inst.*, **229** (1940) 455, 613, and 737.

5.   M. Eigen, *Disc. Faraday Soc.*, **17** (1954) 194; **24** (1957) 25.

6.   J. N. Bradley, *Shock waves in chemistry and physics*, Methuen, London, 1962.

7.   A. G. Gaydon and I. R. Hurle, *The shock tube in high-temperature chemical physics*, Chapman & Hall, London, 1963.

8.   M. Eigen and L. de Mayer, *Z. Elektrochem.*, **59** (1955) 986.

9.   B. Marcandalli, G. Stange, and J. F. Holzworth, *J. Chem. Soc. Faraday Trans. I*, **84** (1988) 2807.

10. R. G. W. Norrish, *Proc. Chem. Soc.*, **1958**, 247.

11. E. D. Becker, *High resolution NMR: theory and chemical applications*, Academic Press, San Diego, 3rd ed., 2000, ISBN 0120846624.

12. T. J. Swift, *Investigation of rates and mechanisms of reactions*, Vol. 6 of *Techniques in chemistry*, G. G. Hammes, ed., Wiley, New York, 3rd ed., 1974, ISBN 0471830968, Part II, Chapter XII.

13. G. R. Fleming, *Chemical applications of ultrafast spectroscopy*, Oxford University Press, 1986, ISBN 0195036441.

14. V. Sundström, ed., *Femtochemistry and femtobiology: ultrafast reaction dynamics at atomic-scale resolution*, Nobel Symposium 101, Imperial College Press, London, 1997, ISBN 1860940390.

15. C. Rullière, ed., *Femtosecond laser pulses: principles and experiments*, Springer, Berlin, 1998, ISBN 3540636633.

16. C. F. Bernasconi, *Relaxation kinetics*, Academic Press, New York, 1976, ISBN 0120929503.

17. Connors (ref. G2), Section 4.2.

18. W. Jennings, E. Mittlefehldt, and P. P. Stremple, *Analytical gas chromatography*, Academic Press, San Diego, 2nd ed., 1997, ISBN 012384357X.

19. H. M. McNair and J. M. Miller, *Basic gas chromatography*, Wiley, New York, 1998, ISBN 0471172618.

20. C. F. Poole and S. K. Poole, *Gas chromatography*, in *Chromatography*, E. Heftmann, ed., Elsevier, Amsterdam, 5th ed., 1992, Vol. A, ISBN 0444882367, Chapter 9.

21. R. P. W. Scott, *Introduction to analytical gas chromatography*, Dekker, New York, 2nd ed., 1998, ISBN 0824700163.

22. D. A. Harris and C. L. Bashford, eds., *Spectrophotometry and spectrofluorimetry*, in *The practical approach series*, Oxford University Press, 1987, ISBN 0947946691.

23. K. Feinstein, *Guide to spectroscopic identification of organic compounds*, CRC Press, Boca Raton, 1995, ISBN 0849394481.

24. W. O. George and P. S. McIntyre, *Infrared spectroscopy*, Wiley, New York, 1987, ISBN 0471913839.

25. B. C. Smith, *Infrared spectral interpretation: a systematic approach*, CRC Press, Boca Raton, 1999, ISBN 0849324637.

26. D. H. Williams and I. Fleming, *spectroscopic methods in organic chemistry*, McGraw-Hill, New York, 5th ed., 1995, ISBN 0077091477.

27. R. V. Kastrup, J. S. Merola, and A. A. Oswald, *Adv. Chem. Ser.*, **196** (1981) 43.

28. G. Dotzlaw and M. D. Weiss, *Chem. Eng. Progr.*, **89**(9) (1993) 42.

29. D. Gibbons and D. A. Lambie, *Radiochemical methods in analysis*, in *Comprehensive analytical chemistry*, C. L. Wilson and D. W. Wilson, eds., Elsevier, Amsterdam, ISBN 0444417354, Vol. II C, 1971, Chapter II.

30. M. F. L'Annunziata, *Radionuclide tracers: their detection and measurement*, Academic Press, London, 1987, ISBN 0124362524.

31. R. M. Smith, *Understanding mass spectra: a basic approach*, Wiley, New York, 1999, ISBN 0471297046.

32. R. K. Harris and B. E. Mann, eds., *NMR and the periodic table*, Academic Press, London, 1978, ISBN 0123276500.

33. J. W. Faller, *Adv. Organometal. Chem.*, **16** (1977) 211.

34. P. S. Pregosin, ed., *Transition metal nuclear magnetic resonance*, Elsevier, Amsterdam, 1991, ISBN 044488176X.

35. U. Weber and H. Thiele, *NMR spectroscopy: modern spectral analysis*, Wiley-VCH, New York, 1998, ISBN 3527288287.

36. R. Jenkins, *X-ray fluorescence spectrometry*, Wiley, New York, 2nd ed. 1999, ISBN 0471299421.

37. D. de Soete, R. Gijbels, and J. Hoste, *Neutron activation analysis*, Wiley, New York, 1972, ISBN 0471203904 (out of print, reprint Books on Demand announced, ISBN 0608176125).

38. C. P. Poole, Jr., *Electron spin resonance: a comprehensive treatise on experimental techniques*, Wiley, New York, 1967, ISBN 047069386X; 2nd ed. reprint Dover, New York, 1997, ISBN 0486694445.

39. J. E. Wertz and J. R. Bolton, *Electron spin resonance: elementary theory and practical applications*, McGraw-Hill, New York, 1972, ISBN 0070694540.

40. C. A. Tolman and J. W. Faller, in *Homogeneous catalysis with metal phosphine complexes*, L. M. Pignolet, ed., Perseus, Cambridge, New York, 1983, ISBN 030641211X, Chapter 2.

41. Hill (ref. G5), p. 29.

42. Levenspiel (ref. G7), Chapter 5.

43. Hill (ref. G5), Section 8.1.

44. Levenspiel (ref. G7), pp. 71-72.

45. Hill (ref. G5), Section 3.1.2.

46. N. R. Draper and H. Smith, *Applied regression analysis*, Wiley, New York, 3rd ed., 1998, ISBN 0471170828.

47. R. D. Cook and S. Weisberg, *Applied regression including computing and graphics*, Wiley, New York, 1999, ISBN 047131711X.

48. S. Chatterjee, A. S. Hadi, and B. Price, *Regression analysis by example*, Wiley, New York, 3rd ed., 2000, ISBN 0471319465.

49. B. S. Gottfried, *Spreadsheet tools for engineers: EXCEL 2000 Version*, McGraw-Hill, New York, 2000, ISBN 0072321660, *Solver* program.

50.   SYSTAT (from Systat, Inc., Evanston, IL, USA).
51.   Hill (ref. G5), p. 66.
52.   W. E. Roseveare, *J. Am. Chem. Soc.*, **53** (1931) 1651.
53.   Connors (ref. G2), p. 52.

Chapter 4

Tools for Reduction of Complexity

Chemical reactions may involve large numbers of steps and participants and thus many simultaneous rate equations, all with their temperature-dependent coefficients. The full set of rate equations is easily compiled as shown in Section 2.4, and to obtain solutions by numerical computation poses no serious problems. With a large number of simultaneous equations, however, it may become too much of a task to verify the proposed network and obtain values for all its coefficients. Therefore, in practice, every available tool must be brought to bear in a concerted effort to reduce the bulk of reaction mathematics, and that without significant sacrifice in accuracy. The present chapter critically reviews the principal tools for such a purpose: the concepts of a rate-controlling step, quasi-equilibrium steps, and quasi-stationary states. Two additional tools that are useful for specific types of reactions, namely, the concept of relative abundance of catalyst-containing species in catalysis and of propagating centers in polymerization as well as the long-chain approximation in chain reactions are also introduced, but will be discussed in more detail later in the context of these reactions (see Sections 8.5.1, 9.3, and 10.4.1).

In essence, the application of any one of the three principal tools allows one or several rate equations to be replaced by algebraic equations which, in turn, can be used to eliminate concentrations of intermediates from the set. In the best of all worlds, mathematics can be reduced to a single rate equation and simple algebraic relationships between the concentrations of the reactants and products. More often, several simultaneous rate equations remain, but the reduced set is nevertheless much easier to handle for network elucidation and more convenient for modeling.

For ease of reference, the present chapter explains the nature of the tools and the errors introduced by their use. Actual applications are a major topic in later chapters.

4.1. Rate-controlling steps

A concept traditionally held in highest esteem, especially by chemists, is that of a rate-controlling step. The idea is that the overall rate is determined by the slowest step in the mechanism, the "bottleneck." For a linear pathway in which one step is much slower than all others, this may allow the set of simultaneous rate equations for all participants to be reduced to one single rate equation of formation of the product or products.

Let us be specific about what is meant when steps are loosely called "slow" or "fast." These terms refer not to rates (number of molecules reacting per unit time and unit volume), but to how soon a given reactant molecule can be expected to react (reaction probability, given by the rate coefficient multiplied with the concentration or concentrations of any co-reactant or co-reactants). For instance, consider a pathway

$$A \longrightarrow K \longrightarrow P \tag{4.1}$$

in which the second step has a much larger rate coefficient than does the first. Just about every molecule of K will then decay to P almost as soon as it is formed from A. The *rates* of the steps $A \longrightarrow K$ and $K \longrightarrow P$ then are practically equal. Yet, the second step is said to be much "faster" than the first because the molecules of K have a much higher *reaction probability* than those of A. Highway traffic provides a good analogy. A bottleneck caused, say, by road work is "slow," the unobstructed road is "fast"—this although only the cars that have passed the bottleneck are actually traveling on the road beyond, so that traffic flow in cars per unit time is no heavier anyplace there than at the bottleneck.

4.1.1. Pathways of irreversible steps

The simplest example of a chemical reaction with rate-controlling step is the pathway 4.1 with very different rate coefficients of the two steps. Solving the rate equations for A, K, and P in a constant-volume batch reaction with only A present at start and without assumptions about the rate coefficients, one obtains for the rate of product formation:

$$r_P \;=\; \frac{k_{AK} k_{KP} C_A^o}{k_{KP} - k_{AK}} \left(\exp(-k_{AK}t) - \exp(-k_{KP}t) \right) \tag{4.2}$$

and for a continuous stirred-tank reactor at steady state with only A in the feed:

$$r_P \;=\; \frac{k_{AK} k_{KP} C_A^o \tau}{(1 + k_{AK}\tau)(1 + k_{KP}\tau)} \tag{4.3}$$

(see Section 5.4 for more detail). In the limiting cases of extremely different rate coefficients and at not too short times or reactor space times, this reduces to:

Case I $(k_{AK} \ll k_{KP})$	*Case II* $(k_{KP} \ll k_{AK})$
batch: $\quad r_P = k_{AK} C_A^o \exp(-k_{AK}t)$	batch: $\quad r_P = k_{KP} C_A^o \exp(-k_{KP}t)$
CSTR: $\quad r_P = k_{AK} C_A^o / (1 + k_{AK}\tau)$	CSTR: $\quad r_P = k_{KP} C_A^o / (1 + k_{KP}\tau)$

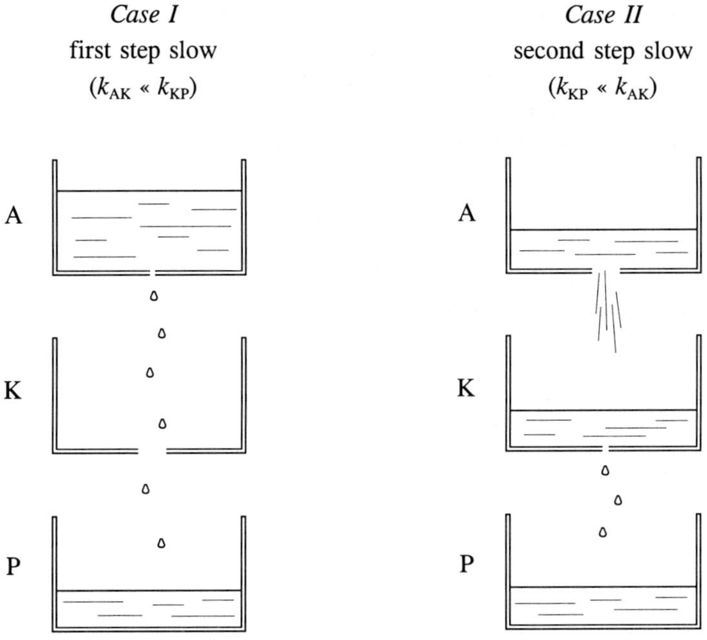

Liquid from container A dribbles right through K into P.
Rate of flow into P is controlled by rate of flow out of A (slow step).

Container A empties quickly into K, from where liquid dribbles slowly into P.
Rate of flow into P is controlled by rate of flow out of K (slow step).

Figure 4.1. Hydrodynamic analogue of batch reaction A $\longrightarrow$ K $\longrightarrow$ P with slow first or second step.

A comparison of the rate equations in the two limiting cases with eqns 3.14 and 3.22 shows them to be the same as for a single step A $\longrightarrow$ P with rate coefficient k_{AK} in Case I and k_{KP} in Case II, that is, with the rate coefficient of the slow step in both cases. A hydrodynamic analogue for the batch reaction is shown in Figure 4.1.

The error incurred with the approximation of a rate-controlling step can be estimated by comparing the approximate rates above with the exact ones given by eqns 4.2 and 4.3. For the two-step reaction 4.1, the relative error—difference between approximate rate and exact rate, divided by the latter—turns out to be

$$
\begin{aligned}
\Delta_{rel} &\cong k_{slow}/k_{fast} & \text{for batch reactor} \\
&\cong 1/(k_{fast}\tau) & \text{for CSTR at steady state}
\end{aligned}
\tag{4.4}
$$

The approximation for the batch reactor at low conversion involves an additional error caused by the omission of the exponential involving the coefficient of the fast step. Provided the rate coefficients differ by at least an order of magnitude, this error is below 1% for reaction times longer than $5/k_{fast}$:

$$\Delta_{rel} < 0.01 \qquad \text{if} \qquad t > 5/k_{fast} \qquad (4.5)$$

The approximation overestimates the rate if based on true rate coefficients (but not, of course, if empirical coefficients in the approximate rate equation are fitted to experimental results).

The argument can be extended to pathways with any number of irreversible steps. In general terms:

If one step in a sequence of irreversible steps is much slower than all others, its rate coefficient alone controls the rate of product formation.

However, this rule is valid only if the steps are sequential and irreversible. Moreover:

- *If the steps have widely different activation energies, rate control may shift from one step to another with change in temperature* (see Section 12.1).

- *If the steps are of different reaction orders, rate control may shift from a high-order step at low conversion to a low-order step at high conversion.* This is because high-order steps slow down more strongly with decreasing reactant concentrations.

In a multistep irreversible reaction, each disregarded fast step introduces an error as discussed above for the two-step pathway: The errors are cumulative.

4.1.2.   *Pathways with reversible steps*

The concept of rate-controlling steps can be formulated more generally for pathways that include reversible steps. The idea here is that the slow reaction of the rate-controlling step gives all others time to approach equilibrium very closely:

If the forward and reverse rate coefficients of a step in the pathway of a reversible reaction are much smaller than all others, the other steps are practically in equilibrium.

In loose parlance, steps that are much faster than others are often said to be in equilibrium. However, as long as the overall reaction proceeds, net conversion in the forward direction must occur through each of its steps: In each, the forward reaction must outweigh the reverse reaction by what amounts to the net conversion of the overall reaction. This precludes ideal equilibrium, in which forward and

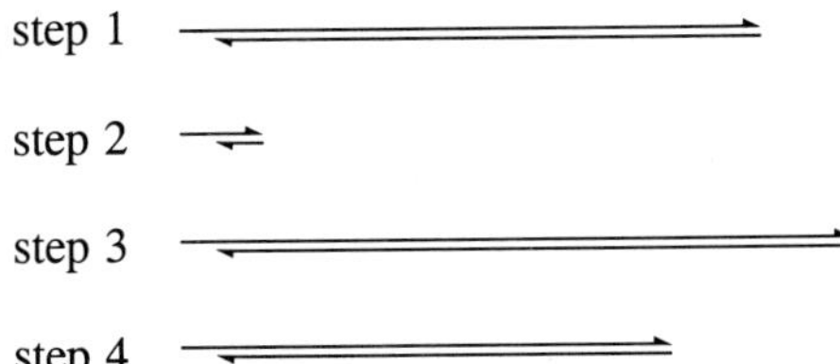

Figure 4.2. Forward and reverse rates of steps in a four-step reaction with rate control by second step (schematic). Forward direction left to right; lengths of arrows indicate magnitudes of rates.

reverse rates are exactly equal. However, if a step is very fast in both directions, its forward and reverse rates differ very little on a *relative* basis, and so can be equated in good approximation. This situation is illustrated in Figure 4.2 for a four-step reaction with rate control by a slow, second step. For precision's sake, the fast steps are said to be in *quasi-equilibrium* (see also next section).

Contrary to what is true for pathways with no reversible steps, fast reversible steps preceding the rate-controlling step *do* affect the rate of product formation. The rate depends on the equilibrium constants of such steps and thus on the ratios of their forward and reverse rate coefficients. Specifically, equilibria favoring the reverse reaction reduce the rate.

As for pathways of irreversible steps, the more general rule allowing for reversibility remains restricted to sequential steps, and rate control may shift to a different step with temperature or concentration. Each quasi-equilibrium step introduces an error into the approximation as will be discussed in the next section.

In its more general form, the concept of the rate-controlling step is often used in catalysis (see Section 8.5.2).

Example 4.1. Nitration of aromatics. The Gillespie-Ingold mechanism [1-3] of nitration of aromatic compounds according to

$$ArH + HNO_3 \longrightarrow ArNO_2 + H_2O$$

is

$$HNO_3 \rightleftarrows H_2NO_3^+ \rightleftarrows NO_2^+ \longrightarrow ArNO_2H^+ \rightleftarrows ArNO_2 \qquad (4.6)$$

where ArH is the aromatic, and HB and B^- are an added strong acid and its anion. (HB may be nitric acid itself, but usually sulfuric acid is added; the step in which $ArNO_2H^+$ is formed may actually involve free radicals as intermediates [4].)

For aromatics of *low reactivity*, the reaction of the nitronium ion with the aromatic is the slow, rate-controlling step. The rate then is

$$r_{ArNO_2} \cong k_{23} C_{NO_2^+} C_{ArH} \qquad (4.7)$$

where k_{23} is the rate coefficient of the slow, third step. (Note that the subsequent steps are irreversible, so that their quasi-equilibrium is shifted entirely to the end product, $ArNO_2$, as though the slow step were producing it directly.) The nitronium ion, NO_2^+, is formed in the first two steps with combined stoichiometry

$$HNO_3 + HB \;\longleftrightarrow\; NO_2^+ + B^- + H_2O$$

The quasi-equilibrium condition (mass-action law) for the reaction with this stoichiometry is

$$\frac{C_{NO_2^+} C_{B^-} C_{H_2O}}{C_{HNO_3} C_{HB}} = K_{02} = \text{const.}$$

Solved for the nitronium ion concentration:

$$C_{NO_2^+} \cong K_{02} \frac{C_{HNO_3} C_{HB}}{C_{H_2O} C_{B^-}} \qquad (4.8)$$

The dissociation equilibrium of the acid HB with dissociation constant K_{HB} is:

$$C_H \cdot C_{B^-} / C_{HB} = K_{HB} \qquad (4.9)$$

and can be used to express the ratio C_{HB}/C_{B-} in terms of the hydrogen ion concentration. Combining eqns 4.7 to 4.9 and taking the concentration of water (the solvent) as constant, one finds:

$$r_{ArNO_2} \cong k_a C_{ArH} C_{HNO_3} C_{H^+} \qquad (k_a \equiv k_{23} K_{02}/K_{HB} C_{H_2O}) \qquad (4.10)$$

Accordingly, the reaction is first order with respect to each the aromatic, undissociated nitric acid, and hydrogen ion.

 On the other hand, aromatics of *high reactivity* react with the nitronium ion much more rapidly than the latter is formed. Here, the formation of that ion becomes the slow, rate-controlling step:

$$r_{ArNO_2} \cong k_{12} C_{H_2NO_3^+} \qquad (4.11)$$

where k_{12} is the rate coefficient of the slow, second step. The protonated nitric acid is formed in the first, fast step, with stoichiometry

$$HNO_3 + HB \;\longleftrightarrow\; H_2NO_3^+ + B^-$$

and quasi-equilibrium condition

$$\frac{C_{H_2NO_3^+} C_{B^-}}{C_{HNO_3} C_{HB}} \cong K_{01} = \text{const.} \qquad (4.12)$$

Solved for the concentration of the protonated nitric acid:

$$C_{H_2NO_3^+} \cong K_{01} C_{HNO_3} C_{HB}/C_{B^-} \tag{4.13}$$

The resulting rate equation in this case is

$$r_{ArNO_2} \cong k_b C_{HNO_3} C_{H^+} \qquad (k_b \equiv k_{12} K_{01}/K_{HB}) \tag{4.14}$$

This rate equation differs from eqn 4.10 in that the reaction is now of zero order with respect to the aromatic. The other orders are the same as before.

For aromatics of intermediate reactivity the rates of the second and third steps may be comparable, so that no single step is rate-controlling. A better tool then is needed to obtain a closed-form rate equation. This case will be examined in Example 4.4 in Section 4.3 and Example 6.1 in Section 6.3.

The postulate of quasi-equilibrium of all steps except a single one that controls the rate is very powerful. It reduces the mathematical complexity of kinetics even of large networks to quite simple rate equations and has become a favorite tool, employed today in a great majority of publications on kinetics of multistep homogeneous reactions, sometimes uncritically. In many cases, a sharp distinction between fast and slow steps cannot be justified. A more general approach that avoids the postulate of a single rate-controlling step and contains the results obtained with it as special cases will be described in Sections 4.3 and 6.3 and widely used in later Chapters.

4.2. Quasi-equilibrium steps

The concept of the rate-controlling step singles out one step as much slower than all others. The concept of a quasi-equilibrium step does the opposite: It singles out one or several steps as much faster than all others. If this is so, the slowness of the other steps gives the fast steps time enough to come essentially to equilibrium:

> If the forward and reverse rate coefficients of one or more steps are much larger than all others, reactants and products of the fast steps are practically in equilibrium.

The use of the equilibrium condition for a fast step can greatly simplify mathematics, as will be seen in various examples in later chapters. Prominent among the fast steps to which the approximation can be applied are dissociation reactions in the gas phase and ionic reactions such as electrolytic dissociation, neutralization, and complex formation, as well as loss, addition, and exchange of ligands in the liquid phase. With few exceptions, such steps are fast compared with other typical chemical transformations.

As discussed in the preceding section and illustrated in Figure 4.2, even a very fast step cannot attain ideal equilibrium as long as the overall reaction proceeds. The term quasi-equilibrium indicates that the forward and reverse rates of the step differ very little on a relative basis, so that the use of the equilibrium condition equating these rates is justified as an approximation.

The quasi-equilibrium approximation can be applied to more than one step. As the preceding section has shown, the postulate of a rate-controlling step implies that all others are in quasi-equilibrium.

The error incurred with the approximation of quasi-equilibrium of a single step is readily estimated. For the first step in a pathway

$$A \longleftrightarrow K \longrightarrow P \tag{4.15}$$

with or without co-reactants or co-products (not shown), the approximation amounts to postulating that the step $K \longrightarrow P$ has only a negligible effect on the concentration of K, that is, $r_{K \to P} \ll r_{K \to A}$. The relative error is

$$\Delta_{\mathrm{rel}} \cong r_{K \to P}/r_{K \to A} \tag{4.16}$$

[stated in terms of rates rather than rate coefficients because the latter do not reflect the effects of possible co-reactants; alternatively, $\Delta_{\mathrm{rel}} \cong \lambda_{KP}/\lambda_{KA}$, with λ coefficients as defined in Section 6.2]. In more complex situations, only the largest of such errors need be considered. If the approximation is applied to several steps, the errors are additive. As in the case of a rate-controlling step, the approximation overestimates the rate if based on the true rate coefficients of the steps.

Example 4.2. The hydrogen-iodide reaction. The reversible formation of hydrogen iodide from hydrogen and iodine in the gas phase, with stoichiometry

$$H_2 + I_2 \longleftrightarrow 2HI$$

nicely follows an empirical rate equation

$$r_{HI} = k_a p_{H_2} p_{I_2} - k_b p_{HI}^2 \tag{4.17}$$

as one would expect for a single-step, reversible reaction that is bimolecular in both directions. Indeed, the reaction had become the classical textbook example of kinetics, even featured in an animated educational movie produced by a Nobel laureate for the American Chemical Society. However, the true mechanism is [5,6]

$$I_2 \longleftrightarrow 2I \overset{H_2}{\longleftrightarrow} 2HI \tag{4.18}$$

The dissociation of I_2 in the first step is very fast, the trimolecular second step is slow. The quasi-equilibrium condition for the first step is

$$p_I^2/p_{I_2} \;\cong\; K_{01}$$

where K_{01} is the dissociation constant of I_2. This expression can be used to eliminate p_I from the rate equation of product formation in the second step:

$$r_{HI} = 2k_{12}p_I^2 p_{H_2} - 2k_{21}p_{HI}^2 \;\cong\; 2k_{12}K_{01}p_{H_2}p_{I_2} - 2k_{21}p_{HI}^2 \qquad (4.19)$$

where k_{12} and k_{21} are the forward and reverse rate coefficients, respectively, of the second step. The rate equation 4.19 is seen to be of the same algebraic form as eqn 4.17 for the single-step mechanism, with $2k_{12}K_{01}$ corresponding to k_a, and $2k_{21}$ to k_b.

Bodenstein, the first to study the reaction, had pointed this out as early as 1898 and had suggested both mechanisms as possibilities [7], but this was long ignored. More than half a century later, the single-step mechanism was finally questioned because the analogous HCl and HBr reactions proceed via quasi-equilibrium dissociation of Cl_2 or Br_2 as the first step. For the HI reaction, such a step had to be even more favored because the I_2 molecule is more strongly dissociated. However, on the basis of the observed concentration dependence of the rate alone, no discrimination between the single-step and two-step mechanisms was possible. Eventually, compelling evidence for the two-step mechanism was provided by Sullivan [5] with experiments in which iodine atoms were produced in the reactor by flash photolysis at temperatures at which thermal dissociation is negligible; he showed that Arrhenius extrapolation of the rate coefficient of the photolytically induced reaction accounted fully for the thermal rates at higher temperatures. Even so, Sullivan's conclusion was questioned on the grounds that the two-step mechanism requires too high an activation energy and therefore should not contribute [8]. However, an examination in terms of molecular orbital theory shows that the single-step mechanism can be ruled out because it violates the Woodward-Hoffmann exclusion rules [9] (see Example 7.8 in Section 7.4). Once again, nature has served notice that to draw conclusions about kinetics from thermodynamic arguments is to skate on thin ice!

Beyond illustrating the application of the quasi-equilibrium approximation, this example also strikingly demonstrates that a mechanism or pathway cannot be deduced with certainty from the rate equation. An empirical rate equation is little more than a symptom. It can disprove a mechanism as incompatible with its algebraic form, but it does not prove a compatible mechanism to be correct. There will always be a number of other potential mechanisms that give the same rate equation in approximations so good that a distinction through experiments that measure the concentration dependence of the rate in a conventional reactor is impossible. If there is any doubt, the experienced kineticist seeks conclusive discrimination by other means (as did Sullivan, see above). Failing this, he applies Occam's razor, settling for the simplest of such rival mechanisms. However, as the example of the hydrogen-iodide reaction has shown, this might be the wrong choice.

Unlike HI, its sisters HCl and HBr are formed by chain reactions (see Chapter 9). The reason for this difference in mechanisms is that the trimolecular step $2X + H_2 \longrightarrow 2HX$ (X = halogen) is too slow for Cl and Br because of the weak dissociation of chlorine and bromine. That step, being second order in halogen atoms, needs a sizeable partial pressure of these to outrun the competing chain mechanism, and only iodine dissociates strongly enough to provide it. Even so, at very high temperatures a chain mechanism takes over [10]. No doubt both the chain and two-step mechanisms play a role in all three hydrogen-halide reactions, the difference lies in which of them dominates. This demonstrates another facet of kinetics: Very often, a multistep reaction proceeds via several pathways in parallel but, fortunately, most of the time one of them is so much faster than the others that it completely dominates the kinetic behavior. Where there is an eight-lane freeway, parallel country roads do not contribute much to traffic flow!

4.3. Quasi-stationary states: the Bodenstein approximation*

The concept of quasi-stationary states is an extremely powerful and widely applicable tool for reduction of mathematical complexity in kinetics. It will be used extensively in the chapters to follow.

If a reaction intermediate X is so unstable that it decomposes practically as soon as it is formed, its concentration necessarily remains quite small. The same must be true for its net rate of formation r_X : If that rate were large and positive, the concentration would rise to large values, which it is known not to do; if that rate were large and negative, the concentration would have to become negative, which it cannot. Accordingly, provided the intermediate is and remains at trace level, its net formation rate r_X is small compared separately with its rates of formation and decay:

> The net rate of formation of an intermediate that is and remains at trace level is negligible compared with its contributing formation and decay rates.

In terms of mathematics, for a pathway

$$A \longleftrightarrow X \longleftrightarrow P \tag{4.20}$$

the rates are

$$\text{net rate:} \quad r_X = k_{AX}C_A + k_{PX}C_P - C_X(k_{XA} + k_{XP})$$
$$\underbrace{\qquad\qquad}_{\text{formation rate}} \quad \underbrace{\qquad\qquad}_{\text{decay rate}}$$

If the intermediate is and remains at trace level:

---

* Suggested independently by Chapman [11] and Bodenstein [12] in 1913, implemented and developed by the latter the same year. Thorough discussions have been presented by Bowen *et al.* [13], Aris [14], and Turányi *et al.* [15], the last of these with copious references to earlier work.

$$|r_X| \ll k_{AX}C_A + k_{PX}C_P, \quad C_X(k_{XA} + k_{XP})$$

net   formation   decay

or

$$r_X = k_{AX}C_A + k_{PX}C_P - C_X(k_{XA} + k_{XP}) \cong 0 \qquad (4.21)$$

as though the net rate r_X of the intermediate were zero. This is called the Bodenstein approximation. In reality, of course, r_X usually is *not* zero, merely negligibly small compared with the formation and decay rates.

Equation 4.21 can be solved for the concentration of the intermediate:

$$C_X \cong \frac{k_{AX}C_A + k_{PX}C_P}{k_{XA} + k_{XP}}$$

The approximation of negligible net formation rate of a trace intermediate can be used to replace a rate equation by an algebraic equation, with which the concentration of the intermediate can be eliminated.

The condition expressed by the Bodenstein approximation $r_X \cong 0$ is often misleadingly called a steady state. It is *not*. It is not a time-independent state, only a state in which a specific variation with time (or reactor space time) is small compared with the others. In fact, some older textbooks applied what they called the steady-state approximation to batch reactions in order to derive the time dependence of the concentrations, unwittingly leading the incorrect presumption of a steady state *ad absurdum*. And a continuous stirred-tank or tubular reactor may, and usually does, come to a true steady state, even if the Bodenstein approximation is and remains inapplicable. [The approximation compares process rates r_i; it is irrelevant for its validity whether or not the reactor comes to a steady state, that is, whether the rates of change, dC_i/dt, become zero.]

The Bodenstein approximation can be applied repeatedly to different trace-level intermediates in succession. Each application removes one rate equation and the concentration of one trace-level intermediate. This makes the Bodenstein approximation especially useful because trace-level intermediates are difficult to detect and their concentrations can rarely be measured accurately.

Example 4.3. Radioactive decay. A long-lived radioisotope A decays; the k intermediates X_1, X_2, etc., are very short-lived; the final product P is stable:

$$A \longrightarrow X_1 \longrightarrow X_2 \longrightarrow \dots \longrightarrow X_k \longrightarrow P$$

Setting the r_i of the intermediates approximately equal to zero:

$$r_i = k_{i-1,i}C_{i-1} - k_{i,i+1}C_i \cong 0 \qquad\qquad (i = 1,\ldots k)$$

(X in indices suppressed here and below) one finds

$$C_A:C_1:C_2:\ldots:C_k \cong (1/k_{A1}):(1/k_{12}):(1/k_{23}):\ldots:(1/k_{kP})$$

that is, the well-known rule that the concentrations of the unstable intermediates are inversely proportional to their decay constants, and so proportional to their half-lives.

The Bodenstein approximation is accurate within reason, provided the intermediate is and remains at trace level, and with the exception of a very short initial time period in which the quasi-stationary state is established [13-15]. It is left to the practitioner to decree how low a concentration must be to qualify as "trace;" the more generous he is, the less accurate will be his results. For the pathway 4.20 with one trace-level intermediate, the error introduced can be estimated in the same way as for rate control by a slow step (see Section 4.1.1):

$$\Delta_{rel} \cong C_X/C_A \qquad\qquad (4.22)$$

(concentration ratio at quasi-stationary state). In repeated applications of the approximation to successive intermediates, the errors are cumulative. In addition, in a batch reaction, time is needed at start to "fill the pipeline" in the approach to quasi-stationary conditions; in a pathway with k trace-level intermediates, this requires a time of about

$$t_{qss} \approx \sum_{i=1}^{k} C_i/(-r_A) \qquad\qquad (4.23)$$

where the C_i are the concentrations of the intermediates under quasi-stationary conditions.   The rate of product formation is overestimated if the Bodenstein approximation is based on the true rate coefficients of the steps.

The error estimates do not apply to chain carriers in chain reactions since these are regenerated by the reaction (see Chapter 9).

The following points should be observed when the Bodenstein approximation is applied:

- *The approximation $r_i \cong 0$ should always be applied to eliminate the concentration C_i of the intermediate X_i, not that of any other intermediate X_{i-1} or X_{i+1}.   [If this rule is not observed, two small quantities may inadvertently be equated to one another.]*

- *If the approximation is applied several times in succession, the easiest way is to begin at the end of the pathway and work one's way toward its start one step at a time.*

- *The approximation is not applicable to short initial time spans and may give inaccurate results in large networks in which quasi-stationary states may be attained at different times in different portions.*

To illustrate the first two points above, the following example shows how the Bodenstein approximation can be applied repeatedly to a reaction that includes reversible steps. Note, however, that the same result can be obtained much more easily with a general formula for "simple" pathways, to be developed in Section 6.3.

Example 4.4. Nitration of aromatics of intermediate reactivity. In Example 4.1 the concept of a rate-controlling step was used to obtain simple rate equations for nitration of aromatics of either low or high reactivity. For aromatics of intermediate reactivity, no single step is rate-controlling. However, if the concentrations of $H_2NO_3^+$, NO_2^+, and $ArNO_2^+$ in the pathway 4.6 remain at trace level—this is a judgment call—the Bodenstein approximation can be applied repeatedly to obtain an explicit, closed-form rate equation.

The Gillespie-Ingold pathway 4.6 can be written

$$
\begin{array}{ccccccccc}
& \text{HB} & & & \text{ArH} & & \text{B}^- & & \\
\text{HNO}_3 & \rightleftharpoons & X_1 & \rightleftharpoons & X_2 & \longrightarrow & X_3 & \longrightarrow & \text{ArNO}_2 \\
& \text{B}^- & & \text{H}_2\text{O} & & & \text{HB} & &
\end{array}
\qquad (4.24)
$$

($X_1 = H_2NO_3^+$, $X_2 = NO_2^+$, $X_3 = ArNO_2H^+$, and indices 0 and 4 will refer to HNO_3 and $ArNO_2$, respectively). The rate of product formation is

$$ r_{ArNO_2} = k_{34} C_3 C_{B^-} \qquad (4.25) $$

(X suppressed in subscripts). The Bodenstein approximation for X_3 is

$$ r_3 = k_{23} C_2 C_{ArH} - k_{34} C_3 C_{B^-} \cong 0 $$

Solved for C_3:

$$ C_3 \cong k_{23} C_2 C_{ArH} / k_{34} C_{B^-} $$

Substitution in eqn 4.25 gives

$$ r_{ArNO_2} \cong k_{23} C_2 C_{ArH} \qquad (4.26) $$

The Bodenstein approximation for X_2 gives

$$ r_2 = k_{12} C_1 - C_2 (k_{23} C_{ArH} + k_{21} C_{H_2O}) \cong 0 $$

Solved for C_2:

$$ C_2 \cong k_{12} C_1 / (k_{23} C_{ArH} + k_{21} C_{H_2O}) \qquad (4.27) $$

Substitution in eqn 4.26 gives

$$ r_{ArNO_2} \cong k_{12} k_{23} C_1 C_{ArH} / (k_{23} C_{ArH} + k_{21} C_{H_2O}) \qquad (4.28) $$

The Bodenstein approximation for X_1 gives

$$r_1 \;=\; k_{01} C_{HNO_3} C_{HB} + k_{21} C_2 C_{H_2O} - C_1(k_{12} + k_{10} C_{B^-}) \;\cong\; 0$$

Solved for C_1:

$$C_1 \;\cong\; (k_{01} C_{HNO_3} C_{HB} + k_{21} C_2 C_{H_2O})/(k_{12} + k_{10} C_{B^-}) \tag{4.29}$$

This equation re-introduces C_2, which must be eliminated with eqn 4.27. Solving again for C_1 after that substitution one finds after considerable rearrangement

$$C_1 \;\cong\; \frac{k_{01} C_{HNO_3} C_{HB} (k_{23} C_{ArH} + k_{21} C_{H_2O})}{k_{12} k_{23} C_{ArH} + k_{10} k_{23} C_{ArH} C_{B^-} + k_{10} k_{21} C_{B^-} C_{H_2O}}$$

Using this to replace C_1 in eqn 4.28 one obtains

$$r_{ArNO_2} \;\cong\; \frac{k_{01} k_{12} k_{23} C_{HNO_3} C_{HB} C_{ArH}}{k_{12} k_{23} C_{ArH} + k_{10} k_{23} C_{B^-} C_{ArH} + k_{10} k_{21} C_{B^-} C_{H_2O}} \tag{4.30}$$

With the equilibrium condition 4.9 for dissociation of the acid HB, the assumption that the concentration of water can be taken as constant, and identification of the participants, eqn 4.30 in a form with fewest empirical coefficients becomes

$$r_{ArNO_2} \;\cong\; \frac{k_a C_{ArH} C_{HNO_3} C_{H^+}}{1 + k_b C_{ArH}(1 + k_c/C_{B^-})} \tag{4.31}$$

where

$$k_a \;\equiv\; \frac{k_{01} k_{12} k_{23}}{k_{10} k_{21} K_{HB} C_{H_2O}} = \frac{K_{01} K_{12} k_{23}}{K_{HB} C_{H_2O}},$$

$$k_b \;\equiv\; \frac{k_{23}}{k_{21} C_{H_2O}}, \qquad k_c \;\equiv\; \frac{k_{12}}{k_{10}}$$

(Ratios of forward and reverse rate coefficients of a step can be replaced by the respective equilibrium constants.)

According to eqn 4.31, the rate is first order each in undissociated nitric acid and hydrogen ion and of an order between zero and one in the aromatic. Moreover, at given pH the rate is of an order between zero and one in the concentration of the anion B^- of the added strong acid.

The rate equations 4.10 and 4.14, derived previously with the assumption of a single rate-controlling step, turn out to be special cases of eqn 4.31, obtained with the Bodenstein approximation: If k_{23} is very small (third step slow), eqn 4.31 reduces to eqn 4.10; if k_{12} and k_{21} are very small (second step slow), it reduces to eqn 4.14. This comparison demonstrates that the Bodenstein approximation leads to results of greater generality.

The derivation of eqn 4.31 shows that the application of the approximation to an intermediate consumed by a reversible step—the second step in the example above—re-introduces the previously eliminated concentration of another intermediate, so that an implicit equation results (in eqn 4.29, replacement of C_2 be eqn 4.27 results in an equation that is implicit in C_1). With each repetition of such an application the amount of algebra required increases steeply and soon gets out of hand. No doubt this is a reason why it has become established practice in kinetics of multistep reactions with trace-level intermediates to postulate a single rate-controlling step instead of using the Bodenstein approximation. However, in many even quite complex cases, a general formula for "simple" pathways, to be given in Section 6.3, allows the rate equation for quasi-stationary conditions to be written down immediately, without need for a lengthy and tedious derivation. This will be demonstrated in Example 6.1 (Section 6.3) by application of the general formula to the nitration pathway 4.24.

Extensive examples of applications of the Bodenstein approximation, with or without the general formula for simple pathways, will be found in later chapters.

4.4. Relative abundance of species in catalysis and polymerization and long-chain approximation

Relative abundance in catalysis and polymerization. In catalysis by species at trace level, the concentration of the free catalyst may not be known. What is known is the total amount of catalyst material, but not how much of it is present as free catalyst rather than bound in the form of reaction intermediates. Therefore, rate equations involving the concentration of free catalyst are of little use. Rather, equations in terms of total amount of catalyst material are needed. If the reaction involves several intermediates, the rate equation in terms of total amount of catalyst becomes quite lengthy, even after the concentrations of the intermediates have been eliminated with the Bodenstein approximation. A major simplification can be achieved with an additional approximation postulating that all but a negligible fraction of the catalyst material is present in the form of one single species, called the most abundant catalyst-containing species (*macs*). This species may be the free catalyst itself or one of the catalyst-containing intermediates. The concept of a *macs* is related to that of the most abundant surface intermediate (introduced and called *masi* by Boudart [16]) in heterogeneous catalysis.

One or several of the reaction intermediates may be known to contain only a negligible fraction of the total catalyst material. A mathematical simplification can then be achieved by omission of the terms stemming from such low-abundance catalyst-containing species (*lacs*).

The approximations obtained with the concept of relative abundance of catalyst-containing species are specific to trace-level catalysis and will be discussed in detail in that context (see Section 8.5.1).

The concept of relative abundance can also be applied in an analogous fashion to propagating centers in ionic polymerization (see Section 10.4.1).

Long-chain approximation. In most chain reactions, a short sequence of steps, once initiated, repeats itself many times until it is terminated. The initiation and termination reactions then contribute very much less to product formation than do the self-repeating steps. The long-chain approximation neglects these minor contributions. It will be taken up in the context of chain reactions (see Section 9.3) and also used in chain-growth polymerization (Sections 10.3 and 10.4).

Summary

The three principal tools for reduction of mathematical complexity of rate equations are the concepts of a rate-controlling step, of quasi-equilibrium steps, and of quasi-stationary states of trace-level intermediates.

If a step in a pathway is much slower than all others, it constitutes a bottleneck that controls the overall rate. Fast, reversible steps in the pathway have enough time to come to quasi-equilibrium. Such equilibria also affect the overall rate. In either case, all steps must be sequential for the rules to apply.

If one or several steps in a pathway or network are much faster than all others, they attain quasi-equilibrium, and their (algebraic) equilibrium conditions can be used to replace rate equations.

Any highly reactive intermediate that is and remains at trace level attains a quasi-stationary state in which its net chemical rate is negligibly small compared separately with its formation and decay rates. This is the basis of the Bodenstein approximation, which allows the rate equation of the intermediate to be replaced by an algebraic equation for the concentration of the intermediate, an equation which can then be used to eliminate that concentration from the set of equations. The approximation can be applied in succession for each trace-level intermediate. It is the most powerful tool for reduction of complexity. It is the basis of general formulas to be introduced in Chapter 6 and widely used in subsequent chapters.

Approximations based on the concept of relative abundance of catalyst-containing species in trace-level catalysis and of propagation centers in ionic polymerization will be discussed in Chapters 8 and 10, respectively; the long-chain approximation in chain reactions will be an important topic in Chapter 9.

Examples include nitration of aromatics, the hydrogen-iodide reaction, and radio-active decay.

References

1.   R. J. Gillespie, E. D. Hughes, C. K. Ingold, D. J. Millen, and R. I. Reed, *Nature*, **163** (1949) 599.

2.   C. K. Ingold, *Structure and mechanism in organic chemistry*, Cornell University Press, Ithaca, 2nd ed., 1969, ISBN 0801404991, Chapter VI.

3.   A. A. Frost and R. G. Pearson, *Kinetics and mechanism: a study of homogeneous chemical reactions*, Wiley, New York, 2nd. ed., 1961, Section 12.E.

4.   C. L. Perrin, *J. Am. Chem. Soc.*, **99** (1977) 5516.

5.   J. H. Sullivan, *J. Chem. Phys.*, **46** (1967) 73.

6.   J. W. Moore and R. G. Pearson, *Kinetics and mechanism, a study of homogeneous chemical reactions*, Wiley, New York, 3rd ed., 1981, ISBN 0471035580, p. 210.

7.   M. Bodenstein, *Z. physik. Chem.*, **13** (1894) 56; **22** (1897) 1; **29** (1898) 295.

8.   K. J. Laidler, *Chemical kinetics*, McGraw-Hill, New York, 2nd ed., 1965, p. 121.

9.   R. Hoffmann, *J. Chem. Phys.*, **49** (1968) 3739.

10. J. H. Sullivan, *J. Chem. Phys.*, **30** (1959) 1292; **36** (1962) 1925; **39** (1963) 3001.

11. D. L. Chapman and L. K. Underhill, *J. Chem. Soc.*, **103** (1913) 496.

12. M. Bodenstein, *Z. physik. Chem.*, **85** (1913) 329.

13. J. R. Bowen, A. Acrivos, and A. K. Oppenheim, *Chem. Eng. Sci.*, **18** (1963) 177.

14. R. Aris, *Am. Sci.*, **58** (1970) 419.

15. T. Turányi, A. S. Tomlin, and M. J. Pilling, *J. Phys. Chem.*, **97** (1993) 163.

16. M. Boudart, *AIChE J.*, **18** (1972) 465.

Chapter 5

Elementary Combinations of Reaction Steps

In his search for the pathway or network of a reaction, the chemist or engineer finds himself in the same position as a physician who must diagnose a patient's illness: He must evaluate symptoms. To do so competently, he must learn to recognize the patterns of kinetic behavior produced by the most common combinations of reaction steps, to be reviewed in the present chapter. Any one of these step combinations might conceivably constitute a complete network, although many such networks would be no more than improbable and uninteresting interconversions of isomers. However, even if embedded in larger networks, the simple combinations to be examined here are apt to produce their characteristic kinetic effects—or can be made to do so by appropriate choice of experimental conditions. In fact, most of the examples to be used for illustration will be of this type, that is, their real networks are combinations of multistep pathways rather than single steps.

This chapter is intended primarily to provide information useful for network elucidation rather than modeling or design. For this reason, emphasis is on behavior in constant-volume batch reactors. Although these are the least common type of reactors in industrial plants, they are advantageous for the purpose at hand. In particular, they make it relatively easy to obtain results at low conversions that may be crucial for a distinction between rival mechanisms (see Section 3.1.1; note that the mathematics of plug-flow tubular and differential recycle reactors is analogous). Also, the focus is on simplicity, on symptoms that will stand out clearly when experimental results are interpreted. Complex or lengthy mathematics has its place in modeling and design, but is of little help in network elucidation.

Most of the step combinations examined in this chapter are covered in advanced texts on kinetics and reaction engineering [G1-G10], with the exception of coupled parallel steps and reactions with fast pre-dissociation (Sections 5.3 and 5.6, respectively) and the equations for continuous stirred-tank reactors. The material is reviewed here for the user's convenience and for ease of reference.

5.1. Reversible reactions

The simplest example of a step combination is a reversible reaction $A \longleftrightarrow P$ composed of a forward step $A \longrightarrow P$ and a reverse step $P \longrightarrow A$. Although such reactions are commonly referred to as single-step reversible, it is more expedient for reaction mathematics to treat them as composed of two separate steps.

5.1.1. *First order-first order reactions*

The most common first order-first order reversible reactions are isomerizations, although many of them are catalytic and involve intermediates. Often they are parts of larger networks. Many other reversible reactions are also first order-first order.

$$\text{stoichiometry:} \qquad A \longleftrightarrow P \qquad\qquad (5.1)$$

$$\text{rate equation:} \qquad -r_A = r_P = k_{AP} C_A - k_{PA} C_P \qquad (5.2)$$

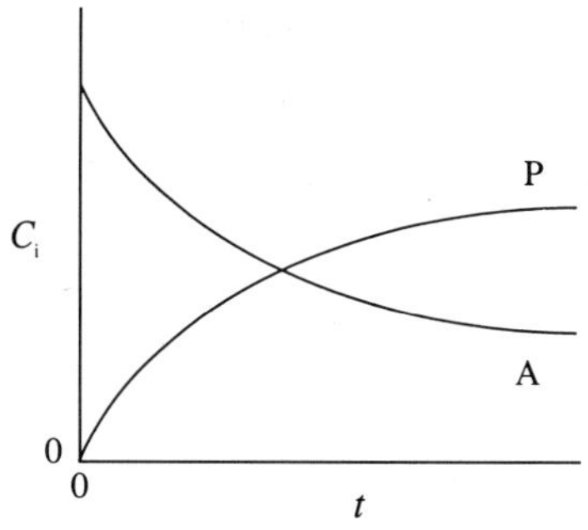

It proves convenient to define new quantities

$$k \equiv k_{AP} + k_{PA}, \qquad C \equiv C_A + C_P$$

In terms of these, the rate equation becomes

$$\text{at time } t: \qquad r_P = k C_A - k_{PA} C$$

$$\text{at equilibrium:} \qquad 0 = k C_A^\infty - k_{PA} C$$

(superscript ∞ indicates equilibrium). Subtraction of these equations from one another yields

$$r_P = k(C_A - C_A^\infty)$$

or, in words:

Figure 5.1. Concentrations vs time for first order-first order reaction A $\longleftrightarrow$ P in batch.

The rate is proportional to the distance from equilibrium. The proportionality factor is the sum of the forward and reverse rate coefficients.

The decrease of distance from equilibrium thus is a first-order process.
     To obtain a more general formula, define:

$$x \equiv \frac{C_i - C_i^\infty}{C_i^0 - C_i^\infty} \qquad (i = A, P) \qquad (5.3)$$

This is the *fractional distance from equilibrium*, varying from $x=1$ at start to $x=0$ at equilibrium and having the same value for A and P. In terms of x, the rate is

$$r_x = -kx \qquad\qquad (5.4)$$

where the process rate r_x is the contribution of the reaction to the rate of change of x.

Specifically for constant-volume batch reactions, in which the reaction rate is the only process rate contributing to the rate of change:

$$\mathrm{d}x/\mathrm{d}t \;=\; r_x \;=\; -kx \qquad \text{and} \qquad \ln x \;=\; -kt \qquad (5.5)$$

A plot of $\ln x$ versus t yields a straight line with slope $-k$ (see Figure 5.2). It makes no difference whether x is calculated with $i = A$ or $i = P$, the lines obtained are identical. Moreover, the method remains applicable even if some product P was already present at zero time.

The individual rate coefficients k_{AP} and k_{PA} can be calculated from k and the equilibrium concentrations C_A^∞ and C_P^∞ with

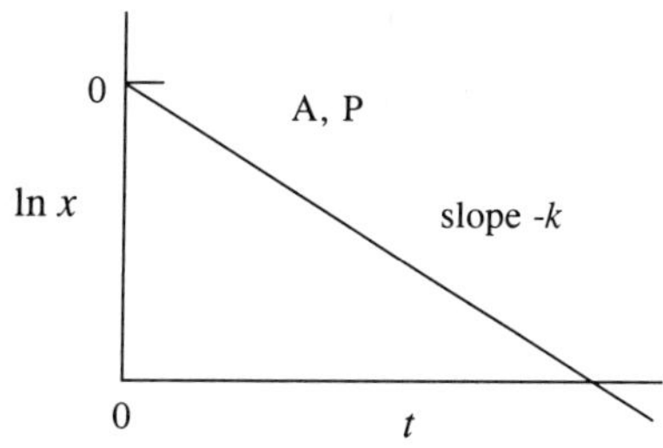

Figure 5.2. First-order plot for reversible reaction in batch order.

$$k_{AP} \;=\; kC_P^\infty/C, \qquad k_{PA} \;=\; kC_A^\infty/C \qquad (5.6)$$

The rate equation 5.5 and the first-order plot remain valid in terms of any physical variable ζ that is a linear function of the concentrations. This is usually true for properties such as refractive index, electric conductivity, specific gravity, and rotation of the plane of polarized light. In all such cases, x is defined as before by eqn 5.3, but with ζ instead of C_i. Equation 5.5 remains valid even if the variable changes its sign in the course of the reaction, as it might with rotation of the plane of polarized light.

Equation 5.5 and the first-order plot also remain valid regardless of volume variations if the fractional distance from equilibrium is defined in terms of amounts N_i instead of concentrations C_i:

$$x \;\equiv\; \frac{N_i - N_i^\infty}{N_i^0 - N_i^\infty} \qquad (i = A, P) \qquad (5.7)$$

The relationship between the characteristic rate coefficient k and the time t_x required to reduce the distance from equilibrium to the fraction x is analogous to that for irreversible first-order reactions (see eqn 3.23):

$$k \;=\; -(\ln x)/t_x \qquad (5.8)$$

In a continuous stirrer-tank reactor at steady state, a first order-first order reversible reaction with no fluid-density variation describes a straight line with slope k and intercept 1 in a plot of $1/x$ versus τ (see Figure 5.3). Again, it makes no difference whether x is calculated from the concentration of A or P or any physical

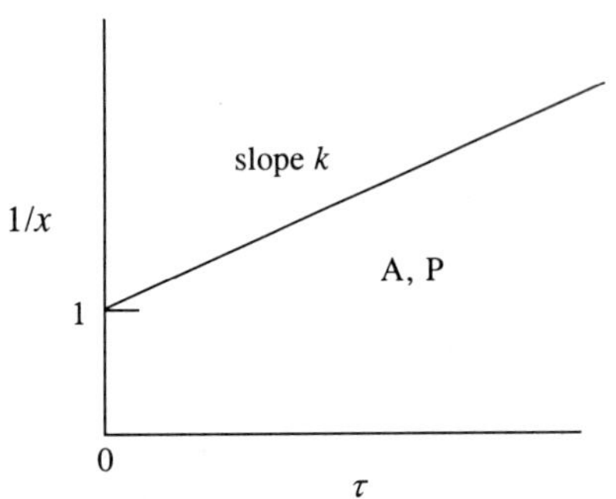

Figure 5.3. First-order plot for reversible reaction in continuous stirred-tank reactor.

property that varies linearly with concentrations. If the reaction involves a fluid-density variation, the same straight-line plot is obtained if x is calculated with eqn 5.7 in terms of amounts rather than concentrations. Moreover, as in a batch reactor, the relationship between the characteristic rate coefficient k and the reactor space time τ_x required to reduced the distance from equilibrium to a fraction x is analogous to that for an irreversible first-order reaction (see eqn 3.25):

$$k = \frac{1 - x}{\tau_x x} \tag{5.9}$$

Example 5.1. Reversible formation of a lactone. Results obtained for the conversion of γ-hydroxybutyric acid into its lactone

$$HO\diagdown\diagup\diagdown COOH \longleftrightarrow \text{(lactone)} = O \;+\; H_2O$$

in aqueous solution at 25° C in a batch reactor and with an initial acid concentration of 0.1823 M are listed in the first two columns of Table 5.1.

The product concentration remained constant within experimental error from 47 to 62 hours. The value at 47 hours, 0.1328 M, can therefore be taken as the equilibrium concentration of lactone, C_P^∞. The corresponding values of x (fractional distance from equilibrium) are listed in the third column. A plot of $\ln x$ versus t is shown in Figure 5.4 (left diagram) and is seen to be nicely linear, indicating that the reaction is first order-first order (the reverse step actually is pseudo-first order as it involves H_2O in large excess as a

Table 5.1. Conversion of γ-hydroxybutyric acid into its lactone in aqueous solution in batch reactor at 25° C [1].

t min	C_P M	$x = 1 - C_P/C_P^\infty$
0	0	1.000
21	0.0241	0.819
36	0373	.720
50	0499	.624
65	0610	.541
80	0708	.467
100	0811	.389
120	0900	.322
160	1035	.221
220	1155	.130
47 h	1328	0
62 h	1326	0

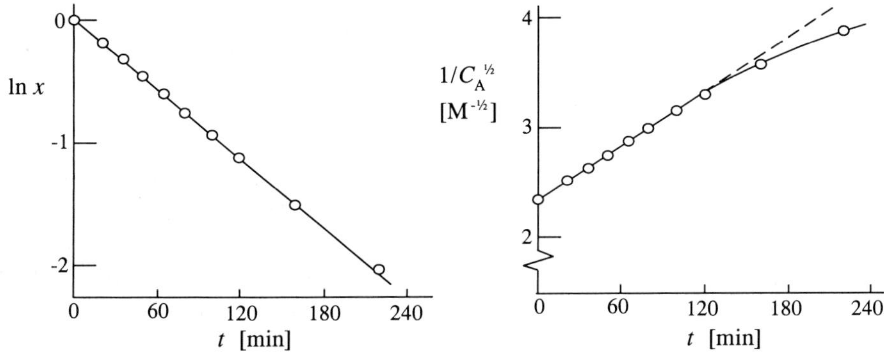

Figure 5.4. Conversion of γ-hydroxybutyric acid into its lactone, plotted as reversible first order-first order reaction (left) and as irreversible reaction of order 1.5 (right).

co-reactant). With $9.42*10^{-3}$ min^{-1} from the slope of the plot in Fig. 5.4, left diagram, the rate coefficients are:

$$k_{AP} = k\,C_P^\infty/C_A^o = 6.84 * 10^{-3} \text{ min}^{-1}, \qquad k_{PA} = k - k_{AP} = 2.58 * 10^{-3} \text{ min}^{-1}$$

The example of this reaction demonstrates another important facet of kinetics. Figure 5.4 shows side by side the experimental data plotted as a first order-first order reversible reaction and as an irreversible reaction of order 1.5. Over a limited conversion range (here about two thirds of the way to equilibrium) the second plot is linear within the scatter of the data points. Although evaluation of the full conversion range leaves no doubt that the reaction is indeed reversible and first order-first order, its rate up to a rather high conversion is approximated surprisingly well by the equation for an irreversible reaction of higher order, in this instance of order 1.5:

$$r_P \cong 0.016\ C_A^{1.5}\ \text{M min}^{-1} \tag{5.10}$$

> Unless results at conversions close to equilibrium are available, a reversible reaction can be mistaken for an irreversible reaction of higher order.

This is a striking example of the dangers of working by recipe with data that do not fully cover the entire conversion range, and of the importance of considering what is chemically reasonable. If the experiment had been terminated at or before 2 hours (about 50% conversion of the acid), eqn 5.10 might have been accepted as adequate, but would have given wrong predictions upon extrapolation to higher conversions. However, a competent chemist would have rejected a reaction order of 1.5 as requiring a kind of mechanism that is highly improbable for a simple ring closure with water loss under mild conditions, while he would gladly have accepted the proposition that the reaction might be first order and reversible.

5.1.2.   First order-second order reactions

Reactions in which a molecule dissociates into two different or equal fragments are very common, although many are so fast that they are practically at equilibrium (e.g., see dissociation of I_2 in the hydrogen-iodide reaction in Section 4.2).

$$\text{stoichiometry:} \qquad A \longleftrightarrow P + Q \tag{5.11}$$

$$\text{rate equation:} \qquad -r_A = r_P = r_Q = k_{AP} C_A - k_{PA} C_P C_Q \tag{5.12}$$

Integration eqn 5.12 for constant-volume batch and $C_P{}^\circ = C_Q{}^\circ = 0$ gives [G8]

$$\frac{f_A^\infty}{2 - f_A^\infty} \ln \frac{f_A^\infty + f_A(1 - f_A^\infty)}{f_A^\infty - f_A} = k_{AP} t$$

where $f_A \equiv 1 - C_A / C_A^\circ$ is the fractional conversion of A. Since this equation is in terms of reactant A, it also holds for reactions of the type $A \longleftrightarrow 2P$.

For a CSTR, the relationship between concentrations and reactor space time is too unwieldy to be of practical use.

5.1.3.   Second order-second order and higher-order reactions

Single-step reactions of this type, with different or identical reactants, are found mostly in textbooks. The mathematics of the HI reaction fits this formalism although the reaction actually proceeds by a different, two-step mechanism (see Section 4.2).

$$\text{stoichiometry:} \qquad A + B \longleftrightarrow P + Q \tag{5.13}$$

$$\text{rate equation:} \qquad -r_A = -r_B = r_P = r_Q = k_{AP} C_A C_B - k_{PA} C_P C_Q \tag{5.14}$$

A solution simple enough to be at least of some use in practice exists only for constant-volume batch with stoichiometric amounts of A and B and no P and Q present initially [G7,G8]:

$$\frac{f_A^\infty}{1 - f_A^\infty} \ln \frac{f_A^\infty - (2f_A^\infty - 1)f_A}{f_A^\infty - f_A} = 2 C_A^\circ k_{AP} t$$

The mathematics of reversible reactions of higher order than second are cumbersome. Rather than struggling with them, the chemist or engineer interested in their kinetics will design his experiments to circumvent this obstacle by determining pseudo-orders or evaluating initial rates (see Section 3.3.2 and 3.3.3).

5.2. Parallel steps

Steps or pathways are called *parallel* (or concurrent) if they originate from the same reactant, but lead to different products. Practically every reaction yields by-products to some extent, and most of these arise from parallel steps at some points in the network. A good understanding of the kinetics of parallel steps is therefore highly important.

5.2.1. Parallel first-order steps

In its literal form, this reaction is only of academic interest because a molecule is unlikely to break up or isomerize irreversibly in two or more different ways. However, situations frequently encountered in practice are those of multistep parallel first-order decomposition reactions and of parallel reactions that involve co-reactants but are pseudo-first order in the reactant A. An example of the first kind is dehydrogenation of paraffins, examples of the second kind include hydration, hydrochlorination, hydroformylation, and hydrocyanation of olefins and some hydrocarbon oxidation reactions. All these reactions are multistep, but the great majority are first order in the respective hydrocarbon, and pseudo-first order if any co-reactant concentration is kept constant.

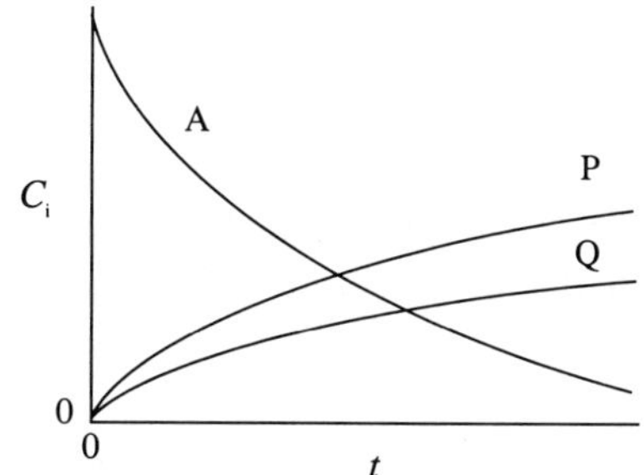

Figure 5.5. Concentrations vs time for parallel first-order steps A ⟷ P and A ⟶ Q in batch.

network:

$$A \longrightarrow \begin{array}{l} P \\ Q \\ \dots \end{array} \qquad (5.15)$$

rate equations:

$$r_A = -kC_A$$

$$r_P = k_{AP}C_A \qquad (5.16)$$

$$r_Q = k_{AQ}C_A$$

$$\dots$$

where

$$k \equiv k_{AP} + k_{AQ} + \dots \qquad (5.17)$$

There may be any number of parallel steps.

According to eqns 5.16, all rates are proportional to the concentration of A and thus, since A reacts to completion, to the distance from the state at complete conversion. Moreover, this being so, the decay or formation rate of each participant is proportional to the *fractional* distance, x, from the state at complete conversion:

$$r_x = -kx$$

where x is given by eqn 5.3, and k by eqn 5.17.

> In a network of irreversible parallel first-order steps, the fractional distance from complete conversion at any conversion level is the same for all participants. The process rate $-r_x$ is proportional to the fractional distance x. The characteristic rate coefficient is the sum of the rate coefficients of all steps.

Since the ratios of the rates remain constant regardless of the conversion level, the concentrations at complete conversion are independent of the reactor type, and so are the distances from that state at any given conversion.

Specifically in constant-volume batch reactors, in which the process rate r_x equals the rate of change dx/dt:

$$dx/dt \;=\; -\,kx$$

or, after integration,

$$\ln x \;=\; -\,kt \qquad (5.18)$$

where x can be expressed in terms of any participant:

$$x \;=\; C_A/C_A^o \;=\; 1 - C_P/C_P^\infty \;=\; 1 - C_Q/C_Q^\infty \;=\; \dots \qquad (5.19)$$

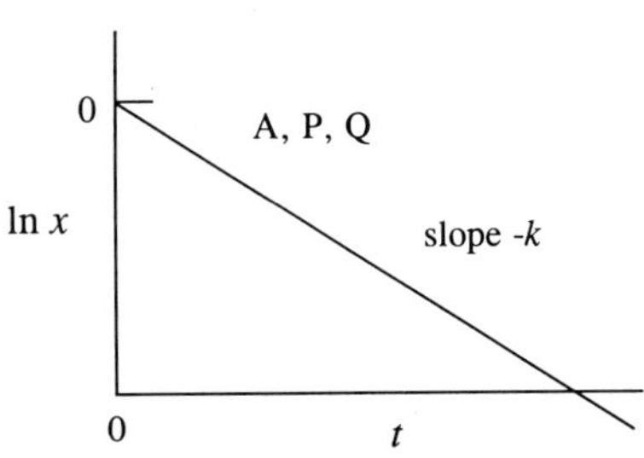

Figure 5.6. First-order plot for parallel steps in batch reactor.

Accordingly, a plot of $\ln x$ versus time gives a straight line with slope $-k$, as in Figure 5.6. It makes no difference whether x is calculated with i = A, P, Q, or any other product, the straight lines obtained are identical. If such behavior is found experimentally, it provides a very strong indication for a network of the type of 5.15 with all reactions first order in A.

The coefficients k_{AP}, k_{AQ}, etc., can be calculated from

$$k_{AP} = (C_P^\infty/C_A^o)k, \qquad k_{AQ} = (C_Q^\infty/C_A^o)k, \qquad \text{etc.} \qquad (5.20)$$

with k obtained from the slope in Figure 5.6 or a corresponding numerical method. Alternatively, the individual coefficients can be calculated from data on incremental conversions:

$$k_{AP} = k\Delta C_P/(-\Delta C_A), \qquad k_{AQ} = k\Delta C_Q/(-\Delta C_A), \qquad \text{etc.} \qquad (5.21)$$

Equations 5.19 assume that no products are present initially. Otherwise one has

$$x = (C_P^\infty - C_P)/(C_P^\infty - C_P^0) = \ldots$$

$$k_{AP} = k(C_P^\infty - C_P^0)/C_A^0, \qquad \text{etc.}$$

It may appear strange that all products give first-order plots with the same slope although their individual formation rate coefficients may differ widely. It is true that the formation rates of products with larger rate coefficients are higher at any time. However, their concentrations at complete conversion are also higher by the same factors. The result is that, in a given time span, the concentration of each product increases by the same *fraction* of the final concentration, regardless of the disparity of the rate coefficients.

In a continuous stirred-tank reactor, eqn 3.20 and the corresponding straight-line plot in Table 3.1 remain valid, and eqn 3.21 for the products becomes

$$C_P = \frac{C_A^0 k_{AP}\tau}{1 + k\tau}, \qquad C_Q = \frac{C_A^0 k_{AQ}\tau}{1 + k\tau}, \qquad \text{etc.} \qquad (5.22)$$

Example 5.2. Hydroformylation of propene [2]. Hydroformylation converts an olefin to an aldehyde of next higher carbon number by addition of carbon monoxide and hydrogen. The reaction is catalyzed by dissolved hydrocarbonyl complexes of transition-metal ions such as cobalt, rhodium, or rhenium. The carbon atom of the carbon monoxide can attach itself to the carbon atom on either side of the olefinic double bond, so that two aldehyde isomers are formed. If the catalyst also has hydrogenation activity, the aldehydes are converted to alcohols and paraffin is formed as by-product. For propene and such a catalyst the (simplified) network is:

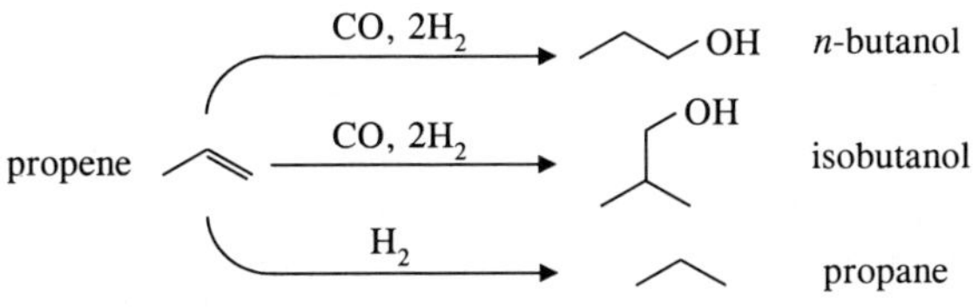

Results obtained in a constant-pressure semi-batch reactor in which synthesis gas was supplied on demand are listed in Table 5.2. The analytical accuracy is said to be within $\pm\ 0.002$ M. [Before alcohol analysis the samples were hydrogenated to convert any residual aldehydes to the corresponding alcohols.] At the resulting constant partial pressures of CO and H_2 in the reactor and with very good gas-liquid mass transfer, the CO and H_2 concentrations in the liquid remain constant and the rate depends only on the propene concentration (see pseudo-orders, Section 3.3.2).

Table 5.2. Hydroformylation of propene in 2-ethylhexanol with phosphine-substituted cobalt hydrocarbonyl catalyst, $HCo(CO)_3Ph$ (Ph = organic phosphine) and synthesis gas of 2:1 H_2-to-CO ratio at 50 atm and 130° C in semi-batch reactor.

time	concentration, M			
min	propene	*n*-butanol	isobutanol	propane
0	.807	0	0	0
10	.679	.107	.009	.010
20	.571	.196	.017	.019
60	.285	.437	.038	.042
120	.101	.592	.051	.055
20 h	< .002	.677	.058	.063

Most homogeneous hydrogenation, hydrohalogenation, halogenation, hydroformylation, and hydrocyanation reactions are first order in the olefinic reactant. A test whether this is the case here suggests itself. A numerical work-up is shown in Table 5.3. The fractional distances from complete conversion are calculated with eqns 5.19 for all participants at all times, and values of the characteristic coefficient k are then obtained for each from eqn 5.18 (global method), with the concentrations at 20 hours taken as C_i^∞.

Table 5.3. Data work-up for hydroformylation of propene. (A = propene, P = *n*-butanol, Q = isobutanol, R = propane.)

t	$x =$	$x = 1 - C_i/C_i^\infty$			$k = -(\ln x)/t * 10^2$, min^{-1}			
min	C_A/C_A°	P	Q	R	A	P	Q	R
10	.8414	.8419	.8448	.8413	1.73	1.72	1.69	1.73
20	.7076	.7105	.7069	.6984	1.73	1.71	1.73	1.79
60	.3532	.3545	.3448	.3333	1.73	1.73	1.77	1.83
120	.1252	.1256	.1207	.1270	1.73	1.73	1.76	1.72

The consistency of the values for k shows that plots of $\ln x$ versus t will be linear and have the same slope for each participant, and indicates strongly that all three reactions are indeed first order in olefin. A network of the general type of 5.15 can therefore be accepted. An average value for k is $1.73*10^{-2}$ min^{-1}. Not unexpectedly, the scatter is largest for the lowest concentrations, where the relative analytical error is greatest. The individual rate coefficients k_{AP}, k_{AQ}, and k_{AR} can be calculated with eqns 5.20. The results are:

$$k_{AP} = \frac{1.73 * 0.677}{0.807} * 10^{-2} = 1.45 * 10^{-2} \text{ min}^{-1},$$

$$k_{AQ} = \frac{1.73 * 0.058}{0.807} * 10^{-2} = 0.12 * 10^{-2} \text{ min}^{-1},$$

$$k_{AR} = \frac{1.73 * 0.063}{0.807} * 10^{-2} = 0.14 * 10^{-2} \text{ min}^{-1}$$

The astute reader will notice that the material balance fails to close within the analytical error: The sum of the final product concentrations leaves more than 1.2% of the initial propene unaccounted for. Factors contributing to this discrepancy are a slight increase in the volume of the liquid phase as C_4 alcohols are formed from C_3 olefin, and a small amount of by-products of higher molecular weights that have escaped the gas-chromatographic analysis.

For use in modeling, the rate coefficients would also have to be corrected for gas-liquid distribution at least of propene and propane. In particular, some propene initially in the gas phase enters the liquid and is converted (making the material-balance discrepancy worse). The analyses of the liquid do not reflect this, and the calculated value of k therefore is slightly too low. Another consequence of the presence of volatiles in the gas phase is that the sum of the partial pressures of CO and H_2 was a little below the 50 atm total pressure, only approaching that level more closely as propene was converted to less volatile products.

Such nuances notwithstanding, the symptom of a network with parallel first-order reactions to all three products emerges quite clearly from this single experiment. Even so, however, the question remains open where exactly the reaction path branches and how the rate depends on the partial pressures of CO and H_2. We shall return to this interesting type of reaction in later examples (Examples 5.3, 6.5, 7.5, 11.2, 12.1, and 12.2).

5.2.2. *Parallel second-order steps*

The simplest case of parallel second-order steps is that of formation of two different dimers of a reactant A, corresponding to the network 5.23 and rate equations 5.24 (see next page). At all times, both products are formed in the same ratio $r_P{:}r_Q = k_{AP}{:}k_{AQ}$, so that the decay of A is an ordinary second-order reaction with rate coefficient $k = k_{AP} + k_{AQ}$. Likewise, the product formations are ordinary second-order reactions. (One could think of the initial amount of A as divided into two portions in the ratio $k_{AP}{:}k_{AQ}$ that react independently of one another and at the same rate, one to P and the other to Q.) All equations and plots for irreversible second-order reactions thus are valid (see Section 3.3.1).

network

$$2A \begin{array}{c} \nearrow P \\ \searrow Q \end{array}$$

(5.23)

rate equations:

$$r_A = -2kC_A^2$$
$$r_P = k_{AP}C_A^2$$
$$r_Q = k_{AQ}C_A^2$$

(5.24)

where

$$k \equiv k_{AP} + k_Q$$

(5.25)

Mathematics becomes more complicated if co-reactants are involved [G10], as for instance in networks such as

However, if the co-reactant is the same in both reactions, as in the network above left, experiments for network identification can be conducted with stoichiometric amounts of the reactants, so that $C_A = C_B$ at all times. The mathematics then reduces to the same simple form as for the network 5.23. With different co-reactants as in the network above right, that technique can also be employed, but the C_B-to-C_C ratio at which equal fractions of these two reactants are consumed would first have to be determined.

5.2.3. *Parallel steps of different orders*

Combinations of parallel steps of different orders are common in practice. The most frequently encountered situation is the formation of a heavy by-product by higher-order dimerization or oligomerization parallel to a first-order main reaction. In other common cases, the side reaction is a first-order decomposition of a reactant parallel to a main reaction of higher order.

network

(5.26)

rate equations:

$$r_{\mathrm{A}} = -2k_{\mathrm{AP}}C_{\mathrm{A}}^2 - k_{\mathrm{AQ}}C_{\mathrm{A}}$$

$$r_{\mathrm{P}} = k_{\mathrm{AP}}C_{\mathrm{A}}^2 \qquad (5.27)$$

$$r_{\mathrm{Q}} = k_{\mathrm{AQ}}C_{\mathrm{A}}$$

Unfortunately, most combinations of steps of different orders lead to rather unwieldy mathematics as far as reactor performance is concerned [G1,G10]. However, one important general facet of such networks warrants special attention and is easily demonstrated with the simplest case, the network 5.26. Of interest are the *yield ratio* of the two products and the *selectivity* of conversion to one of them. For the network 5.26, the instantaneous yield ratio is seen to be

$$Y_{\mathrm{PQ}} = 2r_{\mathrm{P}}/r_{\mathrm{Q}} = 2(k_{\mathrm{AP}}/k_{\mathrm{AQ}})C_{\mathrm{A}} \qquad (5.28)$$

(see definition 1.10 in Section 1.6). The yield ratio (second-order to first-order product) is proportional to the reactant concentration. A consequence is:

Higher reactant concentration favors the
parallel step of higher reaction order.

This qualitative rule is not restricted to the simple network 5.26 and has important consequences for reactor selection in plant design. In a batch or plug-flow tubular reactor, the reactant concentration is at its highest initially—in the batch at start, in the tube at the inlet—and then decreases as conversion progresses to its final level. In contrast, in a well-mixed continuous stirred-tank reactor, the reactant concentration everywhere and at all times is at that final, lowest level because the effluent has the same composition as the reactor contents. Accordingly, if the composition of the initial charge or feed stream and the conversion to be attained are given and if P, formed by the higher-order step, is the desired product, a tubular or batch reactor gives the better selectivity; and if Q is desired, a continuous stirred tank reactor does. [In the first case, if for some overriding reason a stirred tank is the only acceptable reactor configuration, a cascade of stirred tanks in series should be considered because the reactant concentration will be higher at least in its earlier stages.] Also, whatever the reactor type, formation of the product of the higher-order step is more favored at lower final conversion, so there may be a trade-off between selectivity and size of a recycle stream.

For the simple network 5.26 and a reaction with no fluid-density variation, the magnitude of the effect is easily calculated: The cumulative selectivity of conversion to P (moles of A converted to P per mole of A consumed, see definition 1.11) in batch and continuous stirred-tank reactors as a function of fractional conversion, f_{A}, is

$$\text{batch:} \qquad \overline{S}_\text{P} \;=\; 1 \;-\; \frac{1}{f_\text{A}\Phi}\,\ln\frac{1+\Phi}{1+(1-f_\text{A})\Phi} \qquad\qquad (5.29)$$

$$\text{CSTR:} \qquad \overline{S}_\text{P} \;=\; \frac{(1-f_\text{A})\Phi}{1+(1-f_\text{A})\Phi} \qquad\qquad (5.30)$$

where

$$\Phi \;\equiv\; 2k_\text{AP}C_\text{A}^\text{o}/k_\text{AQ} \qquad\qquad (5.31)$$

Results for $\Phi = 2$ are listed in Table 5.4.

fractional conversion $f_\text{A} \equiv 1 - \dfrac{C_\text{A}}{C_\text{A}^\text{o}}$	selectivity to P $\overline{S}_\text{P} = 2C_\text{P}/(C_\text{A}^\text{o} - C_\text{A})$	
	batch	CSTR
0.20	0.642	0.615
0.40	0.612	0.545
0.60	0.574	0.444
0.80	0.524	0.286
0.90	0.491	0.167
0.95	0.472	0.091

Table 5.4. Cumulative selectivity of conversion to higher-order product in first- and second-order parallel steps with no fluid-density variation in batch and CSTR, (calculated for $k_\text{AP}C_\text{A}^\text{o}/k_\text{AQ} = 1$).

Derivation. The instantaneous selectivity to P (moles of A forming P per mole of A reacting) at given conversion level f_A is

$$S_\text{P} \;=\; 2r_\text{P}/(-r_\text{A})$$

(see definition 1.12). With eqns 5.27 and 5.31 this becomes

$$S_\text{P} \;=\; \frac{2k_\text{AP}C_\text{A}}{2k_\text{AP}C_\text{A}+k_\text{AQ}} \;=\; \frac{(C_\text{A}/C_\text{A}^\text{o})\Phi}{1+(C_\text{A}/C_\text{A}^\text{o})\Phi} \;=\; \frac{(1-f_\text{A})\Phi}{1+(1-f_\text{A})\Phi} \qquad (5.32)$$

In a CSTR at steady state, the compositions of reactor content and effluent are the same, so that the cumulative selectivity $\overline{S}_\text{P}$ (moles of A converted to P per mole of A reacted) equals the instantaneous selectivity S_P. In a batch reactor, eqn 5.32 describes the incremental fractional conversion of A to P at the respective conversion level:

$$S_\text{P} \;=\; \frac{2\,\mathrm{d}C_\text{P}}{-\mathrm{d}C_\text{A}} \;=\; \frac{(1-f_\text{A})\Phi}{1+(1-f_\text{A})\Phi}$$

With $\mathrm{d}C_\text{A} = C_\text{A}^\text{o}\,\mathrm{d}(1-f_\text{A})$, integration over $1-f_\text{A}$ from start to final conversion level f_A yields

$$\frac{2(C_P - C_P^o)}{C_A^o} \;=\; -\, \Phi \int_{f_A=0}^{f_A} \frac{(1 - f_A)\,\mathrm{d}(1 - f_A)}{1 + (1 - f_A)\Phi} \;=\; f_A - \frac{1}{\Phi} \ln \frac{1 + \Phi}{1 + (1 - f_A)\Phi}$$

As defined by eqn 1.11, the cumulative selectivity for the case at hand is

$$\overline{S}_P \;=\; 2(C_P - C_P^o)/f_A C_A^o$$

With the integral above this gives eqn 5.29.

This derivation demonstrates that yield ratios and selectivities are more easily obtained by integration over fractional conversion instead of time or reactor space time. Equations for selectivities in other situations with parallel steps of different orders can be derived in the same way, but are more complex.

5.3. Coupled parallel steps

Many reactants in organic chemistry can exist in the form of two or more isomers that give rise to different products. The most typical examples are reactions in which an olefin adds water, halogen, or any compound other than hydrogen under conditions that also promote migration of the double bond. Many reactions of industrial importance belong to this class.

The simplest network of this type is

$$\begin{array}{ccc} A_1 & \longrightarrow & P \\ \updownarrow & & \\ A_2 & \longrightarrow & Q \end{array} \qquad\qquad (5.33)$$

where A_1 and A_2 are the reactant isomers forming products P and Q, respectively. The rate equations are

$$\begin{aligned} r_1 &= k_{21} C_2 - C_1(k_{12} + k_{1P}), & r_P &= k_{1P} C_1, \\ r_2 &= k_{12} C_1 - C_2(k_{21} + k_{2Q}), & r_Q &= k_{2Q} C_2 \end{aligned} \qquad (5.34)$$

("A" is suppressed in the indices to avoid clutter).

The behavior depends strongly on the relative magnitudes of the rate coefficients and the initial isomer ratio. Since all steps are first order, an analytical solution can be given with matrix algebra, but is not easy to interpret. An examination of special cases provides more insight:

Case I:	$k_{12}, k_{21} \gg k_{1P}, k_{2Q}$	isomerization very fast compared with conversion
Case II:	$k_{12}, k_{21} \ll k_{1P}, k_{2Q}$	isomerization very slow compared with conversion
Case III:	$k_{12} \approx k_{21} \approx k_{1P} \approx k_{2Q}$	all rates comparable

Case I. Here, isomerization quasi-equilibrium is established quickly, before any significant conversion to products has taken place. From this short initial time span on the isomers remain in quasi-equilibrium, that is, the fractions of total A present as A_1 and A_2 become constant:

$$\frac{C_1}{C_A} \cong \frac{1}{1 + K_{12}}, \qquad \frac{C_2}{C_A} \cong \frac{K_{12}}{1 + K_{12}} \qquad (5.35)$$

where $C_A \equiv C_1 + C_2$ is the concentration of total A, and K_{12} is the equilibrium constant of isomerization $A_1 \longleftrightarrow A_2$. As a result, all rates can be expressed in terms of the concentration of total A:

$$r_P = \frac{k_{1P}}{1 + K_{12}} C_A, \qquad r_Q = \frac{k_{2Q} K_{12}}{1 + K_{12}} C_A,$$

$$r_A = -(r_P + r_Q) = -\frac{k_{1P} + k_{2Q} K_{12}}{1 + K_{12}} C_A \qquad (5.36)$$

This set is of the same algebraic form as eqns 5.16 for conversion of a single reactant to two products in parallel steps. Because quasi-equilibrium of isomerization is maintained, the reactant acts as a "homogeneous source" despite being split into two isomers, and the mathematics of product formation is the same as for the simpler reaction with network 5.15. Results from constant-volume batch plotted as $\ln x$ versus t for P, Q, and total A give identical straight lines with slope $-k$, as in Figure 5.6, where

$$k \equiv (k_{1P} + k_{2Q} K_{12})/(1 + K_{12}) \qquad (5.37)$$

(see Figure 5.7, top left). If no analytical method to distinguish between the isomers and no information on the very short initial time span before attainment of isomerization quasi-equilibrium were available, the existence of the reactant in the two isomeric forms might go unnoticed.

A good way to determine the rate coefficients k_{1P} and k_{2Q} from experimental results is first to find k from the slope of a first-order plot for the respective reactant or an equivalent numerical method, then to calculate k_{1P} and k_{2Q} from it, the isomerization equilibrium constant K_{12}, and the product concentrations at complete conversion:

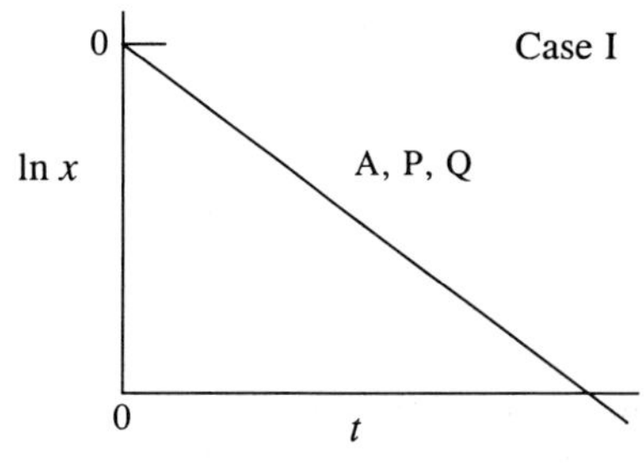

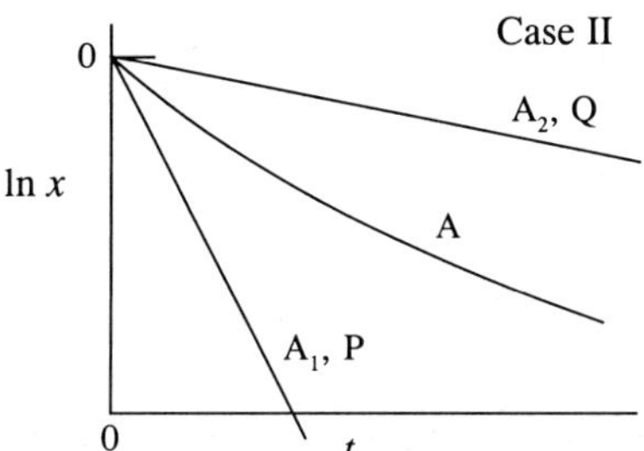

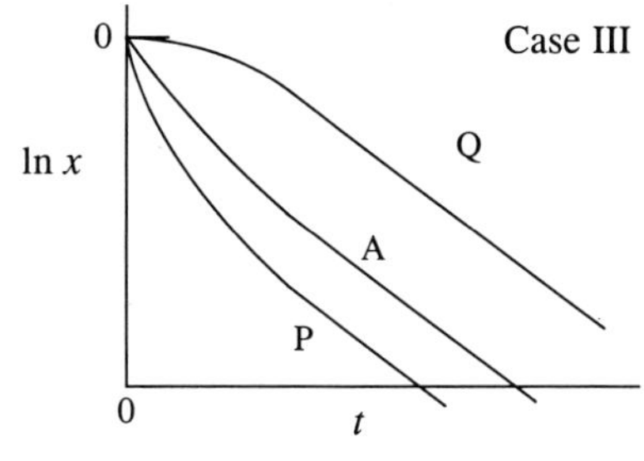

Figure 5.7. First-order constant-volume batch plots for two coupled parallel first-order steps. Top left: very fast isomerization; top right: very slow isomerization; bottom left: comparable rates with $k_{12} > k_{21} > k_{1P} > k_{2Q}$ and initial charge of A_1 (schematic).

$$k_{1P} = k(1 + K_{12})\frac{C_P^\infty}{C_A^o}, \qquad k_{2Q} = k\left[\frac{1 + K_{12}}{K_{12}}\right]\frac{C_Q^\infty}{C_A^o} \qquad (5.38)$$

(granted no products are present at start; otherwise replace the final product concentrations by $C_P^\infty - C_P^o$ and $C_Q^\infty - C_Q^o$). Note that information on the individual isomer concentrations is not needed. The isomerization rate coefficients k_{12} and k_{21} cannot be calculated unless accurate data on the short transient before attainment of isomerization quasi-equilibrium are at hand.

Case II. Here, conversion to products nears completion before any significant isomerization has occurred. The steps $A_1 \longrightarrow P$ and $A_2 \longrightarrow Q$ are uncoupled, proceeding side by side without noticeably affecting one another. First-order plots of $\ln x$ versus t for P and A_1 give identical straight lines with slope $-k_{1P}$, those for Q and A_2 give identical straight lines with different slope $-k_{2Q}$, and that for total A is curved, with a slope that flattens with increasing time and approaches that of the curve for A_2 and Q (see Figure 5.7, top right).

The shape of the plot for total A depends not only on the relative reactivities of the isomers, but also on their relative amounts in the initial charge. If the reactivities differ substantially and the amounts charged are comparable, the early portion of the curve is strongly affected by the fast decay of the more reactive isomer (assumed to be A_1 in Figure 5.7). The late portion, stemming from a time span in which that isomer has all but disappeared, then reflects the slow decay of the less reactive isomer in the absence of the other. Because of the decreasing slope of its first-order plot, the decay of total A resembles a reaction of higher order although the decay rates of the individual isomers are first order.

The fact that the overall behavior of uncoupled reactions occurring side by side resembles that of a reaction of higher order is well-known and important for modeling of simultaneous reactions by lumping. The classic example is that of fundamental kinetic modeling of catalytic cracking by Mobil Oil's research department in the 1960s [3]: Cracking of any single, pure hydrocarbon is approximately first-order (with allowance for catalyst deactivation) [4], but cracking of typical mixtures is well described by a rate equation that is approximately second order in total feed hydrocarbons [3].

Case III. Here, the rates are comparable. This is the most complex, most interesting, and in practice most important situation. The behavior depends strongly on the relative reactivities, isomerization equilibrium, and initial isomer ratio.

Let us say, isomerization is somewhat faster than conversion, and the initial charge is the more reactive isomer A_1 (this is assumed in Figure 5.7, bottom left). Isomerization then produces a steady-state isomer ratio, but not until significant conversion to products has occurred. Loosely said, the system swings from a transient behavior dictated by the initial charge to one resembling that of Case I. At start, the formation rate of P is very high because the parent of P, isomer A_1, is present in high proportion. In contrast, the formation rate of Q is zero initially because the parent of Q, isomer A_2, is absent and must first be formed from A_1 by isomerization. Accordingly, in the first-order plot the curve for P starts with a steep slope that subsequently flattens, and that for Q starts with zero slope and becomes steeper. Because initially all A is present in the form of the more reactive isomer, the curve for total A also starts with a steep slope and then becomes flatter. As time progresses, the isomers asymptotically approach a steady-state distribution that remains constant from then on: The reactant becomes a "homogeneous source." As a result, the curves for P, Q, and total A become parallel straight lines. An experimentally observed behavior of this kind is very strong evidence for coupled, irreversible, parallel steps that are first order in the reactant isomers.

If the initial charge consists of A_2, the less reactive isomer, the opposite behavior results: The curve for Q starts with steeper slope; that for P, with zero slope; and that for total A, with gentler slope. If the initial charge consists of an equilibrium isomer mixture, little shift in isomer distribution occurs during the reaction, and the curves for P, Q, and total A almost coincide, as they do in Case I.

The asymptotically approached constant isomer distribution is not necessarily close to equilibrium. If the conversion rate coefficients differ greatly and the equilibrium fraction of the more reactive isomer is small, the latter becomes depleted because resupply by isomerization cannot quite keep pace with the high consumption rate, and the fraction of A present as the more reactive isomer is significantly smaller than at isomerization equilibrium. Theoretically, the A_1-to-A_2

ratio under reaction conditions can be calculated from the steady-state condition $-r_1/C_1 = -r_2/C_2$ or, if the "drain effect" is large, be estimated with the Bodenstein approximation of quasi-stationary behavior of A_1. In practice, neither method can be recommended. Both require iteration because the values of the coefficients in the equations depend in turn on the magnitude of the drain effect, and both are highly sensitive to the value of the isomerization rate coefficients, usually known only in rough approximation. A better way is to use the coefficient values obtained assuming isomerization equilibrium as initial guess for regression or, if possible, to determine the isomer ratio under reaction conditions directly by analysis.

An observation from everyday life may help to visualize the drain effect. A common type of office hot-water dispenser for instant coffee or tea consists of a container with show glass and stopcock at the bottom of the show glass (see Figure 5.8). It can serve as a hydrodynamic model of our reaction if we drill a small hole into the bottom of the container (or just imagine what would happen if such a hole were there). The water in the container is A_2, that in the show glass is A_1, the water dribbling out of the small hole goes to Q, that draining from the show glass goes to P. If the stopcock and hole are closed, the water level in the show glass is the same as in the container; that corresponds to isomerization equilibrium. However, when the stopcock is opened,

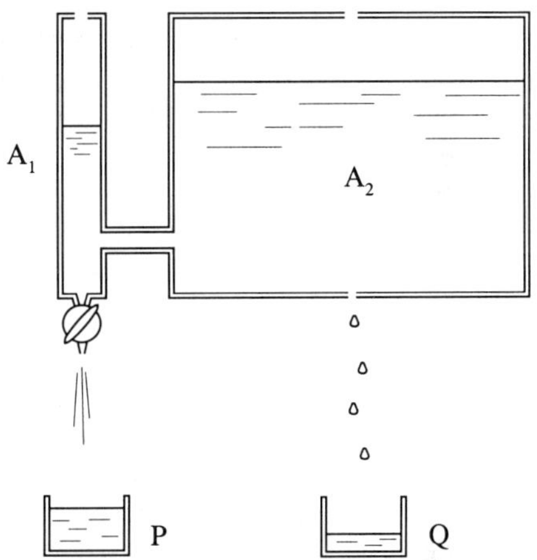

Figure 5.8. Hydrodynamic model for reaction with coupled parallel first-order steps, widely different conversion rate coefficients, and small equilibrium fraction of more reactive isomer.

the level in the show glass falls far below that in the container because resupply from the latter is not fast enough to make up for the quick drain. That the hole in the container allows a small amount of water to dribble out makes little difference. [If we close the stopcock, the level in the show glass overshoots equilibrium and oscillates until settling. This is caused by the inertia of flow through the connection between container and show glass and has no equivalent in reaction-kinetic systems: The hydrodynamic model is not perfect in every detail.]

Just like the approach to isomerization equilibrium in a system without conversion to products, the approach to the steady-state isomer distribution is a first-order process whose characteristic rate coefficient is the sum of the forward and reverse coefficients (see Section 5.1):

$$k_{iso} \equiv k_{12} + k_{21} \tag{5.39}$$

As the steady state is approached, the curves in the first-order plots become parallel straight lines. Accordingly, the fractional-life equation 5.8 can be used to obtain an estimate of k_{iso} from the time required to approach this condition closely. The individual coefficients k_{12} and k_{21} can then be obtained from their sum k_{iso} and the equilibrium constant $K_{12} = k_{12}/k_{21}$:

$$k_{12} = k_{iso}K_{12}/(1 + K_{12}) , \qquad k_{21} = k_{iso}/(1 + K_{12}) \tag{5.40}$$

This is a typical example showing how the kinetic effect of a simple step combination—a first order-first order reversible reaction in the case at hand—can be clearly recognized and evaluated even if the steps are embedded in a larger network.

The coefficient k characteristic for the rate of conversion to products can be determined graphically from the slope of the straight-line portions of the curves in first-order batch plots of $\ln x$ versus t, or numerically from

$$k = - (\Delta \ln x)/\Delta t \tag{5.41}$$

with differences taken from within the range in which the curves are straight and parallel. The conversion rate coefficients k_{1P} and k_{2Q} can then be calculated from k and incremental conversions within that range:

$$k_{1P} \cong \left[\frac{k}{C_1/C_A} \right] \frac{\Delta C_P}{-\Delta C_A} , \qquad k_{2Q} \cong \left[\frac{k}{C_2/C_A} \right] \frac{\Delta C_Q}{-\Delta C_A} \tag{5.42}$$

Here, the isomer fractions C_1/C_A and C_2/C_A are those at steady-state distribution, given by eqns 5.35 unless the drain effect is appreciable and calls for a correction.

Example 5.3. Hydroformylation of 1-pentene with phosphine-substituted cobalt hydrocarbonyl catalyst [2]. Results of hydroformylation of 1-pentene in a semi-batch reactor are shown in Table 5.5.

Since the catalyst and the reaction conditions are similar, one might expect pentene to behave much like propene (see Example 5.2). However, a work-up of the results as for propene in Table 5.3 does not give consistent values of the characteristic coefficient k. Plots of $\ln x$ versus t for pentene, n-hexanol, 2-methyl pentanol, 2-ethyl butanol, and pentane are shown in Figure 5.9. The pattern resembles that of Case III in Figure 5.7 in that the curves start out with different slopes but later become parallel straight lines. This behavior is typical for coupled parallel first-order reactions with comparable rates of isomerization and conversion. Indeed, such a mechanism is eminently plausible because 1-pentene, in contrast to propene, can undergo migration of the double bond from the terminal to an internal position. Attachment of CO to the carbon atom on either side of the double bond would lead to production of n-hexanol and 2-methyl pentanol from 1-pentene (double bond in terminal position), and 2-methyl pentanol and 2-ethyl butanol from 2-pentene (double bond between second and third carbon atom). This suggests the network 5.43, shown on page 102.

Table 5.5. Hydroformylation of 1-pentene in *n*-dodecanol with phosphine-substituted cobalt hydrocarbonyl catalyst and synthesis gas of 2:1 $CO:H_2$ ratio at 50 atm and 150° C in semi-batch reactor.

time	concentration, M				
min	pentene	*n*-hexanol	2-methyl pentanol	2-ethyl butanol	pentane
0	.659	0	0	0	0
10	.233	.350	.030	.0008	.039
20	.210	.368	.032	.002	.041
60	.168	.393	.039	.006	.044
120	.127	.417	.046	.011	.046
180	.098	.436	.051	.015	.048
22 h	< .002	.499	.068	.027	.053

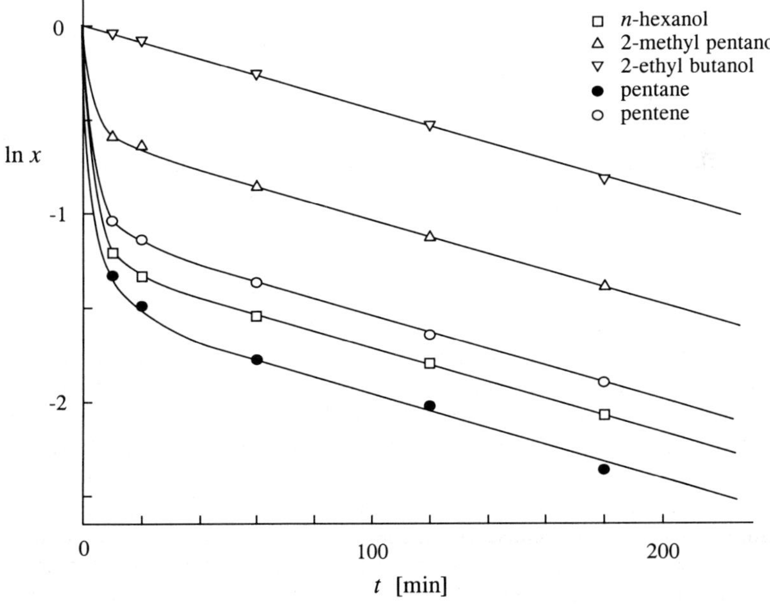

Figure 5.9. First-order batch plots of total olefin and products in hydroformylation of 1-pentene (date from Table 5.5)

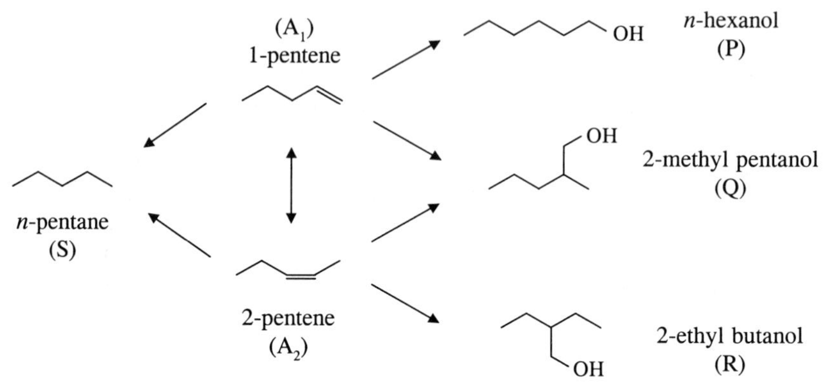

$$(A_1) \text{ 1-pentene} \quad\longrightarrow\quad \text{OH} \quad n\text{-hexanol (P)}$$

$$n\text{-pentane (S)} \qquad\qquad \text{2-methyl pentanol (Q)}$$

$$\text{2-pentene } (A_2) \qquad\qquad \text{2-ethyl butanol (R)} \tag{5.43}$$

[Actually, 2-pentene exists in *cis*- and *trans*-configurations, but this complication is ignored here as relatively unimportant.] Although more complex than the prototype 5.33, this network includes the essential features of coupled parallel first-order reactions and can account for the curves in Figure 5.9.

To estimate the isomerization rate coefficients, eqn 5.8 is applied to the time required for close approach to the straight-line behavior of the first-order curves. Judging this time to be about 25 minutes for a 90% approach to the steady-state isomer distribution, eqn 5.8 yields $k_{\mathrm{iso}} \approx 0.1$ min^{-1}. With this value and an iso-merization equilibrium constant $K_{12} = 20$ at 150°C calculated from thermochemical data [5,6] (with 2-*cis* and 2-*trans* pentene lumped into a single pseudo-component), eqns 5.40 give as rough estimates

$$k_{12} \approx 0.1 \text{ min}^{-1}, \qquad k_{21} \approx 0.005 \text{ min}^{-1}$$

The calculation of the conversion rate coefficients k_{1P}, k_{1Q}, k_{1R}, k_{1S}, k_{2Q}, k_{2R} and k_{2S} (P = n-hexanol, Q = 2-methyl pentanol, R = 2-ethyl butanol, S = pentane) from the limited data available requires some judgment calls. First, the steep initial slope of the plot for pentane indicates that this product is formed practically exclusively from 1-pentene, so that $k_{2S} \cong 0$ and the entire pentane yield is attributable to k_{1S}. Second, for lack of better information, one may tentatively assume that the ratio $k_{1Q}:k_{1P}$ of branched- to straight-chain alcohol formation from 1-olefin is the same as for propene in Example 5.2, namely 0.058:0.677 = 0.086, and to attribute 2-methyl pentanol produced in excess of this to k_{2Q}. Experiments with other initial isomer distributions and sampling at short reaction times would be needed to confirm these two guesses. If they are accepted, a value of $k \cong 0.0045$ min^{-1} from the straight-line slopes of the first-order plots and the data on incremental conversions in the range beyond 60 minutes give the following results with eqns 5.42:

$$k_{1P} \cong 0.059 \text{ min}^{-1}, \qquad k_{1Q} \cong 0.005 \text{ min}^{-1}, \qquad k_{1S} \cong 0.006 \text{ min}^{-1},$$

$$k_{2Q} \cong 0.0007 \text{ min}^{-1}, \qquad k_{2R} \cong 0.0006 \text{ min}^{-1}$$

The values of k_{1P}, k_{1Q}, and k_{1S} for conversion of the highly reactive isomer A_1 still await regression to account for a possible drain effect, which in this case turns out to be rather severe: The corrected coefficients are almost three times as large as the estimates above. Otherwise the fit is fairly reasonable, with the following exception.

The only unsatisfactory feature of the fit is that the first-order plot of the observed results for pentane dips farther down than that for *n*-hexanol, the product whose initial rate should be boosted most strongly by the initial abundance of 1-pentene, its only parent (see Fig. 5.9). This behavior is at odds with any fit based on the network 5.43. A better fit would be obtained with a physically impossible negative value of k_{2S} or inclusion of a mechanistically inconceivable direct pathway from 2-pentene to *n*-hexanol. The most likely explanation of the effect is that mass transfer from the gas phase could not quite keep pace with the extremely high initial rate of consumption of CO and H_2 in the liquid, so that the concentrations of these reactants in the liquid decreased temporarily. The effect would be stronger for CO as the larger molecule, and thereby cause a temporary increase in the H_2-to-CO ratio in the liquid. A higher ratio favors paraffin production at the expense of the alcohols (see Example 7.5 in Section 7.3.2) and so can explain the observed behavior. (For more detail on mass-transfer effects in this reaction, see Section 12.3).

This has been a typical example of how coupled parallel steps or reactions can be recognized in a practical situation, how isomerization embedded in a larger network manifests itself, and how the results of a single batch experiment can yield a wealth of kinetic information and lead to a workable model. However, a good fit to the experimental results of one experiment does not prove the network to be correct, even if backed up by a very plausible mechanism. Confirmation from experiments starting with other isomer compositions is needed, quite apart from the fact that each of the arrows in a network such as 5.43 represent multistep pathways rather than single steps and that the branches may be at nodes along these pathways (for more detail, see Section 7.3.2). In hydroformylation, convincing evidence that the basic structure of the network 5.43 is correct has come from experiments with long-chain olefin isomers [7]: If the starting material is a 1-olefin, the first-order plots for the alcohols as in Figure 5.9 are such that that for the straight-chain product dips deepest, that for the alcohol with the $-CH_2OH$ group at the second carbon atom dips second deepest, etc., and that for the alcohol with the group attached to the middle of the chain dips least; and if the starting material is the isomer with double bond in the center of the molecule (say, the 6-position in dodecene), the sequence is the exact opposite. This confirms the essential features of the network 5.43, namely, that conversion occurs in concert with double-bond migration along the carbon chain and that the alcohols arise from attachment of CO and H_2 to a carbon atom on either side of the double bond.

5.4.   Sequential steps

In almost every chemical reaction, the product or products may react further or decompose. Optimization with respect to yield and product purity then relies on evaluation of *sequential steps* (also called *consecutive steps* or *steps in series*). Other situations involving sequential steps are those in which a substitution or transformation such as chlorination, oxidation, dehydrogenation, etc., occurs successively at different carbon atoms of a reactant molecule, or those involving repeated addition of a reagent to the original reactant, as in ethoxylations of ethanol and other low alcohols to glycol ethers [8,9] (e.g., Union Carbide's *Cellosolve* and *Carbitol* family of solvents) and of long-chain primary alcohols in manufacture of detergents [10]. Lastly, polymerization is a sequential reaction *par excellence* and so important that it has been assigned its own chapter (see Chapter 10).

5.4.1.   Sequential first-order steps

pathway: $A \longrightarrow K \longrightarrow P$ (5.44)

rate equations:

$$r_A = -k_{AK} C_A,$$

$$r_K = k_{AK} C_A - k_{KP} C_K,$$ (5.45)

$$r_P = k_{KP} C_K$$

This is the simplest case of sequential steps. Integration for constant-volume batch with only A present initially gives

$$C_A = C_A^o \exp(-k_{AK}t)$$ (5.46)

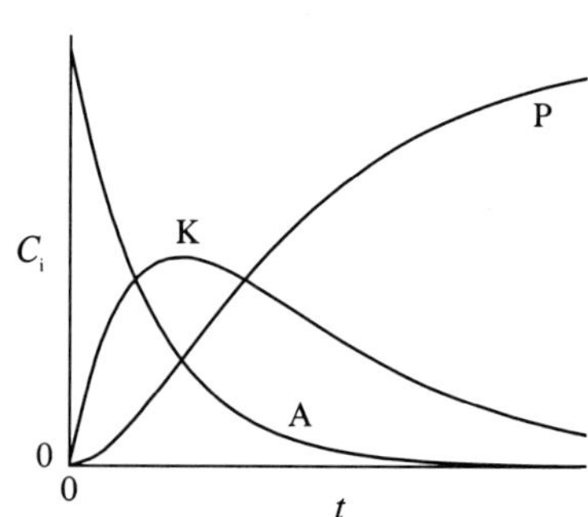

Figure 5.10. Concentrations vs time for first-order sequential steps $A \longrightarrow K \longrightarrow P$ in batch.

$$C_K = C_A^o \frac{k_{AK}}{k_{KP} - k_{AK}} \left(\exp(-k_{AK}t) - \exp(-k_{KP}t) \right)$$ (5.47)

$$C_P = C_A^o \left[1 - \frac{k_{KP} \exp(-k_{AK}t) - k_{AK} \exp(-k_{KP}t)}{k_{KP} - k_{AK}} \right]$$ (5.48)

The concentration of the intermediate K reaches its maximum at the time

$$t_{K_{max}} = \frac{\ln(k_{KP}/k_{AK})}{k_{KP} - k_{AK}}$$ (5.49)

and the height of this maximum is

$$(C_K)_{max} = C_A^o (k_{KP}/k_{AK})^{k_{KP}/(k_{AK} - k_{KP})} \tag{5.50}$$

The latter two equations are useful for determination of the rate coefficients from experimental results.

Equations 5.47 to 5.50 are not applicable if $k_{KP} = k_{AK}$. In that case they must be replaced by

$$C_K = C_A^o kt \exp(-kt) \tag{5.51}$$

$$C_P = C_A^o \left(1 - (1 + kt) \exp(-kt)\right) \tag{5.52}$$

$$t_{K_{max}} = 1/k \tag{5.53}$$

$$(C_K)_{max} = C_A^o/e \tag{5.54}$$

where $k = k_{AK} = k_{KP}$, and $e = 2.71828$ is the basis of natural logarithms.

If K is also present initially, eqn 5.47 contains an additional term $C_K^o \exp(-k_{KP}t)$ on the right-hand side; and eqn 5.48, a term $C_K^o(1 - \exp(-k_{KP}t))$. Also, eqns 5.49 and 5.50 no longer apply. The product P is inert; accordingly, any P present initially has no effect on A and K and only adds to the amount of P formed in the reaction as given by eqn 5.48 and 5.52.

For a continuous stirred-tank reactor, the concentrations as a function of the reactor space time τ are:

$$C_A = C_A^o/(1 + k_{AK}\tau) \tag{5.55}$$

$$C_K = \frac{k_{AK}\tau C_A^o}{(1 + k_{AK}\tau)(1 + k_{KP}\tau)} \tag{5.56}$$

$$C_P = \frac{k_{AK} k_{KP} \tau^2 C_A^o}{(1 + k_{AK}\tau)(1 + k_{KP}\tau)} \tag{5.57}$$

The maximum concentration of the intermediate K is attained at the reactor space time

$$\tau(K_{max}) = (k_{AK} k_{KP})^{-1/2} \tag{5.58}$$

and the height of the maximum is

$$(C_K)_{max} = \frac{(k_{AK}/k_{KP}) C_A^o}{[(1 + (k_{AK}/k_{KP})^{1/2}]^2} \tag{5.59}$$

Often, the pathway consists of more than two steps:

$$A \longrightarrow K_1 \longrightarrow K_2 \longrightarrow \ldots \longrightarrow K_i \longrightarrow \ldots \longrightarrow \ldots \tag{5.60}$$

The equations stated above for A and the first intermediate (now K_1 instead of K) then apply unchanged, and those for P give the sum of the concentrations of all participants beyond the first intermediate. This is because once a molecule has reacted to form K_2, it never reverts to K_1 or A. Mathematics for products beyond K_1 has been developed [8,11-13], but is complex except if all rate coefficients are equal. If so, the concentration of the i'th intermediate is given by [14,15]:

$$C_{K_i} = C_A^o \frac{(kt)^i}{i!} \exp(-kt) \tag{5.61}$$

Example 5.4. Ethoxylation of alcohols [10]. In the presence of base as catalyst and at moderate temperature, liquid alcohols react with gaseous ethene oxide, which is inserted as $-OC_2H_4-$ between the bulk of the molecule and the $-OH$ group. The resulting alcohol successively inserts further $-OC_2H_4-$ blocks to form higher adducts:

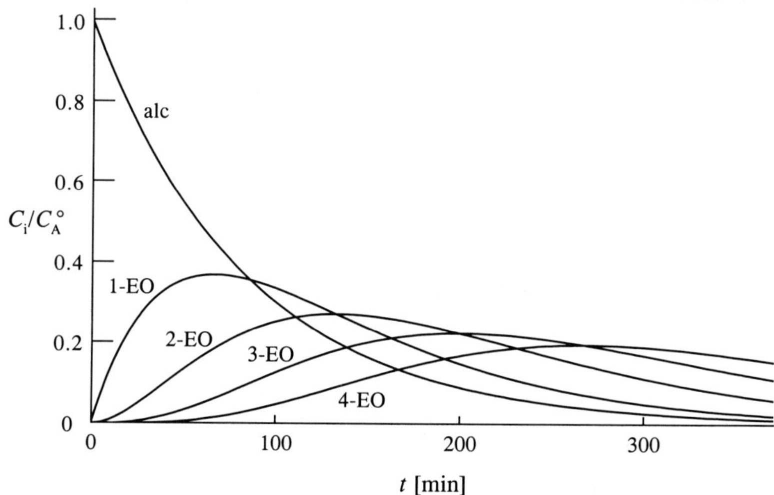

This reaction is important for the manufacture of household detergents, many of which are sulfated ethoxy adducts of straight-chain C_{10} to C_{16} primary alcohols averaging two to six ethoxy blocks per molecule.

Figure 5.11. Concentration histories of alcohol and first four successive ethoxy adducts in batch ethoxylation at constant partial pressure of ethene oxide, calculated for $k_1 = 0.012$ min^{-1} for reaction of alcohol, $k_2 = 0.015$ min^{-1} for reaction of adducts (alc = alcohol; 1-EO, 2-EO, 3-EO, 4-EO = first, second, third, and fourth adduct, respectively).

All steps from the second on amount to insertion of an ethoxy block between a previously inserted block and the $-$OH group, and so have very similar rate coefficients. Usually, the original alcohol reacts at a slightly lower rate. If the reaction is carried out at constant partial pressure of ethene oxide, each insertion including the first is pseudo-first order in the alcohol or ethoxy alcohol reactant. With increasing reaction time in batch, successive adducts reach maximum concentrations and then decay to form higher adducts, as shown for a calculated case in Figure 5.11. The variation in yield structure with reactor space time in a continuous stirred-tank reactor is similar, but with less pronounced concentration maxima.

Often, the first product in a pathway of sequential steps is the desired one. In such cases, the yield ratio of the first product (K) to the subsequent ones (lumped into P) is of special interest. The instantaneous yield ratio is

$$Y_{KP} \equiv r_K/r_P = \frac{k_{AK} C_A}{k_{KP} C_K} - 1 \tag{5.62}$$

At very low conversion, when the concentration of A is still high and little K has as yet been formed, the yield ratio is favorable; with progressing conversion it declines and at some point becomes negative as the decay rate of K starts to outrun the formation rate. Batch and plug-flow tubular reactors give better yields at same conversion than does a continuous stirred-tank reactor. This is because in a batch or tubular reactor the yield ratio is favorable at least initially—in the batch early on, in the tube near the inlet—and deteriorates only as conversion progresses, whereas in the stirred tank it is at the worst, final-conversion level all the time and in all of the reactor because the composition in the latter equals that of the effluent.

The situation is much the same if an early intermediate is the desired product.

If the desired product is the first or an early intermediate in a pathway of sequential steps, batch or plug-flow tubular reactors provide better selectivity than do continuous stirred-tank reactors.

The (cumulative) selectivity for K (fraction of reacted A that is converted to K, see definition 1.11) in a batch reactor is

$$\overline{S}_K = \frac{k_{AK}\left(\exp(-k_{AK}t) - \exp(-k_{KP}t)\right)}{(k_{KP} - k_{AK})(1 - \exp(-k_{AK}t))} \quad \text{if } k_{AK} \neq k_{KP} \tag{5.63}$$

$$\overline{S}_K = \frac{kt \exp(-kt)}{1 - \exp(-kt)} \quad \text{if } k_{AK} = k_{KP} \ (= k) \tag{5.64}$$

and in a continuous stirred-tank reactor:

$$\bar{S}_K \; = \; \frac{1}{1 + k_{KP}\tau} \qquad\qquad (5.65)$$

Equations 5.63 to 5.65 give the selectivity as a function of time or reactor space time. More helpful in practice, however, is the dependence on fractional conversion, f_A. For a batch reactor:

$$\bar{S}_K \; = \; \frac{k_{AK}}{k_{KP} - k_{AK}} \left[\frac{1 - f_A - (1 - f_A)^{k_{KP}/k_{AK}}}{f_A} \right] \qquad \text{if } k_{AK} \neq k_{KP} \qquad (5.66)$$

$$\bar{S}_K \; = \; \frac{1 - f_A}{f_A} \ln\left(\frac{1}{1 - f_A} \right) \qquad \text{if } k_{AK} = k_{KP} \qquad (5.67)$$

and for a continuous stirred-tank reactor:

$$\bar{S}_K \; = \; \frac{1 - f_A}{1 + f_A (k_{KP}/k_{AK} - 1)} \qquad\qquad (5.68)$$

This reduces to $\bar{S}_K = 1 - f_A$ if the two coefficients are equal.

While the selectivity to K decreases monotonically with conversion, the concentration of K is at its maximum in a batch reactor at

$$(f_A)_{K_{max}} \; = \; 1 - \left(\frac{k_{AK}}{k_{KP}} \right)^{\frac{k_{AK}}{k_{KP} - k_{AK}}} \qquad \text{if } k_{KP} \neq k_{AK} \qquad (5.69)$$

$$(f_A)_{K_{max}} \; = \; 1 - 1/e \; = \; 0.6321 \qquad \text{if } k_{KP} = k_{AK} \qquad (5.70)$$

and in a continuous stirred-tank reactor at

$$(f_A)_{K_{max}} \; = \; \frac{1}{1 + (k_{KP}/k_{AK})^{1/2}} \qquad\qquad (5.71)$$

or $(f_A)_{K_{max}} = 1/2$ if the rate coefficients are equal.

Example 5.5. Oxidation of paraffins to secondary alcohols. Alcohols can be produced by oxidation of paraffins with air or oxygen at moderate temperatures (typically 120 to 180° C) in the presence of boric-acid esters or boroxines [16-18]. These intercept the alkyl peroxide, the first oxidation product, preventing it from generating free radicals that would cause further degradation including scission of carbon-carbon bonds and produce aldehydes, ketones, and acids (see also Section 9.6.2). The peroxy borates so formed then are hydrolyzed to yield the alcohol. The carbon atoms at the chain ends are largely immune to oxidation, so the product consists predominantly of isomeric secondary alcohols. The reaction does not stop at the mono-alcohol. Rather, subsequent oxidation at other carbon atoms introduces additional $-OH$ groups:

$$\text{paraffin} \xrightarrow{\frac{1}{2}O_2} \text{mono-alcohol} \xrightarrow{\frac{1}{2}O_2} \text{di-alcohol} \xrightarrow{\frac{1}{2}O_2} \text{tri-alcohol} \xrightarrow{\frac{1}{2}O_2} \dots$$

Since each step consists of the same chemical event, the oxidation of a secondary carbon atom, the rate coefficients of all steps are almost the same. [With each attack the number of still unoxidized secondary carbon atoms in the molecule decreases, and so does the statistical probability of further oxidation; however, for the first few products of a long-chain paraffin this effect remains minor.] At constant partial pressure of oxygen, the steps are pseudo-first order in the respective organic reactant. The dependence of the selectivity to mono-alcohol, usually the desired product, on reactor type and conversion level is dramatic, as the data in Table 5.6 demonstrate.

Table 5.6. Cumulative selectivity to first intermediate in pathway of first-order steps $A \longrightarrow K \longrightarrow \dots$ with equal rate coefficients.

fractional conversion f_A	selectivity to K $C_K/(C_A^\circ - C_A)$	
	batch	CSTR
0.20	0.892	0.800
0.50	0.693	0.500
0.75	0.462	0.250
0.90	0.256	0.100

5.4.2. Sequential steps of other orders

As long as the steps of the sequence are irreversible, the original reactant behaves as in single-step decay, regardless of the reaction orders. Concentration histories of later members in sequences that include steps of orders other than first have been derived for only a few simple cases and are unwieldy even for these [19,20]. Here, numerical solution on a computer is preferable. However, two general rules can be stated:

- If the desired product is the first or an early intermediate, batch and plug-flow tubular reactors provide better selectivity than does a continuous stirred-tank reactor.

- Higher concentration of the original reactant favors the intermediates whose formation rates are of higher reaction orders than are their decay rates.

The first of these rules had already been stated for pathways of first-order steps, the second is a consequence of the stronger concentration dependence of reactions of higher order (see also Section 5.2.3).

The rules have implications for reactor choice and operating conditions in situations in which a desired product undergoes subsequent decay. If the desired reaction is of higher overall order than the decay, selectivity is better in a batch or tubular reactor than in a continuous stirred tank, and in batch or tube it is better at higher charge or feed concentrations. On the other hand, if the decay is of higher order than the desired reaction, the opposite is true.

5.5. Competing steps

The term *competing steps* is used if one and the same component participates as reactant in more than one step of the pathway or network. [The idea is that such steps "compete" with one another for the reactant they have in common; an alternative term is *series-parallel steps*.] The simplest such case is [21]

$$\text{pathway:} \qquad A \longrightarrow K \overset{\displaystyle A}{\searrow} P$$

Of more interest is the slightly more complex pathway with reversible first step:

$$\text{pathway:} \qquad A \longleftrightarrow K \overset{\displaystyle A}{\searrow} P \qquad (5.72)$$

rate equations:

$$r_A = -k_{AK}C_A + k_{KA}C_K - k_{KP}C_AC_K$$

$$r_K = k_{AK}C_A - k_{KA}C_K - k_{KP}C_AC_K \qquad (5.73)$$

$$r_P = k_{KP}C_AC_K$$

Many condensation reactions are of this general type, though the actual mechanism may be more complex (see also Section 8.2.1).

If the intermediate K in the pathway 5.72 remains at trace level, the Bodenstein approximation can be used (see Section 4.3) and gives

$$C_K \cong k_{AK}C_A/(k_{KA} + k_{KP}C_A)$$

so that

$$r_P \cong \frac{k_{AK}k_{KP}C_A^2}{k_{KA} + k_{KP}C_A} \qquad (5.74)$$

Although all exponents are integers, the rate equation is in general not a power law, so the reaction order may vary with conversion. Two special cases are noteworthy:

Case I $(k_{KA} \ll k_{KP}C_A)$	*Case II* $(k_{KA} \gg k_{KP}C_A)$
$r_P \cong k_{AK}C_A$	$r_P \cong (k_{AK}k_{KP}/k_{KA})C_A^2$
reaction is first order in A	reaction is second order in A

> A reaction 2A $\longrightarrow$ P with reversible step A $\longleftrightarrow$ K preceding conversion A + K $\longrightarrow$ P may be first order or second order or have any order between these two, depending on the relative magnitudes of the coefficients.

In Case I, the reverse step K $\longrightarrow$ A is negligible because it is outrun by the much faster second step A + K $\longrightarrow$ P. Here, the slow first forward step controls the rate. In Case II, the step A + K $\longrightarrow$ P is so slow that quasi-equilibrium is established in the first step. Here, the reaction rate is proportional to the rate coefficient of the slow, second step, but is also affected by the equilibrium in the first (equilibrium constant $K_{AK} = k_{AK}/k_{KA}$).

As this examination shows, the order of a reaction with pathway 5.72 depends on the relative magnitudes of k_{KA} and $k_{KP}C_A$. Cases I and II, with first- and second-order behavior, respectively, are the extremes. In between, with the two terms in the denominator of eqn 5.74 of comparable magnitude, the order is between first and second and increases with conversion. It is even possible for the same reactant to have different reaction orders under different conditions, for instance, in different solvents.

The pathway 5.72 is also the simplest that can produce another anomaly. The equilibrium constant K_{AK} may decrease with increasing temperature. If this decrease is stronger than the increase of the rate coefficient k_{KP} and if the behavior is that of Case II, the rate of the overall reaction may decrease with increasing temperature (negative activation energy). This situation will be discussed in more detail in Section 12.1.1.

The pathway 5.72 is the simplest in which a reversible step precedes a competing irreversible one. Such behavior is quite common, especially in catalysis. A non-constant reaction order between one and two with respect to the reactant for which the steps compete is a typical symptom, even though the actual pathway usually is more complex. Two examples—ethyne dimerization and aldol condensation—will be examined later (see Examples 6.4 and 8.2 in Sections 6.4.3 and 8.2.1, respectively).

5.6. Reactions with fast pre-dissociation

A number of reactions involve dissociation of a reactant as the first step. In many such cases, this step is reversible and fast enough compared with subsequent conversion of the dissociation products to be at quasi-equilibrium. Typical examples are gas-phase reactions of halogens or hydrogen halides.

The simplest network of this type is

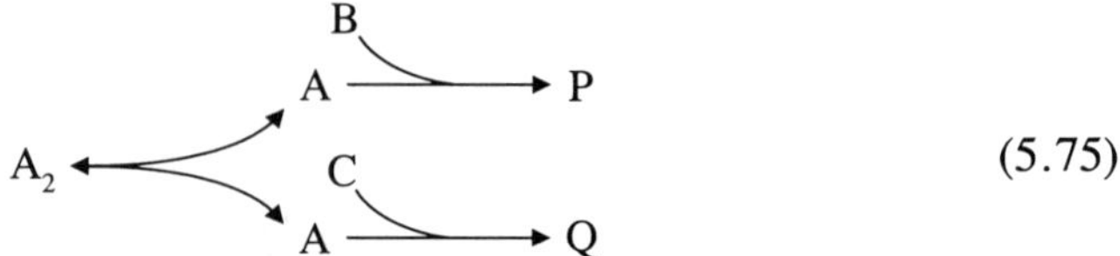

$$(5.75)$$

With the quasi-equilibrium condition

$$p_A^2/p_{A_2} \;\cong\; K_d \qquad \text{so that} \qquad p_A \;\cong\; (K_d p_{A_2})^{1/2}$$

(K_d = dissociation constant of A_2) the rate equation becomes

$$r_P \;=\; k_{AP} p_A p_B \;\cong\; k_{AP} p_B (K_d p_{A_2})^{1/2} \qquad\qquad (5.76)$$

If conversion of A is reversible, a term for the reverse step appears:

$$r_P \;\cong\; k_{AP} p_B (K_d p_{A_2})^{1/2} - k_{PA} p_P \qquad\qquad (5.77)$$

Conversion of the dissociation product may occur via a more complex pathway rather than a single step, or may involve other co-reactants or co-products. However, as long as that conversion is first order in A, the square-root factor $(K_d p_{A_2})^{1/2}$ in the rate equation remains unchanged.

The reaction may start with dissociation of a reactant into unequal fragments. The rate equation then becomes complex unless dissociation equilibrium is highly unfavorable so that the dissociation products remain at trace level:

$$(5.78)$$

If so, the formation rates of P and Q are necessarily equal by virtue of the overall stoichiometry, $A \longrightarrow P + Q$ (amounts present as intermediates remain negligible). This leads to a simple rate equation:

$$r_P \;\cong\; r_Q \;\cong\; (K_d k_{XP} k_{YQ} p_A p_B p_C)^{1/2} \qquad\qquad (5.79)$$

If conversion of the dissociation products is reversible, an explicit rate equation can still be obtained, but involves the root of a quadratic equation.

Derivation of eqn 5.79. From the equality of the formation rates of P and Q

$$r_P \;=\; k_{XP} p_X p_B \;=\; r_Q \;=\; k_{YQ} p_Y p_C$$

one finds

$$p_Y = p_X k_{XP} p_B / k_{YQ} p_C$$

Using this relationship to replace p_Y in the quasi-equilibrium condition

$$p_X p_Y / p_A \cong K_d$$

and solving for p_X one obtains

$$p_X \cong (K_d p_A k_{YP} p_C / k_{XP} p_B)^{1/2}$$

Substitution of this expression for p_X in the rate equation $r_P = k_{XP} p_X p_B$ gives eqn 5.79.

Characteristic of the rate equations 5.76 and 5.79 is their one-half order with respect to the dissociating reactant, in the case of eqn 5.79 with respect to the co-reactants B and C as well. This is an exception to the rule that a reasonably simple mechanism does not give a rate equation with fractional exponents. Conversely, an observed, conversion-independent order of one half is an indication that the reaction might involve fast pre-dissociation.

On the other hand, a power-law rate equation with integer reaction orders cannot be taken as evidence against fast pre-dissociation. An example is the hydrogen-iodide reaction, which involves fast pre-dissociation although the reaction orders are integers (see Example 4.2 in Section 4.2). Here, conversion $2I + H_2 \longrightarrow 2HI$ of the dissociation products is second order in these, so that the square-root factor is squared, making the reaction first order in I_2.

Positive or negative fractional exponents of one half or integer multiples of one half are also common in rate equations of chain reactions, where, however, they are caused by binary termination steps rather than fast pre-dissociation (see Chapter 9).

Apart from chain reactions, the most common occurrence of fractional exponents of one half or integer multiples of one half is in heterogeneous catalysis. For example, in hydrogenation on Group VIII transition metals, hydrogen is adsorbed as atoms, $H_2 \longleftrightarrow 2H(ads)$, producing such behavior.

5.7. General solution for first-order networks

An elegant, general solution for first-order networks has been provided in a classic publication by Wei and Prater [22].* In essence, the mathematics are developed for a reaction system with any number of participants that are all connected with one another by direct first-order pathways. For example, in a system with five participants, each of these can undergo four reactions, for a total of twenty first-order steps. Matrix methods are used to obtain concentration histories in constant-volume batch reactions, and a procedure is described for determination of all rate coefficients from such batch

---

* An abstract mathematical solution without the concept of reaction paths and no guidance for determination of coefficients from experiments had been given earlier by Matsen and Franklin [23].

experiments. The method is clearly presented in the original publication and in several advanced texts on kinetics [22,24,25], to which the reader wishing to apply it is referred for details. Only a brief outline is given here.

In principle, all the reactions in a completely interconnected network are coupled with one another. The key to obtaining explicit solutions is to uncouple them. Wei and Prater achieve this by defining and identifying "pseudo-components" which are combinations of real components in ratios such that they do not interconvert into one another. If the amounts of the real components are divided up among the pseudo-components in this way, the latter react independently of one another, that is, their reactions are mathematically uncoupled. This allows the rate coefficients to be determined from experiments and provides for a simpler description of the reaction behavior.

The determination of the rate coefficients from experimental results is best described with an example. The simplest case is that of interconversion of three isomers:

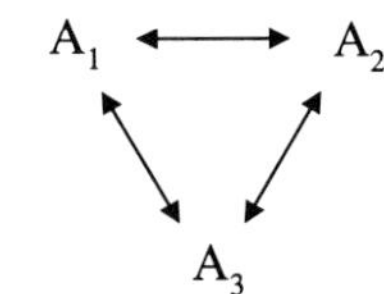

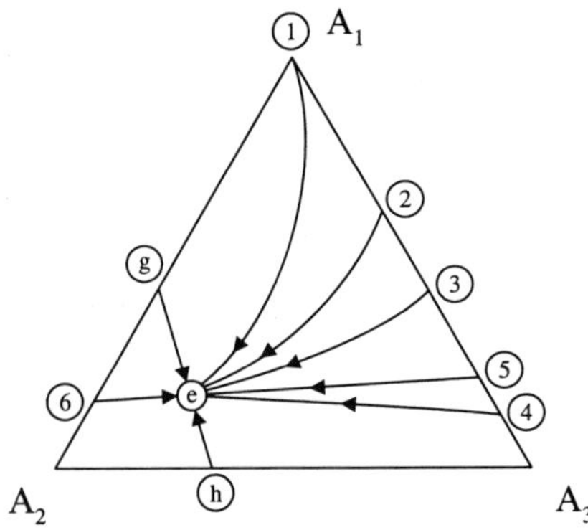

Figure 5.12. Reaction paths in a three-component, first-order reaction system (from Froment and Bischoff [25], schematic).

Figure 5.12 shows histories of several batch experiments, plotted in a composition diagram (triangular diagram with the isomer mole fractions as coordinates). The composition variation with time appears as a curved path. The path of each experiment ends up at the equilibrium point, e. A first experiment starts with pure isomer A_1 (point 1) and describes a curved path, and so do experiments starting with mixtures of A_1 and A_3 corresponding to points 2, 3, and 4. Determination of the rate coefficients calls for identification of the point from which a *straight-line path* originates. By trial and error, point 5 is found to meet that requirement. For confirmation, an experiment is started with an A_1-A_2 mixture corresponding to point 6, on the linear extension of the straight-line path from 5 to e, and its history is found to follow that extension as required. In a like manner, the end points g and h of a second straight-line path across the diagram are identified. Matrix manipulations can then be used to calculate the values of the six rate coefficients from the compositions 5, 6, g, and h and the rate at which the composition point of a system moves from one of these points toward the equilibrium point. [The points of origin of the straight-line paths are the pseudo-components.]

The method is noteworthy for its generality and mathematical elegance and for the insight into reaction behavior it has contributed. However, it is more often quoted than applied in practice. The reason is not mathematical complexity; today, the required matrix operations are easily performed on a computer. Rather, as the authors themselves pointed out, a relatively large number of experiments must be performed and evaluated. Also, if the number of participants is large, the very generality of the method becomes a disadvantage because many straight-line paths in a multidimensional space must be determined, and the possibility of converging on a false solution becomes real. Lastly, the method, although general in all other respects, has one limitation in that it implies a fully reversible system with unique equilibrium composition. Two examples may illustrate simple cases in which the method fails to give complete results. Figure 5.13 shows typical reaction paths of a three-component system with sequential, irreversible steps A $\longrightarrow$ K $\longrightarrow$ P (see also Section 5.4). There is only one straight-line reaction path, from K to P. Any initial mixture containing A produces a curved path. Here, the method yields the coefficient k_{KP} (from the rate of advance along that path) and the information that the coefficients k_{PK}, k_{PA}, k_{KA}, and k_{AP} must be zero or close to zero, but no value of k_{AK}.

The second example is that of a network with coupled parallel steps:

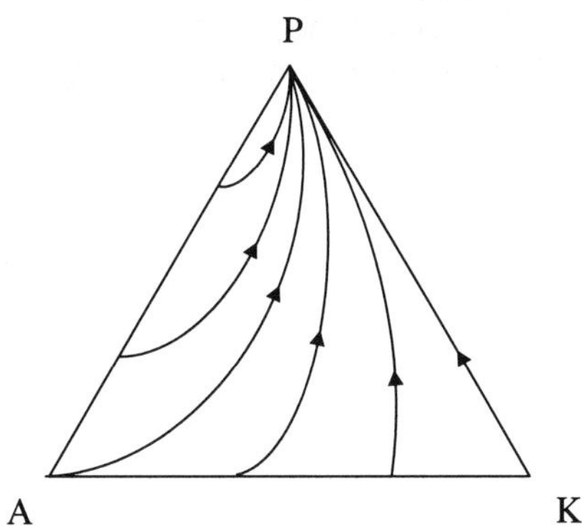

Figure 5.13. Reaction paths in a three-component, first-order reaction system with pathway A $\longrightarrow$ K $\longrightarrow$ P (schematic).

$$A_1 \longrightarrow P$$
$$\updownarrow \qquad\qquad\qquad (5.33)$$
$$A_2 \longrightarrow Q$$

(see Section 5.3). Here, too, only one straight-line reaction path can be found, namely, that for initial mixtures containing the isomers A_1 and A_2 in their steady-state ratio (this mixture, called a "homogeneous source" in Section 5.3, is a pseudo-component as defined here). Unless isomerization is very fast compared with conversion, there is no unique equilibrium point because the relative amounts of P and Q at complete conversion depend on the initial isomer distribution. The point of origin of the single straight-line path and the rate of progress along that path are insufficient for determination of the four non-zero rate coefficients of the network.

The method is at its best when applied to interconversion of not too large a number of isomers such as the butenes. For networks involving combinations of multistep pathways, techniques to be described in the next chapters are usually easier to handle.

Summary

This chapter surveys the kinetic behavior of the following step combinations:

reversible steps,

parallel steps (concurrent steps),

coupled parallel steps,

sequential steps (consecutive steps, steps in series),

competing steps (series-parallel steps), and

reactions with fast pre-dissociation.

In first order-first order reversible reactions, the rate of approach to equilibrium is proportional to the fractional distance from equilibrium, measured in terms of any quantity that is a linear function of the concentrations. The same rule holds true for any participant in reactions with first-order parallel steps.

If not carried to high conversion, the rate of a reversible reaction may be indistinguishable from that of an irreversible reaction of higher order.

In reactions with parallel steps of different reaction orders or with sequential steps, selectivities depend on the reactor type. Batch and plug-flow tubular reactors give higher selectivities to the product formed by the parallel step of higher order, or to the first product in a step sequence, than do continuous stirred-tank reactors.

Coupled parallel steps are an important combination not covered in any standard texts, and are therefore examined in more detail. Typical examples are isomerization in concert with conversion of the isomers to different products. If isomerization is very fast compared with conversion, the isomers are at quasi-equilibrium and act as "homogeneous source," producing a kinetic behavior like that of a single reactant. If isomerization is very slow compared with conversion, the reactions of the different isomers are essentially uncoupled. If the rates of isomerization and conversion are comparable, a more complex behavior ensues. Most interesting is the case with isomerization being somewhat faster than conversion. The isomer distribution then approaches a steady state (not necessarily close to equilibrium), and from then on the isomers act as homogeneous source.

Reactions in which a step competes with an earlier reversible step are the simplest that may show features not found in single-step reactions: The rate equation may not be a power law, the apparent reaction order may be fractional and vary with conversion, and the activation energy may be negative (lower rate at higher temperature).

Reactions with fast pre-dissociation may, but need not, lead to fractional reaction orders of one half or integer multiples of one half or non-power law rate equations involving such exponents.

The various combinations may show their characteristic symptoms, or be made to do so by choice of experimental conditions, even if they are composed of multistep pathways rather than single steps or if they are embedded in larger networks.

A brief overview of the Wei-Prater general mathematical solution for arbitrary networks consisting entirely of reversible first-order steps is provided.

The presentation is geared toward recognition of symptoms of typical step combinations in observable kinetic behavior as an essential facet of network elucidation.

Examples include reversible lactone formation, hydroformylation of propene and 1-pentene, ethoxylation of aliphatic alcohols, and oxidation of paraffins to secondary alcohols.

References

General references

G1. S. W. Benson, *The foundations of chemical kinetics*, McGraw-Hill, New York, 1960, updated and corrected reprint, Krieger, Melbourne, 1982, ISBN 0898741947, Chapters III and V.

G2. J. J. Carberry, *Chemical and catalytic reaction engineering*, McGraw-Hill, New York, 1976, ISBN 0070097909, Section 2-9.

G3. N. H. Chen, *Process reactor design*, Allyn & Bacon, Boston, 1983, ISBN 0205079032, Sections 3.12 to 3.15.

G4. K. A. Connors, *Chemical kinetics: the study of reaction rates in solution*, VCH Publishers, New York, 1990, ISBN 3527218223, Section 3.1.

G5. J. H. Espenson, *Chemical kinetics and reaction mechanisms*, McGraw-Hill, New York, 2nd ed., 1995, ISBN 0070202605, Chapters 3, 4, and 5.

G6. C. G. Hill, Jr., *An introduction to chemical engineering kinetics and reactor design*, Wiley, New York, 1977, ISBN 0471396095, Chapter 5.

G7. O. Levenspiel, *The chemical reactor omnibook*, OSU Book Stores, Corvallis, 1989, ISBN 0882461648, Chapters 2 and 8.

G8. J. W. Moore and R. G. Pearson, *Kinetics and mechanism: a study of homogeneous chemical reactions*, Wiley, New York, 3rd. ed., 1981, ISBN 0471035580, Chapter 8.

G9. J. I. Steinfeld, J.S. Francisco, and W. L. Hase, *Chemical kinetics and dynamics*, Prentice-Hall, Englewood Cliffs, 2nd ed., 1999, ISBN 0137371233, Chapter 2.

G10. Z. G. Szabó, in *The theory of kinetics*, Vol. 2 of *Comprehensive chemical kinetics*, C. H. Bamford and C. F. H. Tipper, eds., Elsevier, Amsterdam, 1969, ISBN 0444406743, Chapter 1.

Specific References

1. N. M. Emanuel' and D. G. Knorre, *Kurs khimicheskoi kinetiki (gomogennye reaktsii)*, Izdatel'stvo Vysshaya Shkola, Moscow, 2nd ed., 1969; English translation *Chemical kinetics—homogeneous reactions*, Wiley, New York, 1973, ISBN 0706513185, pp. 167-169.

2. F. G. Helfferich, unpublished, 1965.

3.   V. W. Weekman, Jr., and D. M. Nace, *AIChE J.*, **16** (1970) 397.

4.   D. M. Nace, *I&EC Prod. Res. Develop.*, **8** (1969) 24.

5.   Hill (ref. G6), Appendix A.

6.   *Handbook of chemistry and physics*, 80th ed., D. R. Lide, ed. CRC Press, Boca Raton, 1999, ISBN 0849304806, Section 5, pp. 40 and 47.

7.   F. G. Helfferich, *Technical Report 203-65*, Shell Development Company, 1965.

8.   B. Weibull and B. Nycander, *Acta Chem. Scand.*, **8** (1954) 847.

9.   J. E. Logsdon, *Ethanol*, in *Kirk-Othmer, Encyclopedia of chemical technology*, 4th ed., J. I. Kroschwitz and M. Howe-Grant, eds., Wiley, New York, Vol. 9, 1994, ISBN 0471526770, p. 846.

10.   J. L. Lynn, Jr., and B. H. Bory, *Surfactants*, in *Kirk-Othmer, Encyclopedia of chemical technology*, 4th ed., J. I. Kroschwitz and M. Howe-Grant, eds., Wiley, New York, Vol. 23, 1997, ISBN 0471526924, p. 501-502.

11.   E. Abel, *Z. physik. Chem.*, **56** (1906) 558.

12.   Szabó (ref. G10), Section 1.IV.3.

13.   A. E. R. Westman and D. B. DeLury, *Can. J. Chem.*, **34** (1956) 1134.

14.   P. J. Flory, *J. Am. Chem. Soc.*, **62** (1940) 1561.

15.   Benson (ref. G1), Section III.6.

16.   A. N. Bashkirov, *Dokl. Akad. Nauk SSSR*, **118** (1958) 149.

17.   F. Broich and H. Grasemann, *Erdöl Kohle Erdgas Petrochem.*, **18** (1965) 360.

18.   N. Kurata and K. Koshida, *Hydrocarbon Proc.*, **57**(1) (1978) 145.

19.   Benson (ref. G1), Section III.7.

20.   Szabó (ref. G10), Section 2.IV.

21.   M. Frenklach and D. Clary, *I&EC Fundamentals*, **22** (1983) 433.

22.   J. Wei and C. D. Prater, *Adv. Catal.*, **13** (1962) 203; *AIChE J.*, **9** (1963) 77.

23.   F. A. Matsen and J. L. Franklin, *J. Am. Chem. Soc.*, **72** (1950) 3337.

24.   M. Boudart, *Kinetics of chemical processes*, Prentice-Hall, Englewood Cliffs, 1968, Section 10.2.

25.   G. F. Froment and K. B. Bischoff, *Chemical reactor analysis and design*, Wiley, New Work, 2nd. ed., 1990, ISBN 0471510440, Sections 1.4 and 1.7.

Chapter 6

Practical Mathematics of Multistep Reactions

This chapter addresses the establishment of practical mathematics of kinetics for given pathways and networks. The complementary problem of establishing pathways or networks from observed kinetic behavior will be taken up in the next chapter. The discussion of special aspects of catalysis, chain reactions, and polymerization is deferred to Chapters 8 to 10.

In principle, a multistep reaction with any network is fully described by the complete set of rate equations of all participants, compiled as shown in Section 2.4. However, if the reaction is complex, the experimental work required to verify the assumed mechanism and to determine all its coefficients and their activation energies can get out of hand. A reduction of complexity then is imperative, and is also desirable for both a better understanding of reaction behavior and more efficient numerical modeling. The present chapter describes ways of achieving this.

The key elements of this approach are

- the use of pseudo-first order rate coefficients,

- a reduction of the number of simultaneous rate equations by use of the Bodenstein approximation, and

- the establishment of general equations for rates and yield ratios applicable to pathways, network segments, and networks regardless of their number of steps, locations of nodes, and points of co-reactant entries and co-product exits.

6.1. Simple and non-simple pathways and networks

For convenience, a distinction is made between "simple" and "non-simple" pathways or networks. A pathway, network, or any portion of one of these is called *simple* if it meets both of the following criteria:

- *all intermediates are and remain at trace level,* and

- *no step* (forward or reverse) *involves two or more molecules of intermediates as reactants.**

---

* Networks which meet this second condition have been called "linear" [1,2]. This term is avoided here because of the seeming self-contradiction if a structure with branches is declared linear.

The first of these conditions ensures that the Bodenstein approximation of quasi-stationary behavior (see Section 4.3) can be used for all intermediates, the second guarantees that the algebra is linear. If both conditions are met, explicit equations or algorithms for rates and yield ratios of all reactants and products can be given, regardless of the actual complexity of the network.

To be sure, a simple pathway or network, or any portion of these, may be of any size and topology and contain any number of steps with higher molecularities as long as the co-reactants or co-products are not themselves intermediates:

<table>
<tr><td>allowed steps</td><td>disallowed steps</td></tr>
</table>

$$X_j + A \longrightarrow X_k, \qquad\qquad 2X_j \longrightarrow X_k,$$
$$X_j \longrightarrow X_k + Q, \qquad\qquad X_j + X_\ell \longrightarrow X_k,$$
$$X_j + A + B \longrightarrow X_k \qquad\qquad 2X_j + A \longrightarrow X_k$$

where A, B, and Q are bulk reactants and products, and the X_i are intermediates.

A non-simple network can be broken at the offending point or points into piecewise simple portions (see Section 6.5).

The notation X_i will remain reserved for trace-level intermediates, with sequential indices. In indices of concentrations, rates, rate coefficients, etc., X is suppressed for simplicity. For instance, the concentration of X_j is given as C_j; the formation rate of X_j, as r_j; and the rate coefficient of a step $X_j \longrightarrow X_k$, as k_{jk}. For consistency in indices of sums and products, rate coefficients of steps $A \longrightarrow X_1$ are given as k_{01}, those of steps $X_{k-1} \longrightarrow P$ as $k_{k-1,k}$ (P = end product). Intermediates that are not, or do not remain, at trace level are designated K, L, etc.

6.2. Pseudo-first order rate coefficients

To use linear algebra as much as possible, pseudo-first order rate coefficients are introduced for steps with co-reactants. Each such coefficient is the product of the actual, higher-order rate coefficient and the concentration of the respective co-reactant (or concentrations of the co-reactants, if there are several) raised to the power that corresponds to the respective stoichiometric number. For example, for a step

$$X_j + B \longrightarrow X_k$$

the rate contribution

$$(-r_j)_{j \to k} = (r_k)_{j \to k} = k_{jk} C_j C_B$$

is written instead

$$(-r_j)_{j \to k} = (r_k)_{j \to k} = \lambda_{jk} C_j$$

where

$$\lambda_{jk} \equiv k_{jk}C_B \qquad \text{(for } X_j + B \longrightarrow X_k) \qquad (6.1)$$

This convention is used for all forward and reverse steps with more than one reactant (provided no more than one molecule of intermediates functions as reactant). For steps without co-reactant, e.g., $X_j \longrightarrow X_k$ or $X_j \longrightarrow X_k + Q$, the λ coefficient is identical with the actual, first-order k coefficient:

$$\lambda_{jk} \equiv k_{jk} \qquad \text{(for } X_j \longrightarrow ...) \qquad (6.2)$$

With this convention, and provided no step involves two or more molecules of intermediates as reactants, any reaction with or without co-reactants and co-products and with actual rate coefficients k_{jk} becomes mathematically formally equivalent to one composed exclusively of pseudo-first order steps with pseudo-first order rate coefficients λ_{jk}.*

As an example, a pathway

$$B \qquad\qquad A$$
$$A \rightleftharpoons X_1 \rightleftharpoons X_2 \rightleftharpoons X_3 \longleftrightarrow P$$
$$Q$$

is equivalent to the pseudo-first order pathway

$$A \longleftrightarrow X_1 \longleftrightarrow X_2 \longleftrightarrow X_3 \longleftrightarrow P$$

with coefficients

$$\lambda_{01} \equiv k_{01}C_B, \qquad \lambda_{12} \equiv k_{12}, \qquad \lambda_{23} \equiv k_{23}C_A, \qquad \lambda_{34} \equiv k_{34},$$

$$\lambda_{10} \equiv k_{10}, \qquad \lambda_{21} \equiv k_{21}C_Q, \qquad \lambda_{32} \equiv k_{32}, \qquad \lambda_{43} \equiv k_{43}.$$

6.3. General formula for simple pathways

The keystone in the reduction of mathematical complexity is a general rate equation for simple (linear) pathways with arbitrary number of steps and with or without co-reactants and co-products [6-9].

For the arbitrary simple pathway

$$A \longleftrightarrow X_1 \longleftrightarrow X_2 \longleftrightarrow ... \longleftrightarrow X_{k-1} \longleftrightarrow P \qquad (6.3)$$

with or without co-reactants and co-products (not shown) and, by definition, with all intermediates X_i at trace level, the full set of rate equations is

---

* The pseudo-first order rate coefficients λ_{jk} used here were first introduced by Christiansen [3] as "reaction probabilities" ω_j. Equivalent quantities are also standard in generalized treatments of chain reactions [4,5].

$$r_A \;=\; -\lambda_{01} C_A \,+\, \lambda_{10} C_1$$

$$r_i \;=\; \lambda_{i-1,i} C_{i-1} \,-\, (\lambda_{i,i-1} + \lambda_{i,i+1}) C_i \,+\, \lambda_{i+1,i} C_{i+1} \qquad (i = 1,\dots,k-1)$$

$$r_P \;=\; \lambda_{k-1,k} C_{k-1} \,-\, \lambda_{k,k-1} C_P$$

(Indices 0 and k in λ coefficients are used for A and P, respectively, to avoid complications in later formulas with sums and products.) If the end members A and P additionally act as co-reactants and co-products, the respective rates r_A or r_P must be replaced by $(1/n_A)r_A$ or $(1/n_P)r_P$ to account for the stoichiometry. After elimination of the concentrations of all the intermediates by repeated application of the Bodenstein approximation, the set of rate equations can be reduced to the single rate equation

$$-r_A \;\cong\; r_P \;\cong\; \Lambda_{0k} C_A \,-\, \Lambda_{k0} C_P \qquad\qquad (6.4)$$

where

$$\Lambda_{0k} \;\equiv\; \frac{\displaystyle\prod_{i=0}^{k-1} \lambda_{i,i+1}}{D_{0k}}, \qquad\qquad \Lambda_{k0} \;\equiv\; \frac{\displaystyle\prod_{i=0}^{k-1} \lambda_{i+1,i}}{D_{0k}}, \qquad (6.5)$$

$$D_{0k} \;\equiv\; \sum_{j=1}^{k} \left[\prod_{i=1}^{j-1} \lambda_{i,i-1} \prod_{i=j}^{k-1} \lambda_{i,i+1} \right] \qquad\qquad (6.6)$$

(products Π to be taken as unity if lower limit exceeds upper).

As eqn 6.4 shows:

A multistep simple pathway

A $\longleftrightarrow$ X$_1$ $\longleftrightarrow$ ... $\longleftrightarrow$ P

with or without co-reactants or co-products can be reduced to a single, pseudo-first order step

A $\longleftrightarrow$ P

The forward and reverse *segment coefficients* Λ_{0k} and Λ_{k0} of that pseudo-single step, however, are not in general concentration-independent. Given by eqns 6.5 and 6.6, they are functions of the true rate coefficients k_{ij} of all steps and of the concentrations of any co-reactants and co-products, but are independent of the concentrations of the intermediates. The problem of accounting for the effects of co-reactants and co-products has not been solved, but has been deferred to a time when it is more easily taken care of.

Equations 6.5 and 6.6 for the segment coefficients look rather formidable, but are in fact quite simple and very easy to remember and apply. The numerator of Λ_{0k}, the forward coefficient, is the product of all forward λ coefficients; similarly, the numerator of Λ_{k0}, the reverse coefficient, is the product of all reverse λ coefficients. The denominator D_{0k}, common to both segment coefficients, is easily obtained with the following recipe:

- *Construct a square matrix of order* k, *with elements 1 along the diagonal, with forward λ coefficients of the* m'*th step in the* m'*th column in all rows above the diagonal, and with reverse λ coefficients of the* m'*th step in the* m'*th column in all rows below the diagonal; then obtain D_{0k} as the sum of the products of the elements in each row.*

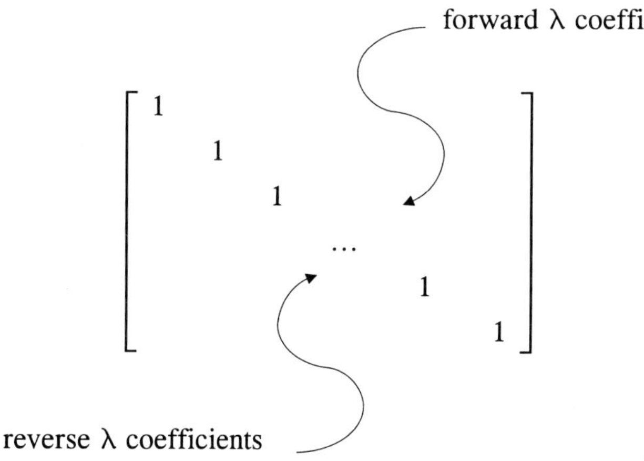

In detail, the matrix is:

$$
\begin{bmatrix}
1 & \lambda_{12} & \lambda_{23} & \cdots & \lambda_{k-2,k-1} & \lambda_{k-1,k} \\
\lambda_{10} & 1 & \lambda_{23} & \cdots & \lambda_{k-2,k-1} & \lambda_{k-1,k} \\
\lambda_{10} & \lambda_{21} & 1 & \cdots & \lambda_{k-2,k-1} & \lambda_{k-1,k} \\
 & & & \cdots & & \\
\lambda_{10} & \lambda_{21} & \lambda_{32} & \cdots & 1 & \lambda_{k-1,k} \\
\lambda_{10} & \lambda_{21} & \lambda_{32} & \cdots & \lambda_{k-1,k-2} & 1
\end{bmatrix}
\tag{6.7}
$$

For example, for a pathway with three intermediates (k=4) and with D_{0k} compiled as described, the procedure yields

One of the CO ligands is then inserted between metal and alkyl carbon to form a cobalt acyl (X_5), which is hydrogenated to a dihydride (X_6). Finally, the latter splits into aldehyde (2-methyl butanal) and the CO-deficient hydrocarbonyl (cat') to complete the cycle. At very high CO pressures, the cobalt acyl reversibly adds another CO ligand to form a catalytically inactive tetracarbonyl-acyl species (Y). All members of the cycle are at trace level; the tetracarbonyl catalyst is not, nor is the tetracarbonyl acyl at very high CO pressures. Ligand dissociation cat $\longleftrightarrow$ cat' + CO and olefin addition cat' + ole $\longleftrightarrow$ X_1, involving only coordinative bonds, can safely be taken as fast, so that cat, cat', olefin, and X_1 are at quasi-equilibrium (shown in a box in the network 6.9). Under reaction conditions, formation of the carbon-carbon bond ($X_4 \longrightarrow X_5$) is irreversible, but all preceding steps are reversible. With only a slight modification to be discussed farther below, this is the mechanism originally postulated by Heck and Breslow [10,11] and still found in current texts such as the most recent edition of Kirk-Othmer's encyclopedia [12].

If for the time being the side reaction to the tetracarbonyl acyl is disregarded, the pathway from π-complex to aldehyde in the network 6.9 can be written

$$
X_1 \xrightarrow{\ H_2\ } X_2 \underset{H_2}{\rightleftarrows} X_3 \xrightarrow{\ CO\ } X_4 \longrightarrow X_5 \xrightarrow{\ H_2\ } X_6 \xrightarrow{\ cat'\ } \text{ald}
$$

Application of eqns 6.4 to 6.6 to this pathway gives

$$
r_{\mathrm{ald}} \cong \frac{\lambda_{12}\lambda_{23}\lambda_{34}\lambda_{45}\cancel{\lambda_{56}}\cancel{\lambda_{67}}C_1 \;-\; \lambda_{21}\lambda_{32}\lambda_{43}\cancel{\lambda_{54}}\cancel{\lambda_{65}}\cancel{\lambda_{76}}\overrightarrow{C}_{\mathrm{ald}}}{\left(\begin{array}{l}\lambda_{23}\lambda_{34}\lambda_{45}\cancel{\lambda_{56}}\cancel{\lambda_{67}} \;+\; \lambda_{21}\lambda_{34}\lambda_{45}\cancel{\lambda_{56}}\cancel{\lambda_{67}} \;+\; \lambda_{21}\lambda_{32}\lambda_{45}\cancel{\lambda_{56}}\cancel{\lambda_{67}} \\ +\; \lambda_{21}\lambda_{32}\lambda_{43}\cancel{\lambda_{56}}\cancel{\lambda_{67}} \;+\; \cancel{\lambda_{21}\lambda_{32}\lambda_{43}\lambda_{54}\lambda_{67}} \;+\; \cancel{\lambda_{21}\lambda_{32}\lambda_{43}\lambda_{54}\lambda_{65}}\end{array}\right)} \tag{6.10}
$$

As in the previous example, terms containing zero-valued reverse coefficients (here λ_{54} and λ_{76}) are struck out, and so are the forward coefficients that then cancel. The condition of quasi-equilibrium of cat, cat', olefin, and π-complex (X_1)

$$
\frac{C_1 p_{\mathrm{CO}}}{C_{\mathrm{cat}} C_{\mathrm{ole}}} \cong K_1 = \text{const.}
$$

(validity of Henry's law is assumed) gives

$$
C_1 \cong \frac{K_1 C_{\mathrm{cat}} C_{\mathrm{ole}}}{p_{\mathrm{CO}}}
$$

With this substitution, replacement of the λ by k coefficients multiplied with co-reactant or co-product concentrations, and collection of terms, eqn 6.10 becomes

$$
r_{\mathrm{ald}} \cong \frac{K_1 k_{12} k_{23} k_{34} k_{45} C_{\mathrm{ole}} C_{\mathrm{cat}} p_{\mathrm{H_2}}}{(k_{23} k_{34} k_{45} + k_{21} k_{34} k_{45}) p_{\mathrm{CO}} + (k_{21} k_{32} k_{45} + k_{21} k_{32} k_{43}) p_{\mathrm{H_2}}} \tag{6.11}
$$

After division of numerator and denominator by the second term of the latter, a very simple rate equation with only two phenomenological coefficients is obtained:

$$ r_{ald} \cong \frac{k_a C_{ole} C_{cat}}{1 + k_b p_{CO}/p_{H_2}} \qquad (6.12) $$

where

$$ k_a \equiv \frac{K_1 k_{12} k_{23} k_{34} k_{45}}{k_{21} k_{32}(k_{45} + k_{43})} \,, \qquad k_b \equiv \frac{(k_{23} + k_{21}) k_{34} k_{45}}{k_{21} k_{32}(k_{45} + k_{43})} $$

Equation 6.12 is in very good agreement with the observed dependence on total pressure and H_2-to-CO ratio. In fact, an empirical rate equation of this form was established by Martin [13] well before the mechanism had been elucidated.

The effect of a "dead-end" side reaction such as that to the tetracarbonyl acyl will be examined in detail in Section 8.7. Suffice it for now to say that accumulation of cobalt in the form of this inactive species reduces the rate, which, nevertheless, still remains given by an equation of the form of eqn 6.12.

The key intermediates in the mechanism, the cobalt alkyls and acyls, are firmly established by independent experiments and have been synthesized [11]; the trihydride has not. The original Heck-Breslow mechanism [10,11] lacked that intermediate, and so do the networks found in current texts. Heck and Breslow postulated rate control by hydrogenation of cobalt acyl. Various minor modifications, most of them concerning rate control, were suggested by others [15-19]. However, neither Heck and Breslow's original mechanism nor any of these modifications can account for the observed rate behavior according to the Martin equation 6.12* or the fact that double-bond migration in the olefin, presumably occurring via the cobalt alkyl (X_3) but hardly the acyl dihydride (X_6), is accelerated by an increase in H_2 pressure as much as is conversion to aldehydes. Moreover, all must make the unlikely assumption that, under reaction conditions, CO insertion with carbon-carbon bond formation is reversible as otherwise the rate would be independent of hydrogen pressure. Barring new evidence to the contrary, the network 6.9 that includes the elusive and still controversial trihydride [14] appears to be the most likely candidate.

The example was chosen to illustrate pathway reduction and the resulting simplification of mathematics. So as not to distract from that message, complications encountered in practice with this particular reaction were disregarded. These include isomerization of 2-*cis*- to 2-*trans*- and 1-butene, conversion of 1-butene to 2-methyl butanal and *n*-pentanal, and aldol condensation of the latter (see also Example 11.1 in Section 11.2).

---

* This is true even if it is admitted that intermediates may contain significant fractions of total cobalt. Christiansen mathematics (see Section 8.4) shows that all then appearing additional denominator terms of the rate equation contain as factors a zero-value reverse coefficient of an irreversible step or the olefin concentration and must therefore be negligible (since the reaction is first order in olefin, the denominator may not include additive terms containing its concentration as a factor).

In this example, the unreduced network (without the side reaction to the tetracarbonyl-acyl) has twelve rate equations, one for each participant, and fourteen coefficients. Four rate equations can be replaced by stoichiometric constraints, namely, the carbon skeleton, cobalt, hydrogen, and CO balances. This still leaves eight rate equations with ten coefficients. Pathway reduction has reduced this to a single rate equation, eqn 6.12, with two coefficients.

The rate equation 6.12 was derived without use of any short-cuts. With rules to be shown in Section 7.3.1, algebra could have been reduced. Specifically, the steps following the irreversible formation of the cobalt acyl (X_5) could have been omitted (Rule 7.12) and the pairs of steps from X_1 to X_3 and from X_3 to X_5 been consolidated each into a single step (Rule 7.24). This would have reduced the mathematics of conversion of the π-complex to the aldehyde to that of a pathway with only two instead of six steps.

The general formula 6.4 to 6.6 for simple pathways is so important that a closer examination of its properties is warranted. First, we can see that it meets the requirement of thermodynamic consistency (Section 2.5.1): Setting $r_P = 0$ for equilibrium one finds

$$\prod_{i=0}^{k-1} \lambda_{i,i+1} C_A \;=\; \prod_{i=0}^{k-1} \lambda_{i+1,i} C_P$$

The forward λ coefficients contain the concentrations of all co-reactants as factors; the reverse λ coefficients, those of all co-products (the concentrations include those of A and P if these species additionally appear as co-reactants or co-products). Accordingly, after replacement of the λ coefficients by means of eqns 6.1 and 6.2:

$$\prod_{i=0}^{k-1} \frac{k_{i,i+1}}{k_{i+1,i}} \;=\; \text{const.} \;=\; \frac{C_P^{n_P} C_Q^{n_Q} \dots}{C_A^{n_A} C_B^{n_B} \dots} \tag{6.13}$$

where A, B, ... are the reactants and P, Q, ... are the products. This is in accordance with the mass-action law, so that the requirement of thermodynamic consistency is met.

Incidentally, eqn 6.13 also shows that, for multistep pathways, the thermodynamic equilibrium constant equals the ratio of the product of the forward rate coefficients to that of the reverse rate coefficients, or a power thereof.

Moreover, eqn 6.13 can be applied to a reversible catalytic reaction in which the catalyst acts as a co-reactant in an early step and is restored as a co-product in a later one. Its concentration, C_{cat}, then appears in both the numerator and denominator on the right-hand side of eqn 6.13, and so cancels, in accordance with the requirement that the presence of a catalyst does not affect equilibrium. In the rate equation developed from eqns 6.4 to 6.6, C_{cat} appears as co-factor in both terms of the numerator. This makes the rate first-order in the catalyst, provided C_{cat} does not also appear in some terms of denominator, as it will unless the catalyst reacts in the first step and is reformed in the last, as is usually true.

Lastly, the rate equation of a reversible reaction A ⟷ ... ⟷ P with linear simple pathway must have a symmetry property: In principle, one could declare P instead of A as the reactant, and A instead of P as the product. The rate equation then obtained must be the "mirror image" of the original one. Specifically, eqns 6.4 to 6.6 must be invariant to the exchange of all indices i for k − i (with $0 \le i \le k$), including A for P. Equation 6.4 is seen to have this symmetry property when the Λ coefficients are expressed in terms of the λ coefficients.

Derivation of equation 6.4. The general formula 6.4 to 6.6 for simple pathways can be derived in several different ways. First, the equation is a special case of a much more complex formula for catalytic reactions, derived by Christiansen [3] and later independently by King and Altman [20] (see Section 8.4). A proof along these lines, but specifically for the simple pathway 6.3, is as follows [9].

A necessary corollary of the Bodenstein approximation in a pathway is that the net rates of conversion are the same for all steps (an intermediate with higher formation than decay rate would not remain at trace level). In matrix form this condition is:

$$M * \begin{Bmatrix} 1/r \\ C_1/r \\ C_2/r \\ \dots \\ C_{k-1}/r \end{Bmatrix} = \begin{Bmatrix} 1 \\ 1 \\ 1 \\ \dots \\ 1 \end{Bmatrix}$$

Here, r is the net forward rate of each step:

$$r \equiv \lambda_{i,i+1} C_i - \lambda_{i+1,i} C_{i+1} \qquad (i = 0,\dots,k-1)$$

and M is a square matrix of order k given by

$$M \equiv \begin{bmatrix} \lambda_{01} C_0 & -\lambda_{10} & 0 & 0 & \dots & 0 & 0 \\ 0 & \lambda_{12} & -\lambda_{21} & 0 & \dots & 0 & 0 \\ 0 & 0 & \lambda_{23} & -\lambda_{32} & \dots & 0 & 0 \\ & & & \dots & & & \\ 0 & 0 & 0 & 0 & \dots & \lambda_{k-2,k-1} & -\lambda_{k-1,k-2} \\ -\lambda_{k,k-1} C_k & 0 & 0 & 0 & \dots & 0 & \lambda_{k-1,k} \end{bmatrix}$$

(Note indices 0 and k indicate A and P, respectively.) With Cramer's rule one finds:

$$1/r = \det M^\circ / \det M \qquad (6.14)$$

A network may be non-simple because an intermediate builds up to higher than trace concentration, or because an intermediate (at trace or higher concentration) reacts with itself. In both these cases, the network is cut into portions of formation and decay of that intermediate (Cases I and II, respectively, in Table 6.1).

If an intermediate (at trace or higher concentration) reacts with another intermediate, the network is cut into portions of formation and decay of the these intermediates and, if both are on the same pathway within the network, the portion of formation of one from the other (Cases III and IV in Table 6.1).

Table 6.1. Examples of break-up of non-simple reduced networks into piecewise simple portions (co-reactants and co-products of other steps not shown, and other parts of network assumed to be simple).

case	non-simple reduced network	simple portions
I	$A \leftrightarrow K \nearrow^{P} \searrow_{Q}$	$A \leftrightarrow K \qquad K \nearrow^{P} \searrow_{Q}$
II	$A \leftrightarrow X_j \nearrow^{X_j} \rightarrow P$	$A \leftrightarrow X_j \qquad 2X_j \leftrightarrow P$
III	$A \leftrightarrow X_j \qquad B \leftrightarrow X_k \searrow\nearrow P$	$A \leftrightarrow X_j \qquad B \leftrightarrow X_k \qquad X_j + X_k \leftrightarrow P$
IV	$A \leftrightarrow X_j \leftrightarrow X_k \nearrow^{X_j} \rightarrow P$	$A \leftrightarrow X_j \qquad X_j \leftrightarrow X_k \qquad X_j + X_k \leftrightarrow P$
V	$A \leftrightarrow X_j \nearrow^{X_k \leftrightarrow P} \searrow_{X_\ell \leftrightarrow Q}$	$A \leftrightarrow X_k + X_\ell \qquad X_k \leftrightarrow P \qquad X_\ell \leftrightarrow Q$

Loss of simplicity may or may not result from a step in which an intermediate dissociates into two others. If the step is reversible, its reverse and thus the network are non-simple. The latter then is broken into portions of formation and decay of the dissociation products (Case V in Table 6.1). On the other hand, if the step is irreversible, it does not offend simplicity. Mathematics then is the same as that of a fictitious (and stoichiometrically inconsistent) network in which two *parallel* steps with identical rate equations replace the single dissociation step:

$$\ldots \leftrightarrow X_j \Big\langle \begin{array}{l} X_k \leftrightarrow \ldots \\[4pt] X_\ell \leftrightarrow \ldots \end{array} \qquad \text{is replaced by} \qquad \ldots \leftrightarrow X_j \Big\langle \begin{array}{l} X_k \leftrightarrow \ldots \\[4pt] X_\ell \leftrightarrow \ldots \end{array}$$

If the networks in Table 6.1 contain other non-simple features, the affected portion would have to be subdivided accordingly.

As a rule, a reduction to a single, explicit rate equation (plus algebraic equations for stoichiometric constraints and yield ratios) is not achieved. Rather, the equations for the end members of the piecewise simple network portions must be solved simultaneously. Nevertheless, The concentrations of all trace-level intermediates that do not react with one another have been eliminated by this procedure and, in many cases of practical interest, the reduction in the number of simultaneous rate equations and their coefficients is substantial.

The reduction in mathematical complexity of non-simple networks is mainly of value in modeling. For network elucidation, a better approach usually is to study as many portions of the network separately, say, by experiments that use synthesized intermediates as starting materials or by appropriate lumping, as the following example will illustrate (see also Section 7.3.3).

Example 6.5.   Olefin hydroformylation with phosphine-substituted cobalt hydrocarbonyl catalyst [30]. The pathway 6.9 of olefin hydroformylation with the "oxo" catalyst, $HCo(CO)_4$, has been shown in Example 6.2 in Section 6.3. For phosphine-substituted catalysts, $HCo(CO)_3Ph$ (Ph = organic phosphine), the pathway olefin $\longrightarrow$ aldehyde is essentially the same. However, these catalysts also promote hydrogenation of aldehyde to alcohol (Examples 7.3 and 7.4) and of olefin to paraffin (Example 7.5). Moreover, straight-chain primary aldehydes under the conditions of the reaction undergo to some extent condensation to aldol, which is subsequently dehydrated and hydrogenated to yield an alcohol of twice the carbon number (e.g., 2-ethyl hexanol from *n*-butanal; see Section 11.2). The entire reaction system is

$$
\begin{array}{ccccc}
\text{olefin} & \xrightarrow{\;+\;H_2,\,CO\;} & \text{aldehyde} & \xrightarrow{\;+\;H_2\;} & \text{alcohol} \\[4pt]
{\scriptstyle +\,H_2}\Big\downarrow & & {\scriptstyle +\,\text{aldehyde}}\Big\downarrow & & \\[8pt]
\text{paraffin} & & \text{aldol} & \xrightarrow[{-\;H_2O}]{\;+\;2H_2\;} & \text{heavy alcohol}
\end{array}
\qquad (6.43)
$$

All arrows represent multistep, irreversible pathways, and aldehyde and possibly aldol build up to higher than trace concentrations. The network is non-simple for that reason and also because the aldehyde, an intermediate, acts in addition as a co-reactant in the pathway to the aldol.

The network can be cut into four piecewise simple portions:

$$\begin{array}{ll}
\text{olefin} \longrightarrow \text{aldehyde,} & \text{aldehyde} \longrightarrow \text{alcohol,} \\
\downarrow & \text{2 aldehyde} \longrightarrow \text{aldol,} \\
\text{paraffin} & \text{aldol} \longrightarrow \text{heavy alcohol}
\end{array} \tag{6.44}$$

(H_2 and CO co-reactants and H_2O co-product not shown). Since aldol dehydration and hydrogenation is fast, the portion from aldehyde to heavy alcohol can be treated without much detriment as a single, simple portion, at least if heavy-alcohol formation is a minor side reaction. The two pathways from olefin to aldehyde and paraffin cannot be separated because they must be expected to go through a common intermediate.

Since the portions are irreversible, none of them feeds back into a preceding one. Accordingly, the portions are not coupled except through the concentrations of their common members (aldehyde and aldol).

The network 6.43 still fails to include olefin isomerization, the distinction between isomeric products, and the effects of catalyst equilibria. The last of these will be examined in Section 8.2.2. A more complete network for the total reaction with a typical industrial catalyst and including isomerization will be discussed in Section 11.2.2.

In a number of reactions of practical interest, the offending non-simple step is a reversible dissociation of a reactant or intermediate, as in Case V in Table 6.1. Often, such a step is fast compared with the others and thus is at quasi-equilibrium. If so, the quasi-equilibrium approximation (see Section 4.2) can greatly simplify mathematics, in some instances even lead to an explicit rate equation. This has been discussed in Section 5.6.

Another case in point is chain reactions (see Chapter 9). As a rule, these are non-simple because their termination step involves two intermediates as reactants. Nevertheless, explicit rate equations can often be derived with the Bodenstein and long-chain approximations. The classical example is that of the hydrogen-bromide reaction (see Section 9.3).

An extensive, though abstract treatment of networks based on graph theory and including non-simple configurations has been provided in two recent books by Russian authors [1,2].

Summary

This chapter describes practical, simplified mathematics for ordinary multistep reactions, excluding for the time being catalysis, chain reactions, and polymerization.

The concept of "simplicity" of a pathway or network is introduced. For a pathway to be "simple," all its intermediates must be and remain at trace level, and no step may involve two or more molecules of intermediates as reactants. The first condition ensures applicability of the Bodenstein approximation to all intermediates, the second guarantees mathematics to be linear. In addition, pseudo-first order rate coefficients λ_{jk} are introduced, defined as the products of the real rate coefficients k_{jk} and the concentrations of any co-reactants.

The principal tool for simplifying mathematics is network reduction: Any simple pathway or simple network segment between nodes or between a node and an end member can be reduced to a pseudo-single, reversible, pseudo-first order step with first-order "segment coefficients" that are functions of all the real rate coefficients and the concentrations of any co-reactants and co-products, but are independent of the concentrations of the intermediates. An easily remembered general formula for these coefficients allows the rate equation for any simple segment to be written down immediately, regardless of the number of steps and participation of co-reactants and co-products. The procedure does not eliminate the complications of co-reactant and co-product involvement, but postpones it from the hard task of establishing a rate equation to simple substitutions in that equation.

For simple pathways, a single, explicit rate equation in terms of the concentrations of the bulk participants is immediately obtained. For simple networks of any degree of complexity, rate behavior is given by simultaneous rate equations for the network end members. An algorithm for compilation of these rate equations is provided.

For elucidation of all but quite small networks, rate equations are too unwieldy even if the network is simple. Here, the much more compact yield ratio equations are a better tool. These can also profitably be used to replace rate equations in modeling.

Many reactions of practical interest have non-simple pathways or networks, i.e., the concentration of an intermediate rises above trace level or a (forward or reverse) step involves two or more molecules of intermediates as reactants. If the majority of the steps meet the simplicity conditions, a significant reduction in mathematical complexity can still be achieved by cutting the network into piecewise simple portions at the offending steps. In some other instances, the quasi-equilibrium or long-chain approximations can be invoked in order to obtain explicit rate equations although the network is non-simple. On the other hand, if a majority of steps in a network are non-simple, the tools described here are of little use.

Practical examples include nitration of aromatics, olefin hydroformylation with cobalt hydrocarbonyls and phosphine-substituted hydrocarbonyls as catalysts, and ethyne dimerization.

References

1.   G. S. Yablonskii, V. I. Bykov, A. N. Gorban', and V. I. Elokhin, *Kinetic models of catalytic reactions*, Vol. 32 of *Comprehensive chemical kinetics*, R. G. Compton, ed., Elsevier, Amsterdam, 1991, ISBN 0444888020.
2.   O. N. Temkin, A. V. Zeigarnik, and D. Bonchev, *Chemical reaction networks: a graph-theoretical approach*, CRC Press, Boca Raton, 1996, ISBN 0849328675.
3.   J. A. Christiansen, *Adv. Catal.*, **5** (1953) 311.

4. A. A. Frost and R. G. Pearson, *Kinetics and mechanism, a study of homogeneous chemical reaction*, Wiley, New York, 2nd. ed., 1961, Chapter 10.

5. J. J. Carberry, *Chemical and catalytic reaction engineering*, McGraw-Hill, New York, 1976, ISBN 0070097909, Section 2-12.

6. F. G. Helfferich, *Detergent-range alcohols via hydroformylation; V. Kinetics*, Technical Report 203-65, Shell Development Co., 1965.

7. F. Wilkinson, *Chemical kinetics and reaction mechanisms*, Van Nostrand-Reinhold, New York, 1979, ISBN 0442302487, Chapter 3.

8. F. G. Helfferich, *J. Phys. Chem.*, **93** (1989) 6676.

9. J.-M. Chern and F. G. Helfferich, *AIChE J.*, **36** (1990) 1200.

10. D. S. Breslow and R. F. Heck, *Chem. & Ind.*, **1960**, 467.

11. R. F. Heck and D. S. Breslow, *J. Am. Chem. Soc.*, **83** (1961) 4023.

12. E. Billig and D. R. Bryant, *Oxo reaction*, in *Kirk-Othmer, Encyclopedia of chemical technology*, 4th ed., J. I. Kroschwitz and M. Howe-Grant, eds., Wiley, New York, Vol. 17, ISBN 047152686X, 1996, p. 902.

13. A. R. Martin, *Chem. & Ind.*, **1954**, 1536.

14. J. E. Mahler, unpublished, 1963.

15. J. Falbe, *Carbon monoxide in organic synthesis* (translated from German), Springer, Berlin, 1970, Section I.2.

16. M. van Boven, N. H. Alemdaroglu, and J. M. L. Penniger, *I&EC Prod. Res. Dev.*, **14** (1975) 259.

17. N. H. Alemdaroglu, J. M. L. Penniger, and E. Oltay, *Monatshefte Chem.*, **107** (1976) 1153.

18. J. P. Collman, L. S. Hegedus, J. R. Norton, and R. G. Finke, *Principles and applications of organotransition metal chemistry*, University Science Books, Mill Valley, 2nd ed., 1987, ISBN 0935702512, p. 622.

19. G. W. Parshall and S. D. Ittel, *Homogeneous catalysis: the application and chemistry of catalysis by soluble transition metal complexes*, Wiley, New York, 1992, ISBN 0471538299, Section 10.3.

20. E. L. King and C. Altman, *J. Phys. Chem.*, **60** (1956) 1375.

21. Parshall and Ittel (ref. 19), p. 197.

22. O. N. Temkin, G. K. Shestakov, and Yu. A. Treger, *Acetylene chemistry, reaction mechanisms, and technology*, Khimiya, Moscow, 1991 (in Russian).

23. Temkin *et al.* (ref. 2), p. 267.

24. Temkin *et al.* (ref. 2), Chapter 5.

25. *Landolt-Börnstein, New Series, Radical reaction rates in liquids*, H. Fischer, ed., Springer, Berlin, Part II Vol. 13 (5 subvolumes), 1983-1985, ISBN 0387126074, 0387132414, 0387117253, 0387121978, 0387136762; Vol. 18 (5 subvolumes), 1994-1997, 3540560548, 3540560556, 3540560564, 3540603573, 3540560572.

26. D. L.Baulch, C. J. Cobos, R. A. Cox, C. Esser, P. Frank, T. Just, J. A. Kerr, M. J. Pilling, J. Troe, R. W. Walker, and J. Warnatz, *J. Phys. Chem. Ref. Data*, **21** (1992) 411.

27. D. L. Baulch, C. J. Cobos, R. A. Cox, P. Frank, G. Hayman, T. Just, J. A. Kerr, T. Murrells, M. J. Pilling, J. Troe, R. W. Walker, and J. Warnatz, *J. Phys. Chem. Ref. Data*, **23** (1994) 847.

28. W. G. Mallard, F. Westley, J. T. Herron, R. F. Hampson, and D. H. Frizzell, *NIST chemical kinetics data base: Windows Version 2Q98*, National Institute of Standards and Technology, Gaithersburg, MD, 1998.

29. *Handbook of chemistry and physics*, D. R. Lide, ed.-in-chief, CRC Press, Boca Raton, 80th ed., 1999-2000, ISBN 0849304806, Section 5, pp. 505 to 519.

30. F. G. Helfferich and P. E. Savage, *Reaction kinetics for the practical engineer*, Course #195, AIChE Educational Services, New York, 7th ed., 1999, Section 8.2.

Chapter 7

Network Elucidation

Network elucidation starts with the identification of the participants in the reaction: reactants, products, known intermediates, and possibly catalysts and any other species that affect the rate. From there on approaches differ.

Access to powerful computers has made it possible in principle to begin with a general network in which each participant in the reaction is connected to all others by reversible steps that conform with stoichiometry (*maximum model* [1]). Values of the rate coefficients of this network are then fitted to experimental results. Steps with very small coefficients are dropped, and the values of the remaining coefficients are refined. The entire procedure is easy to learn and can be automated. However, in a reaction of any degree of complexity, the number of possible combinations of reaction steps is staggering, and this "blind" approach not only requires a lot of calculation, but also is in danger of converging on a false optimum. Moreover, there may be unknown intermediates. On the other hand, if the reaction is relatively simple, theoretical chemistry, experience, and common sense almost always allow a few plausible mechanisms to be singled out from improbable or sterically or energetically impossible ones, making the comprehensive screening of all a waste of time. For these reasons, maximum models are rarely advantageous, and no further space will be given to them here.

In most cases, a more cost-effective and reliable approach is to undertake a preliminary sorting by reaction orders and, if called for, product ranks and then proceed in either of two ways:

- *compile empirical rate equations from extensive experimental results over a wide range of conditions; then find a plausible mechanism and network that can produce such rate equations*, or

- *compile all networks that make sense from the point of view of theoretical chemistry and experience (stereochemistry, thermodynamics, molecular orbital theory, selection and exclusion rules, analogies with other reactions of the same type or with same catalyst, etc.); then devise bench-scale experiments to discriminate most effectively between these rival networks.*

Which of the two alternatives is more promising will differ from case to case. Usually, the first approach is preferable if a large amount of reliable quantitative kinetic data is already available; the second, if the kinetic behavior of the reaction is still largely unknown and the reaction engineer has a say in the design of the kinetic experiments yet to be carried out. Several examples of compilation of rate equations for postulated networks have already been given in Chapters 4, 5, and 6. Examples of deduction of networks from empirical rate equations will be provided in the present chapter. Regardless of what approach is taken, the ultimately postulated network and its mathematics should be validated by verification of their predictions, preferably counterintuitive ones, under conditions not previously studied (see also Chapter 11).

Large networks are common in industrial practice. Of course, the larger the network, the harder is its elucidation. The best approach to large networks is to break them down as much as possible into portions that can be studied separately and independently. In particular, non-trace intermediates can usually be synthesized and used as reactants in kinetic studies. Also, sometimes, side reactions or product decomposition can be suppressed in laboratory studies by additives or by structural modification of the reactants in a way that will not affect the pathway of the reaction of principal interest (e.g., see Example 6.5 in Section 6.5). For this reason, emphasis in this chapter is on the analysis of relatively small and primitive networks.

7.1. Order and rank

Once the identities of the participants in a reaction have been established, a logical next step toward proposing and testing networks is the determination of reaction orders and, in the case of reactions with many products, of product ranks. The present section examines these preliminaries.

7.1.1. Reaction orders

Reaction orders and experimental techniques to establish them are discussed in great detail in texts on kinetics and reaction engineering (see general references in Chapter 3). A brief survey concentrating on practical aspects of equipment and data evaluation has been given in Sections 3.1 and 3.3.

In the context of network elucidation, the determination of reaction orders is a preliminary step whose results are intended mainly for orientation. Note that constant, integer reaction orders may not exist in the case at hand. A fractional and possibly varying reaction order, while reflecting ignorance of the mechanism and true rate equation, is nevertheless a key symptom that can prove important in the subsequent work to establish the latter. On the other hand, just because orders are apt to be variable, any effort to determine their precise values under the conditions of the experiment is a waste of time: All the information sought at this stage is a range—say, an order between plus one and plus two—and, if the order is variable, in which direction it changes with the concentrations of the participants. Often, plots for guessed reaction orders and conclusions from any curvatures that become apparent are all that is needed. Examples in the present chapter will illustrate such methods.

Most reactions of practical interest have orders with respect to several participants. The methods to determine individual orders in such cases are therefore especially important.

It is also important to realize that, in multistep reactions in principle, the rate equation even of only the forward reaction may involve the concentrations of any participants, not only those of the reactants. This includes catalysts, products, and "silent partners" whose presence affects the rate although they are not catalysts nor are formed or consumed by the reaction and so do not appear in the stoichiometric equation. The determination of reaction orders can therefore not remain restricted to reactants, even if the reaction is irreversible.

Unusual reaction orders are found in autocatalysis, that is, in product-promoted and reactant-inhibited reactions, the former with positive apparent order with respect to a product, the latter with negative apparent order with respect to a reactant (see Section 8.9). An example of a product-promoted reaction is acid-catalyzed ester hydrolysis. An example of a reactant-inhibited reaction has already been encountered, namely, olefin hydroformylation, whose order with respect to CO is negative (see eqn 6.12 in Section 6.3). Such behavior is also not uncommon in enzyme catalysis ("substrate-inhibited" reactions in biochemistry lingo). Examples from heterogeneous catalysis are hydrogenation over nickel under certain conditions and oxidation of CO over some ceramic catalysts. A reaction having an order with respect to a silent partner—CO in a homogeneous hydrogenation—will be examined in some detail later in this chapter (see Examples 7.3 and 7.4).

The autocatalytic nature of a reaction can have profound consequences for reaction engineering and can even be design-limiting, as will be discussed in more detail in Section 12.3.

7.1.2. *Ranks and Delplot* *

The *rank* (primary, secondary, etc.) of an intermediate or product of a multistep reaction reflects the provenance of the species (see also Section 1.5). A *primary* species is formed directly from the original reactant or reactants; a *secondary* species, from a primary one; etc. The formation from a species of next lower rank may involve more than one step, but only if all but one of these are very fast.

---

* The rank is a recent concept, developed and formalized at the University of Delaware [2,3], based in part on earlier work by Myers and Watson [4]. The name "Delplot" for the plots to determine it alludes to this origin. The concept and procedure have not yet found their way into standard texts on reaction engineering.

Rank-ordering of products on the basis of experimental results is one of the tools that can be used in elucidation of the network of a reaction about which very little is known as yet. Such ranking often involves judgment calls, especially if rates along different branches of the network differ greatly. Moreover, a sharp experimental distinction between higher ranks on the basis of experimental results can be arrived at only in relatively simple cases, particularly if the network includes higher-order steps, and is impossible if some or all of the steps are reversible. Ranking is essentially a qualitative tool and, in practice, is rarely carried beyond a distinction between primary and higher-rank participants.

Ranking requires detailed information on behavior at very low conversions, best obtained with batch reactors.

Identification of primary participants. The distinction between primary products or intermediates and those of higher ranks is easy and usually unambiguous. By definition, primary products or intermediates arise directly and exclusively from the original reactant or reactants, initially present at finite concentrations; in contrast, products or intermediates of higher ranks arise from participants whose initial concentrations are zero. As a result, the initial formation rates of primary partici-pants are finite, those of participants of higher rank are zero. An examination of the concentration histories allows this distinction to be made. (A participant may appear to be primary although formed in two or more steps, namely, if all but one of these are very fast.)

This principle is formalized and sharpened in the so-called *Delplot* [3], in which selectivities $\overline{S}_i \equiv y_i / f_A$ are plotted versus f_A , where y_i is the yield of participant i and f_A is the fractional conversion of the original reactant (or limiting reactant) A:

$$y_P \equiv \frac{(N_P - N_P^o)/n_P}{N_A^o/n_A} \tag{1.6}$$

$$f_A \equiv 1 - N_A/N_A^o \tag{1.4}$$

(see Section 1.6). The plots are extrapolated to zero conversion. Primary partici-pants, having finite initial formation rates, give plots with finite and positive inter-cepts; participants of higher rank, having zero initial formation rates, give plots with zero intercepts. This is true regardless of reactor type, reaction orders of the steps, and whether or not these are reversible. Moreover, among several primary participants, those with higher initial formation rates produce larger intercepts.

Primary participants give Delplots with finite intercepts.
Participants of higher ranks give Delplots with zero intercepts.

Example 7.1. Delplots of a hypothetical reaction [3]. A simple example may illustrate the principle of the Delplot. Consider a hypothetical reaction in which A is transformed into K, P, and Q, with network

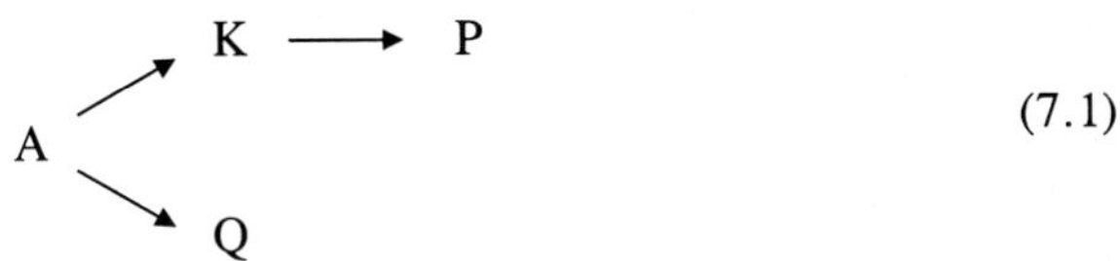

$$\tag{7.1}$$

The corresponding Delplots are shown in the upper diagram of Figure 7.1. The plots for K and Q have finite intercepts while that for P has a zero intercept. This allows K and Q to be identified as primary participants, and P as one of higher rank.

The plots also allow the question to be settled whether P is formed from K or Q. The ordinate of the plots is the selectivity $\overline{S}_i$ to the respective product i (see eqn 1.11). The selectivity to an intermediate in a pathway of sequential steps decreases with progressing conversion (see Section 5.4). The curve for K in Figure 7.1, top diagram does so, that for Q does not, indicating that P is formed from K rather than Q.

Distinction between higher ranks. A distinction between higher ranks of participants, difficult to make by mere inspection of concentration histories, is possible but not always unambiguous.

For reactions in which all steps are first order and irreversible, the procedure is as follows. For each intermediate or product i, the quantity $y_i/(f_A)^R$ is plotted versus f_A

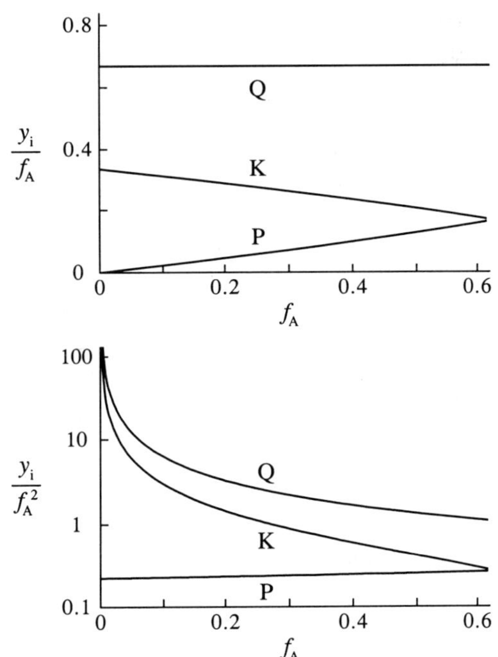

Figure 7.1. Delplots for K, P, and Q in batch reaction with network 7.1 and rate coefficients $k_{AK} = 1$, $k_{KP} = 4$, $k_{AQ} = 2$ (arbitrary units). Top: first-rank plots; bottom: second-rank plots. (Adapted from Bhore *et al.* [3].)

successively with $R = 1, 2$, etc. ("R-rank Delplots") and extrapolated to zero conversion. Second-rank plots (i.e., with $R = 2$) give finite intercepts for secondary participants, zero intercepts for participants of higher ranks, and diverging intercepts for primary participants. In general terms, R-rank Delplots give finite intercepts for participants of rank R, zero intercepts for participants of ranks higher than R, and diverging intercepts for participants of ranks lower than R. Second-rank Delplots for K, P, and Q in the reaction with network 7.1 are shown in the lower diagram of Figure 7.1.

For reactions that include irreversible steps of higher reaction orders, the same procedure is used. However, a distinction between the *Delplot rank* and *network rank* of a participant must now be made. The Delplot rank R is related to the intercepts as for networks with first-order steps only and is obtained by inspection of the plots. The network rank N indicates the provenance: $N = 1$ for primary participants, $N = 2$ for secondary participants, etc. The Delplot rank of the participant L of a step $n_K K \longrightarrow L$ (n'th order in K) is related to the Delplot rank of its parent K by

$$R_L \;=\; 1 + n_K R_K \tag{7.2}$$

More generally, for a step $n_{K_1}K_1 + n_{K_2}K_2 + \ldots + n_{K_n}K_n \longrightarrow n_L L + \ldots$:

$$R_L \;=\; 1 + \sum_{i=1}^{n} n_{K_i} R_{K_i} \tag{7.3}$$

For example, if the step $K \longrightarrow P$ in the network 7.1 were replaced by a second-order step $2K \longrightarrow P$, the Delplot rank of P would become $1 + 2*1 = 3$, although the network rank of P as a secondary product is 2.

This procedure introduces uncertainties because the deduction of the network rank from the Delplot rank requires some prior knowledge about the network. For example, suppose the Delplot ranks of K and L are 1, that of M is 2, and that of N is 3, that M is likely to be formed from K, and N from K and L. If so, K and L could react directly to N in a second-order step, and K to M in a separate first-order step (Case I). Instead, K could form M, which then reacts with L in a two-step reaction that is first order in M and zero order in L (Case II, with $X + L \longrightarrow N$ much faster than $M \longrightarrow X$):

Case I   *Case II*

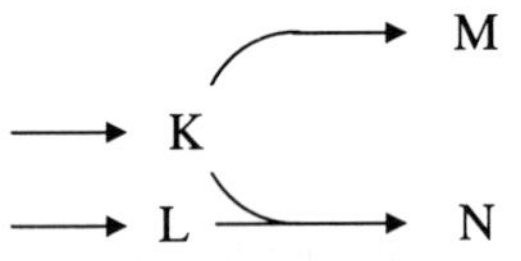

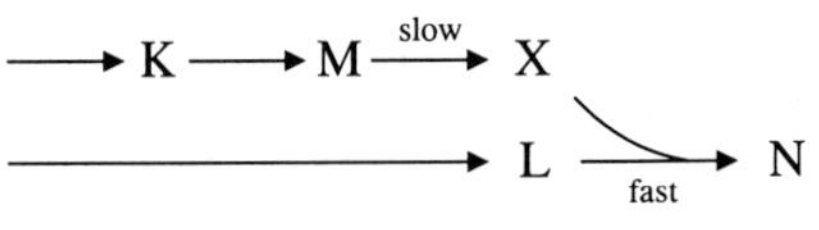

The Delplots cannot distinguish between these two possibilities.

To date, no procedures have been worked out for networks that include reversible steps.

Application. Delplots are of greatest utility in the early stages of exploration of complex reactions that yield large numbers of products. To date they have mostly been applied to oxidation and pyrolysis reactions of organic substances, reactions in which most or all steps are irreversible. An example is shown for illustration.

Example 7.2. Pyrolysis of n-pentadecylbenzene [5]. Upon pyrolysis, *n*-pentadecyl-benzene decomposes to about sixty different products. The major ones are toluene, styrene, *n*-tridecane, 1-*n*-tetradecene, and ethylbenzene. First-rank Delplots of experimental results are shown in Figure 7.2.

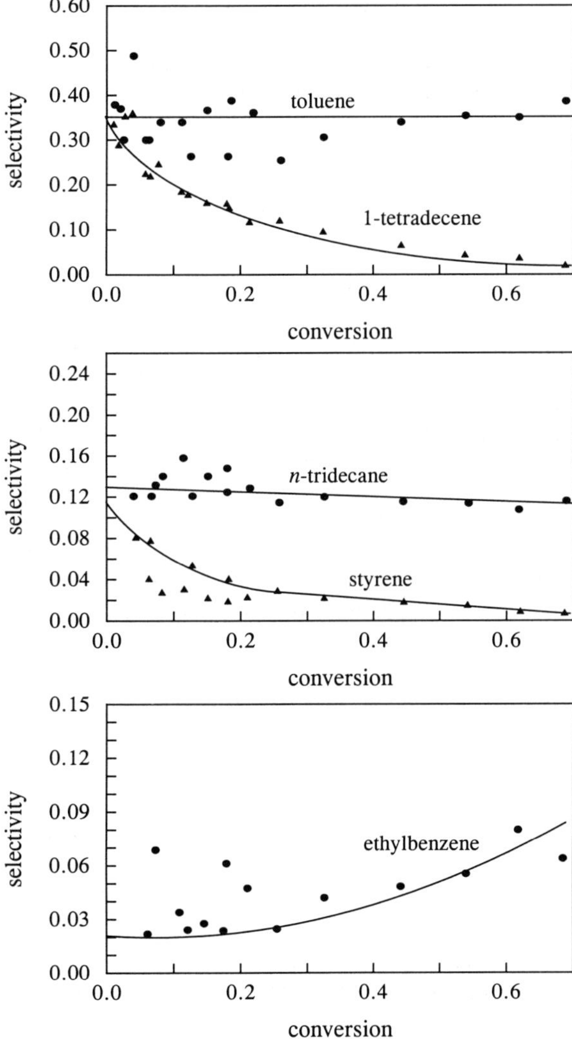

Figure 7.2. First-rank Delplots of major decomposition products in pyrolysis of *n*-pentadecylbenzene at 400°C (from Savage and Klein [5]).

All plots with the possible exception of that of ethylbenzene show finite intercepts, indicating that the products arise directly from *n*-pentadecylbenzene. This suggests a network

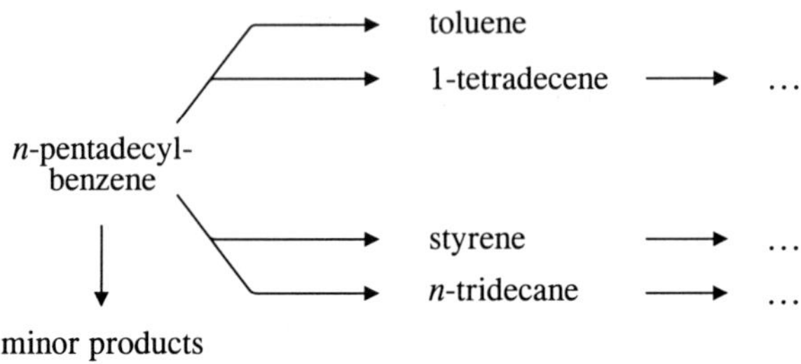

With a network with these primary steps, concentration histories in agreement with observation were indeed obtained [5].

7.2. "One-plus" rate equations

For mathematical convenience and economy of effort, rate equations in network elucidation and modeling are best written in terms of the minimum necessary number of constant "phenomenological" coefficients, which may be combinations of rate coefficients of elementary steps. This not only simplifies algebra and increases clarity, but also lightens the experimental burden: fewer coefficients, fewer experiments to determine them and their temperature dependences.

Rate equations of product formation usually contain additive terms in the denominator if the pathway or network includes reverse steps. The number of phenomenological coefficients can then be reduced by one if numerator and denominator are divided by one of the terms. The result is a "one-plus" rate equation, with a "1" as the leading term in the denominator. (Exception: This procedure is superfluous if all terms in the denominator consist only of coefficients, or of coefficients multiplied with the same concentration or concentrations, so that they can be combined to give a true power-law rate equation.)

One-plus equations are quite common in many fields of science and technology, the most notable being the Langmuir adsorption isotherm [6].

One-plus rate equations play a key role in network elucidation. Perhaps the most difficult step in that endeavor is the translation of a mathematical description of experimental results into a correct network of elementary reaction steps. The observed behavior can usually be fitted quite well by a traditional power law with empirical, fractional exponents, at least within a limited range of conditions. This has indeed been standard procedure in times past. However, such equations are highly unlikely to result from a combinations of elementary steps. Their acceptance may be expedient, but as far as network elucidation is concerned they are a dead end. In contrast, one-plus rate equations *can* result from step combinations, and their establishment therefore is an important stepping stone in the course of network elucidation, as the present chapter will demonstrate.

> If fitting a power law requires fractional exponents,
> a one-plus rate equation with integer exponents should
> be tried instead.

Moreover, being more likely to reflect the true mechanism, the one-plus rate equation is also more likely to remain valid upon extrapolation to still unexplored ranges of conditions.

There is one important exception: Certain types of chain reactions and reactions involving dissociation produce exponents of one half or integer multiples of one half in power-law or one-plus rate equations (see Sections 5.6, 9.2, and 10.3.1). Such exponents should be accepted if found not to vary with conversion and if there is good reason to believe that a mechanism of this kind may be operative.

7.2.1. *Types of one-plus rate equations*

For an example of the simplest type of one-plus rate equation, let us return to the reaction

$$A \longleftrightarrow X \overset{A}{\searrow} P \qquad (5.72)$$

with rate equation

$$r_P \cong \frac{k_{AX} k_{XP} C_A^2}{k_{XA} + k_{XP} C_A} \qquad (5.74)$$

Dividing numerator and denominator by the first term of the latter one obtains

$$r_P = \frac{k_a C_A^2}{1 + k_b C_A} \qquad (7.4)$$

with the two phenomenological coefficients

$$k_a \equiv k_{AX} k_{XP}/k_{XA} \qquad \text{and} \qquad k_b \equiv k_{XP}/k_{XA}$$

The number of coefficients, three in eqn 5.74, has been reduced to two.

Another, already encountered example of a one-plus rate equation is that of olefin hydroformylation (see Example 6.2 in Section 6.3). Here, the rate equation after cancellations but before reduction was

$$r_{\text{ald}} \;\cong\; \frac{K_1 k_{12} k_{23} k_{34} k_{45} C_{\text{ole}} C_{\text{cat}} p_{H_2}}{(k_{23} k_{34} k_{45} + k_{21} k_{34} k_{45}) p_{\text{CO}} + (k_{21} k_{32} k_{45} + k_{21} k_{32} k_{43}) p_{H_2}} \qquad (6.11)$$

and contained seven rate coefficients of steps and one equilibrium constant.
Collection of terms and division of numerator and denominator by the second term
of the latter gave

$$r_{\text{ald}} \;\cong\; \frac{k_a C_{\text{ole}} C_{\text{cat}}}{1 + k_b p_{\text{CO}}/p_{H_2}} \qquad (6.12)$$

with only two phenomenological coefficients

$$k_a \;\equiv\; \frac{K_1 k_{12} k_{23} k_{34} k_{45}}{k_{21} k_{32}(k_{45} + k_{43})}, \qquad\qquad k_b \;\equiv\; \frac{(k_{23} + k_{21}) k_{34} k_{45}}{k_{21} k_{32}(k_{45} + k_{43})}$$

The example of the hydroformylation reaction demonstrates another point.
The denominator of a rate equation obtained from the general formula (eqns 6.4 to
6.6) may contain several terms that involve the same combination of concentrations
of co-reactants. Such terms can, of course, be lumped. Thus, in the original
hydroformylation rate equation 6.10, the first two denominator terms both contained
the co-factor p_{CO}, the third and fourth both contained p_{H2}. In eqn 6.11, obtained by
lumping the terms of these pairs, the number of denominator terms had been
reduced from four to two. The one-plus form 6.12 was then obtained by division
by the second of these two, actually the sum of the third and fourth terms in the
original equation.

Even after such lumping, the denominator may be left with more than two
terms. The one-plus equation will then have more than two phenomenological
coefficients.

7.2.2. *Establishment of one-plus rate equations from experimental data.*

The principle of establishing a one-plus rate equation and the values of its
phenomenological coefficients is very simple. If the reaction is irreversible and
found to be of an order between zero and one with respect to a participant i, the
simplest one-plus equation contains the respective concentration C_i (or p_i) as a factor
in the numerator and in some but not all terms of the denominator. More generally,
if the order is between n (positive integer) and $n + 1$, the simplest equation contains
the factor C_i^{n+1} in the numerator and C_i in some but not all terms of the
denominator. Many other combinations are possible, but less likely. For instance,
an order between zero and plus one might also result from a numerator with factor
C_i^2 and a denominator with C_i in some terms and C_i^2 in the others. Occam's razor
suggests the best policy: to try the simplest option first.

The approach for negative reaction orders is similar. For example, the simplest equation giving an order between zero and minus one contains the respective concentration in one or more terms of the denominator, but not in the numerator.

To summarize the forms of some of the simplest one-plus equations:

order	factor in numerator	factor in some but not all denominator terms
between 0 and $+1$:	C_i	C_i
between 0 and -1:	$-$	C_i
between $+n$ and $+n+1$:	C_i^{n+1}	C_i
between 0 and $-n$:	$-$	C_i^n

As an example, the four simplest one-plus rate equations for an irreversible reaction of first order in A and orders between zero and plus one in B and C are:

$$r_P = \frac{k_a C_A C_B C_C}{1 + k_b C_B + k_c C_C}, \qquad r_P = \frac{k_a C_A C_B C_C}{1 + k_b C_B C_C}$$

$$r_P = \frac{k_a C_A C_B C_C}{1 + k_b C_B + k_c C_B C_C}, \qquad r_P = \frac{k_a C_A C_B C_C}{1 + k_b C_B C_C + k_c C_C} \tag{7.5}$$

A convenient procedure for testing a tentative one-plus equation and obtaining values of its coefficients is to look for a straight-line plot. A plot of the rate versus a concentration or power of a concentration is nonlinear because of the additive terms in the denominator. However, the reciprocal of the rate is given by additive terms with different concentration dependences and so lends itself to such a purpose. For example, if the rate is

$$r_P = \frac{k_a C_A}{1 + k_b C_A}$$

its reciprocal is

$$\frac{1}{r_P} = \frac{1}{k_a C_A} + \frac{k_b}{k_a}$$

so that a plot of $1/r_P$ versus $1/C_A$ gives a straight line with slope $1/k_a$ and intercept k_b/k_a if the coefficients are constant. If the plot of the experimental results does not give a straight line, a different one-plus form may have to be tried.

The following example illustrates the procedure step by step as applied to a moderately complex reaction.

Example 7.3. One-plus rate equation for hydrocarbonyl-catalyzed hydrogenation of aldehyde [7]. Homogeneous liquid-phase hydrogenation of aldehydes to alcohols

$$\text{aldehyde} \xrightarrow{\quad H_2 \quad} \text{alcohol} \tag{7.6}$$

is catalyzed by dissolved phosphine-substituted cobalt hydrocarbonyl, $HCo(CO)_3Ph$, where Ph is a tertiary organic phosphine. The reaction requires the presence of CO as well as H_2 in order to keep the catalyst stable. Table 7.1 lists experimental results of hydrogenation of 2-ethylhexanal at different aldehyde concentrations and partial pressures of H_2 and CO.

Table 7.1. Rates of hydrogenation of 2-ethylhexanal in *n*-dodecanol at 165°C, measured in a continuous stirred-tank reactor [8] (concentration of catalyst, $HCo(CO)_3Ph$, same in all runs).

run	aldehyde concentration	partial pressure atm		rate
	M	of H_2	of CO	M min^{-1}
1	0.100	80	80	$1.83*10^{-3}$
2	.109	80	40	3.90
3	.105	40	40	3.33
4	.101	20	40	2.42
5	.100	80	20	7.20
6	.111	40	20	6.87
7	.129	20	20	6.18
8	.136	10	20	4.44
9	.050	80	40	1.81
10	.051	40	20	3.17
11	.061	20	20	2.90
12	.020	40	20	1.20
13	.024	20	20	1.16
14	.010	40	20	0.62

Most homogeneous hydrogenation reactions are first order in the organic reactant. A cursory inspection of the results suggest that this may be the case here, too. To test this hypothesis, a tentative rate equation is written:

$$r_{alc} = k_{app}(p_{H_2}, p_{CO}) C_{ald}$$

whose apparent first-order rate coefficient, k_{app}, still is a function of the partial pressures of H_2 and CO. For each run, $k_{app} = r_{alc}/C_{ald}$ is calculated. Comparison of the values (fifth column in Table 7.2) shows them to be the same within experimental error for runs with different aldehyde concentrations but same conditions with respect to H_2 and CO (runs 2 and 9 at 80 atm H_2 and 40 atm CO; runs 6, 10, 12, and 14 at 40 and 20 atm; runs 7, 11, and 13 at 20 and 20 atm). Thus, k_{app} is independent of the aldehyde concentration, i.e., the reaction is first order in aldehyde as expected.

Table 7.2. Work-up of data on aldehyde hydrogenation in Table 7.1 [7].

run	C_{ald}	p_{H2}	p_{CO}	$k_{app} = r_{alc}/C_{ald}$	r_{alc} M min^{-1}	
	M	atm	atm	min^{-1}	observed	calculated
1	0.100	80	80	$1.83*10^{-2}$	$1.83*10^{-3}$	$1.81*10^{-3}$
2	.109	80	40	3.58	3.90	3.95
3	.105	40	40	3.17	3.33	3.24
4	.101	20	40	2.40	2.42	2.41
5	.100	80	20	7.20	7.20	7.24
6	.111	40	20	6.19	6.87	6.86
7	.129	20	20	4.79	6.18	6.16
8	.136	10	20	3.26	4.44	4.46
9	.050	80	40	3.62	1.81	1.81
10	.051	40	20	6.22	3.17	3.15
11	.061	20	20	4.75	2.90	2.91
12	.020	40	20	6.00	1.20	1.24
13	.024	20	20	4.83	1.16	1.15
14	.010	40	20	6.20	0.62	0.62

To establish the order with respect to CO, values of k_{app} at same partial pressure of H_2 but different partial pressures of CO are compared. Within each of these sets (runs 1, 2, 5, and 9 at 80 atm H_2; runs 3, 6, 10, 12, and 14 at 40 atm; runs 4, 7, 11, and 13 at 20 atm) the coefficient is seen to vary in good approximation in inverse proportion to the CO pressure. The order with respect to CO thus is minus one within experimental error.

In a like manner, to establish the order with respect to H_2, values of k_{app} at same partial pressure of CO but different partial pressures of H_2 are compared. At both 40 atm CO (runs 2, 3, 4, and 9) and 20 atm CO (runs 5, 6, 7, 8, and 10 to 14) the coefficient is found to increase with increasing pressure of H_2, but clearly less than in proportion to that pressure: The order with respect to H_2 thus is between zero and plus one.

The simplest one-plus equation with the established apparent reaction orders and first order in catalyst is

$$r_{alc} = \frac{k_a C_{ald} p_{H_2} C_{cat}}{p_{CO}(1 + k_b p_{H_2})} \qquad (7.7)$$

To test it, eqn 7.7 is rearranged into a form with reciprocal rate on the left-hand side and simplest possible additive terms on the right:

$$\frac{C_{ald} C_{cat}}{r_{alc} p_{CO}} = \frac{1}{k_a p_{H_2}} + \frac{k_b}{k_a} \qquad (7.8)$$

If eqn 7.7 reflects the data correctly, a plot of the left-hand side of eqn 7.8 versus the reciprocal partial pressure of H_2 will give a straight line. Figure 7.3 shows that to be true.

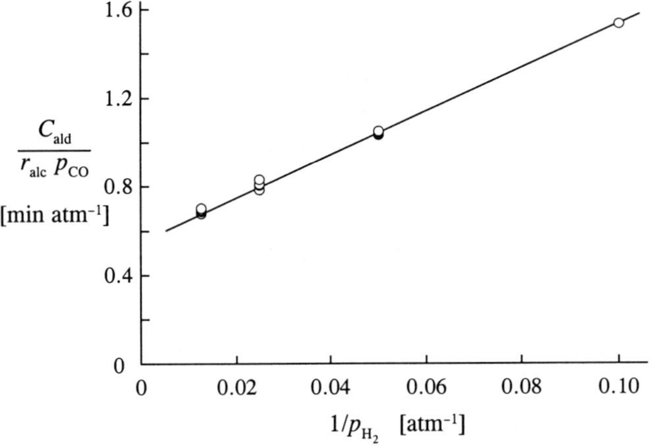

Figure 7.3. Plot to test rate equation 7.7 (C_{cat}, same in all runs, not included in calculation) [7].

According to eqn 7.8, the slope of the straight line in Figure 7.3 is $1/k_a$ and the intercept is k_b/k_a. Evaluation of slope and intercept yields

$$k_a = (0.105 \text{ min}^{-1})/C_{cat}, \qquad k_b = 0.060 \text{ atm}^{-1}$$

The last two columns of Table 7.2 show a comparison of observed rates with those calculated with eqn 7.7 and these coefficients. The agreement is excellent. Even so, eqn 7.7 should not be viewed as established beyond doubt until a plausible mechanism leading to it has been found and predictions made with it have proved correct. The search for a mechanism will be described in Example 7.4 in the next section.

A fit to the experimental results may require a one-plus rate equation with three or more terms in the denominator (e.g., see eqns 7.5). If so, the coefficients can be determined by linear regression or, long-hand, by cross-plotting. Say, the one-plus equation is

$$ r_P \;=\; \frac{k_a C_A C_B}{1 + k_b C_A + k_c C_B} $$

One of several possible graphical procedures then is to group the rate data into different sets, each with different C_A but same (constant) C_B. The rate equation is rearranged to

$$ \frac{C_B}{r_P} \;=\; \frac{k_d}{C_A} + \frac{k_b}{k_a} \qquad \text{where} \qquad k_d \;\equiv\; \frac{1 + k_c C_B}{k_a} = \text{const. at const.} C_B $$

and k_d and k_b/k_a are obtained from slope and intercept of a plot of C_B/r_P versus $1/C_A$. The k_d values from the different sets are then plotted versus C_B to obtain k_c/k_a and $1/k_a$ from slope and intercept. The only still unknown coefficient, k_b, can now be calculated from the previously determined ratio k_b/k_a.

7.3. Relationships between network properties and kinetic behavior

At the very basis of chemical kinetics as we know it is the knowledge of how the mechanics of a molecular event is reflected in observable kinetic behavior: the knowledge that the spontaneous decay or rearrangement of a molecule occurs at a rate that is proportional to the concentration of the species, that the formation of a product by collision of two molecules does so at a rate proportional to the concentrations of both reactants, etc. To every chemist and chemical engineer this has been self-evident for as long as he or she can remember. Unfortunately, this core of elementary knowledge tells us only what happens in single-step reactions and proves woefully inadequate when we face the complications of real-life chemistry. Efficient handling of kinetic problems in practice, and especially of network elucidation, calls for a broadening of that basis to include multistep reactions. The present section addresses this problem with the deduction of additional rules [7,9].

While covering additional ground, the set of rules in the present section still leaves important areas of kinetics of homogeneous reactions untouched. Three such areas—trace-level catalysis, chain reactions, and polymerization—will be examined in the next three chapters. A third, kinetics of reaction with periodic or chaotic behavior, is beyond the scope of this book.

The discussion in this section will require familiarity with the material in the previous chapter or frequent reference to it.

Table 7.3. Algebraic forms of segment coefficient Λ_{kP} of segment with stoichiometry $X_k + B \longrightarrow P$ and different configurations (from Helfferich and Savage [7]).

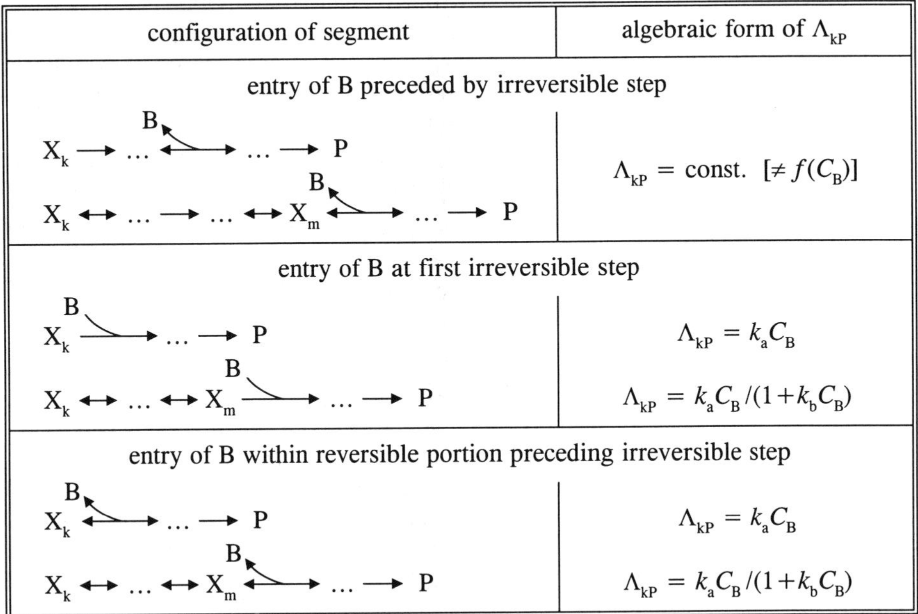

$$\text{olefin} \begin{cases} \xrightarrow{\;+H_2,\ CO\;} \text{aldehyde} \xrightarrow{\;+H_2\;} \text{alcohol} \\[2mm] \xrightarrow{\;+H_2\;} \text{paraffin} \end{cases} \qquad (7.30)$$

    The catalyst differs from the "oxo" catalyst, $HCo(CO)_4$, only in that one CO ligand is replaced by a tertiary organic phosphine, and rates obey the Martin equation 6.12 (see Example 6.2 in Section 6.3). Therefore, it seems a safe bet that the olefin-to-aldehyde pathway is the same as with the oxo catalyst (network 6.9). Also, almost certainly, paraffin arises from an intermediate along that pathway rather than from a different olefin-catalyst adduct. Logical candidates for this node intermediate are the π-complex (X_1 in network 6.9), the trihydride (X_2), and the cobalt phosphino-di- and tricarbonyl alkyls (X_3 and X_4, respectively) as paraffin can no longer be expected to form once irreversible carbonylation to cobalt acyl (X_5) has occurred. Granted these premises and the additional simplest assumption that paraffin forms from the respective node intermediate in either a single step or in steps that can be consolidated with Rule 7.24 into a single step, four networks are in contention. They are shown on next page with their yield ratio equations. To avoid clutter, only the pathway portions from π-complex to cobalt acyl and paraffin are shown (since carbonylation to $X_4 \longrightarrow X_5$ is irreversible, subsequent steps have no effect).

Table 7.4. Trial networks for hydroformylation with paraffin by-product formation.

(I)

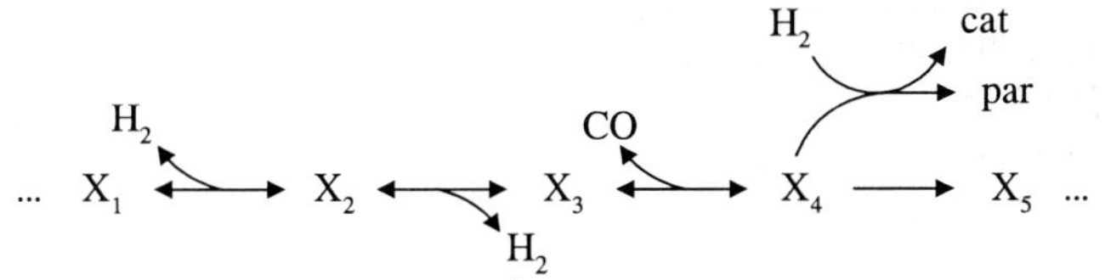

$$Y_{\text{par/ald}} = \Lambda_{4\text{par}}/\Lambda_{4\text{ald}} = (k_{4\text{par}}/k_{45})p_{\text{H}_2} \tag{7.31}$$

(II)

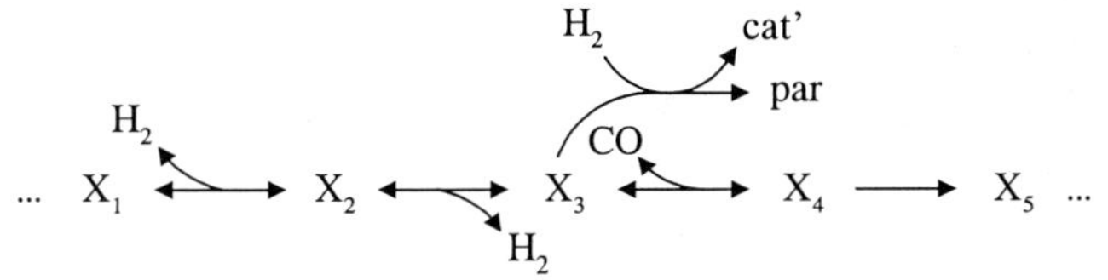

$$Y_{\text{par/ald}} = \frac{\Lambda_{3\text{par}}}{\Lambda_{3\text{ald}}} = \frac{k_{3\text{par}}(k_{45} + k_{43})}{k_{34}k_{45}} \left[\frac{p_{\text{H}_2}}{p_{\text{CO}}}\right] \tag{7.32}$$

(III)

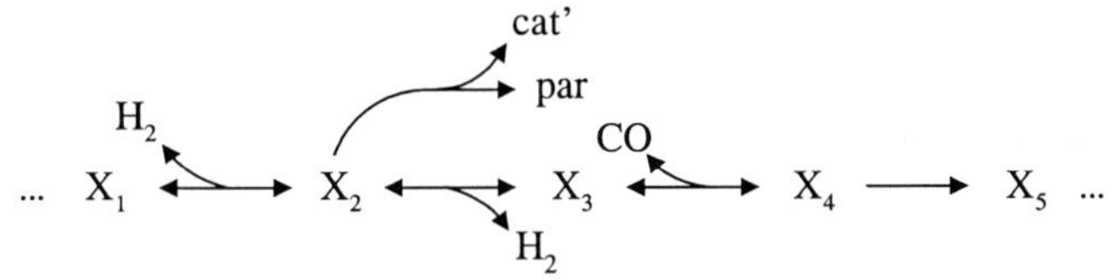

$$Y_{\text{par/ald}} = \frac{\Lambda_{2\text{par}}}{\Lambda_{2\text{ald}}} = \frac{k_{2\text{par}}(k_{34}k_{45}p_{\text{CO}} + (k_{32}k_{43} + k_{32}k_{43})p_{\text{H}_2})}{k_{23}k_{34}k_{45}p_{\text{CO}}}$$

$$= \frac{k_{2\text{par}}}{k_{23}} + \frac{k_{2\text{par}}k_{32}(k_{45} + k_{43})}{k_{23}k_{34}k_{43}} \left[\frac{p_{\text{H}_2}}{p_{\text{CO}}}\right] \tag{7.33}$$

(IV)

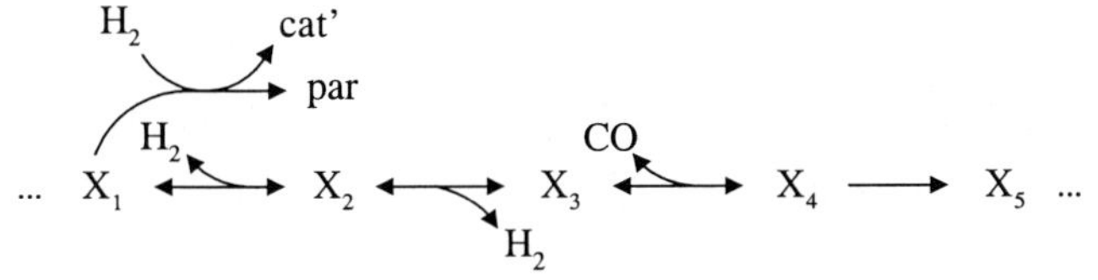

$$Y_{\text{par/ald}} = \frac{\Lambda_{1\text{par}}}{\Lambda_{1\text{ald}}} = \frac{k_{1\text{par}}(k_{23}k_{34}k_{45}p_{\text{CO}} + k_{21}k_{34}k_{45}p_{\text{CO}} + k_{21}k_{32}k_{45}p_{\text{H}_2} + k_{21}k_{32}k_{43}p_{\text{H}_2})}{k_{12}k_{23}k_{34}k_{45}p_{\text{CO}}}$$

$$= \frac{k_{1\text{par}}(k_{23} + k_{21})}{k_{12}k_{23}} + \frac{k_{1\text{par}}k_{21}k_{32}(k_{45} + k_{43})}{k_{12}k_{23}k_{34}k_{43}} \left[\frac{p_{\text{H}_2}}{p_{\text{CO}}}\right] \tag{7.34}$$

The Λ coefficients of the segments leading to paraffin and aldehyde are obtained from the general formula 6.4 to 6.6 in terms of λ coefficients, which then are replaced by the true rate coefficients multiplied by co-reactant concentrations where called for.

In network I, paraffin is formed by hydrogenation of the tricarbonyl alkyl, X_4. Here, the segment coefficients are

$$\Lambda_{4\,par} = \lambda_{4\,par} = k_{4\,par}p_{H_2}, \qquad\qquad \Lambda_{4\,ald} = \lambda_{45} = k_{45}$$

In network II, paraffin is formed by hydrogenation of the dicarbonyl alkyl, X_3, and the segment coefficients are

$$\Lambda_{3\,par} = \lambda_{3\,par} = k_{3\,par}p_{H_2}, \qquad\qquad \Lambda_{3\,ald} = \frac{\lambda_{34}\lambda_{45}}{\lambda_{45} + \lambda_{43}} = \frac{k_{34}k_{45}p_{CO}}{k_{45} + k_{43}}$$

In network III, paraffin is split off from the trihydride, X_2, and the segment coefficients are

$$\Lambda_{2\,par} = \lambda_{2\,par} = k_{2\,par}$$

$$\Lambda_{2\,ald} = \frac{\lambda_{23}\lambda_{34}\lambda_{45}}{\lambda_{34}\lambda_{45} + \lambda_{32}\lambda_{45} + \lambda_{32}\lambda_{43}} = \frac{k_{23}k_{34}k_{45}p_{CO}}{k_{34}k_{45}p_{CO} + k_{32}(k_{45} + k_{43})p_{H_2}}$$

Lastly, in network IV, paraffin is formed by hydrogenation of the π-complex, X_1, and the segment coefficients are

$$\Lambda_{1\,par} = \lambda_{1\,par} = k_{1\,par}p_{H_2}$$

$$\Lambda_{1\,ald} = \frac{\lambda_{12}\lambda_{23}\lambda_{34}\lambda_{45}}{\lambda_{23}\lambda_{34}\lambda_{45} + \lambda_{21}\lambda_{34}\lambda_{45} + \lambda_{21}\lambda_{32}\lambda_{45} + \lambda_{21}\lambda_{32}\lambda_{43}}$$

$$= \frac{k_{12}k_{23}k_{34}k_{45}p_{H_2}p_{CO}}{(k_{23} + k_{21})k_{34}k_{45}p_{CO} + k_{21}k_{32}(k_{45} + k_{43})p_{H_2}}$$

The yield ratio equations, obtained as ratios of the segment coefficients according to eqn 6.38, are shown with the respective networks in Table 7.4 (preceding page). For network I, the yield ratio is proportional to the partial pressure of H_2. In the other three cases, the yield ratio is seen to depend only on the H_2-to-CO ratio, not on total pressure at same H_2-to-CO ratio. However, the dependence on that ratio differs. For network II it is of the form

$$Y_{par/ald} = k_a(p_{H_2}/p_{CO}) \tag{7.35}$$

whereas for networks III and IV:

$$Y_{par/ald} = k_a + k_b(p_{H_2}/p_{CO}) \tag{7.36}$$

This suggests experiments in which the yield ratio is measured at different total pressures and H_2-to-CO ratios. Results of such experiments are shown in Figure 7.4, in which the yield ratio is plotted versus the H_2-to-CO ratio. The yield ratio is seen to be independent of total pressure at same H_2-to-CO ratio, to vary linearly with that ratio, and to remain finite when extrapolated to a zero value of that ratio; that is, eqn 7.36 is obeyed but eqns 7.31 and 7.35 are not. The agreement with eqn 7.36 is good confirmation of the initial assumptions, and rules out networks I and II. A dis-

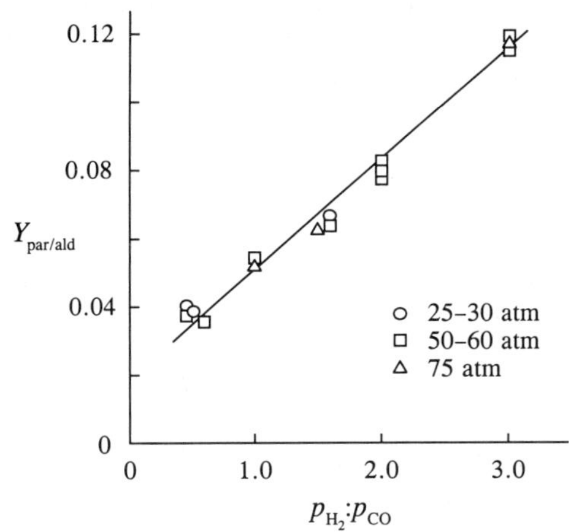

Figure 7.4. Molar paraffin-to-aldehyde yield ratios in hydroformylation of *n*-dodecene catalyzed by $HCo(CO)_3Ph$ at 185 °C as function of H_2-to-CO ratio at different total pressures [11].

crimination between networks III and IV is not possible on the basis of the information at hand. Although the exact location of the node in the network has remained ambiguous, eqn 7.36 appears reliable as it represents all experimental results and can be explained with an eminently plausible mechanism.

7.3.3. *Non-simple pathways and networks*

Many reactions in industrial practice have non-simple networks. Their variety is so great that standard recipes for elucidation cannot be stated: What works in one case will not in another. Only some strategies that might be useful can be suggested. Some of the more common ones will be shown in this section and be illustrated with examples.

Non-simplicity is caused by intermediates whose concentrations rise above trace level, or by steps in which two or more molecules of intermediates function as reactants. Non-simplicity caused by the first of these possibilities usually becomes apparent immediately, when the known participants in a reaction are sorted into reactants, products, intermediates, and possibly catalysts and silent partners. Where this is not so, say, because the number of participants is very large—not uncommon in hydrocarbon processing and combustion—, Delplot rank ordering can help to distinguish intermediates from end products (see Section 7.1.2). Non-simplicity caused by reactions of trace intermediates with one another may not be apparent at the outset, only to turn up as the mechanism becomes clearer. If so, the kineticist will have to cross that bridge when he comes to it.

In practice, many reaction systems involve non-trace intermediates, but no obvious non-simple reactions of intermediates. A good strategy in such situations is to cut the overall reaction network into portions at the non-trace intermediate or intermediates (see Section 6.5), then reduce the portions as described for simple networks in Section 6.4.1. Network reduction makes it unnecessary to keep track of trace intermediates (except those reacting in a non-simple manner) and so obviates much of the hard work: Trace intermediates are the more troublesome ones in network elucidation because they are difficult or impossible to detect, identify, analyze for, or synthesize, tasks that usually do not pose problems with intermediates that rise above trace level. Often, the network portions will turn out to be "piecewise simple" (see Section 6.5). If not, further cutting at additional non-simple steps is called for when these become apparent.

If the pathway or segment of a portion producing a non-trace intermediate is irreversible, no subsequent portion of the overall network feeds back into it. As a rule, this allows the subsequent portion or portions to be studied independently by using the separately synthesized non-trace intermediate as starting material. It also allows the portion yielding the non-trace intermediate to be studied independently: For this purpose, all subsequent intermediates and products are lumped with the intermediate produced by the portion (i.e., the concentrations are added) to obtain the total production of the portion. Alternatively, before analysis, all intermediates are converted to end products, and only these then need to be analyzed for and lumped.

Piecewise simple portions feeding into others may, of course, be reversible. This complicates network elucidation significantly. Often, however, such back-reactions can be blocked by an additive, the omission of a catalyst or co-catalyst, or some other experimental stratagem. Alternatively, the intermediate produced by the portion can be trapped in some fashion; this allows at least the forward reaction through the portion to be studied without interference.

The following example illustrates such strategies with a relatively complex network.

Example 7.6.   Olefin hydroformylation with phosphine-substituted cobalt hydrocarbonyl catalyst [7]. The overall reaction system of olefin hydroformylation with a phosphine-substituted cobalt hydrocarbonyl catalyst to produce alcohol, paraffin, and a heavy alcohol has been shown in Example 6.5 (Section 6.5):

$$
\begin{array}{ccccc}
 & \xrightarrow{+\ H_2,\ CO} & & \xrightarrow{+\ H_2} & \\
\text{olefin} & & \text{aldehyde} & & \text{alcohol} \\
\end{array}
$$

$$
\begin{array}{ll}
+\ H_2 \ \downarrow & +\ \text{aldehyde} \ \downarrow \\
\end{array}
\qquad (6.43)
$$

$$
\text{paraffin} \qquad\qquad \text{aldol} \xrightarrow[-\ H_2O]{+\ 2H_2} \text{heavy alcohol}
$$

Aldehyde and aldol are the only non-trace intermediates. If the network is cut at these, the portions are

$$\text{olefin} \longrightarrow \text{aldehyde}, \qquad\qquad \text{aldehyde} \longrightarrow \text{alcohol},$$
$$\downarrow \qquad\qquad\qquad\qquad\qquad 2\text{ aldehyde} \longrightarrow \text{aldol}, \qquad (6.44)$$
$$\text{paraffin} \qquad\qquad\qquad\qquad \text{aldol} \longrightarrow \text{heavy alcohol}$$

These can be investigated separately as follows.

The study of the first portion, olefin to aldehyde and paraffin, can be conducted under normal reaction conditions. For the mathematical evaluation, aldehyde and all products arising from it—alcohol, aldol, and heavy alcohol—are lumped into one single pseudo-species (counting aldol and heavy alcohol each as two aldehyde molecules). This is possible because, in the network, the aldehyde marks a point of no return: Once a molecule has been converted to aldehyde, it can only remain aldehyde or react on to alcohol, aldol, or heavy alcohol, but not revert back to olefin or paraffin.

Aldol condensation is catalyzed by base, but not by the cobalt catalyst. This allows the pathway from aldehyde to alcohol to be studied separately with aldehyde as starting material and under conditions that preclude aldol condensation. For example, condensation is minimal in the absence of a base or with an aldehyde whose carbon skeleton has a branch adjacent to the aldehyde group.

The pathway from aldehyde to aldol can be studied in the presence of a base and absence of the cobalt catalyst. If need be, the last pathway, from aldol to heavy alcohol, can be studied separately under normal reaction conditions with synthesized aldol as starting material.

Actually, the network 6.43 also contains a step in which two molecules of intermediates react with one another, namely, the step of entry of the second aldehyde molecules along the pathway from aldehyde to aldol. However, the cutting of the network at the aldehyde had removed that non-simplicity, too.

If a pathway or network should turn out to contain a step in which two or more molecules of the same or different intermediates react with one another, it can be cut at the offending species into piecewise simple portions as discussed in Section 6.5. However, it will rarely be possible to study these portions independently because, more often than not, the respective trace intermediates cannot be synthesized for use as starting materials.

If many or even a majority of the steps are non-simple, the network reduction methods described here are of little use in network elucidation. This is typically the case in hydrocarbon pyrolysis and combustion, where reactions of free radicals with one another are common. Fortunately, an extensive data base of rate coefficients and activation energies of reaction steps of species in this field of chemistry has been compiled over the years and can be of help in network elucidation [12-16].

7.4. Other criteria and guidelines

The rules and regularities developed in the preceding section and their adaptations
to catalysis in Section 8.6 provide a good deal of information about what features
a pathway or network may have, and definitely cannot have. As is generally true
in kinetics, they allow possibilities to be ruled out as incompatible with observed
behavior, but cannot serve to prove a compatible mechanism to be correct. Other
considerations, well covered in standard texts, are called for to narrow the field.
A brief survey is given here.

Stereochemistry. On the most primitive level, mere steric considerations can often
suggest plausible features of a network. Common sense tells us that conversion
must take place at some reactive group or configuration of the reactant molecule,
most likely with no or only minimal changes elsewhere.

　　　Catalytic hydrocyanation of mono-olefinic compounds may serve as a very
simple example. It stands to reason that HCN is likely to add to the olefinic double
bond, H going to one of the latter's carbon atoms, and CN to the other:

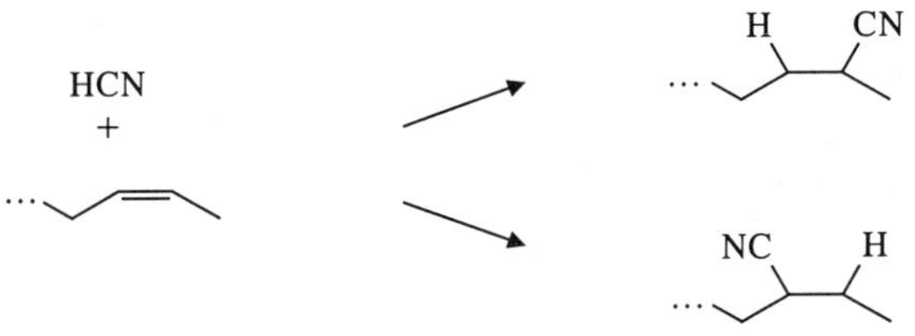

$$(7.37)$$

That this is indeed so has been confirmed by experiments with pentenenitrile
isomers and deuterium labeled hydrogen cyanide (DCN) [17]:

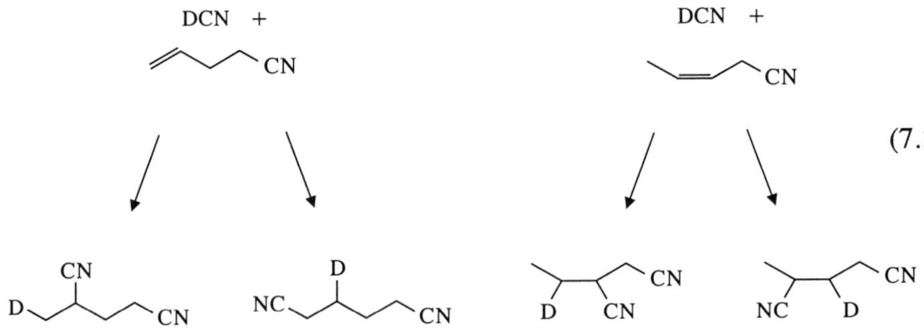

$$(7.38)$$

(Arrows represent multistep pathways;   for details of these, see Example 8.7 in
Section 8.5.4)

Under typical hydrocyanation conditions, double-bond migration along the carbon chain of the olefinic reactant is relatively slow, so that reactant isomerization remains unimportant. This makes the example above almost trivial. A similar but more complicated situation is encountered in hydroformylation of olefins (see Example 6.2 in Section 6.3). Here, the CO carbon atom can be expected to attach itself to either of the two double-bonded carbon atoms:

$$(7.39)$$

Depending on what catalyst is used, double-bond migration may be so fast that the product isomer distribution depends only little on what olefin isomer or isomer mixture was used as starting material. This, of course, makes it harder to identify the olefin isomer parents of any given product isomer, so much so that at one time a single, common intermediate was postulated from which all product isomers were said to arise [18]. In contrast, if CO attachment to either double-bonded carbon atom is accepted, the (simplified) network to be expected for a straight-chain mono-olefin is

$$(7.40)$$

(H$_2$ and CO reactants not shown;   arrows represent multistep pathways;   for mechanistic details, see Example 6.2 in Section 6.3). This basic network structure of coupled parallel steps has been verified as described in Section 5.3, in refutation of the earlier postulate of a common intermediate.

Molecularities. An equally elementary criterion is the fact that a great majority of reaction steps are uni- or bimolecular; trimolecular steps are rare and slow, and steps of still higher molecularities are unheard of (see Section 2.1). A trimolecular forward or reverse step in a postulated mechanism calls for an explanation why its reactants are not consumed by bimolecular steps before they have a chance to undergo the trimolecular one (e.g., see the Example 7.8 farther below). No mechanism involving a forward or reverse step of even higher molecularity should ever be considered.

Thermodynamics. Despite its name, thermodynamics is a science of equilibrium, not of dynamics. It compares energies of states of matter, and such a procedure by itself does not allow rates and mechanisms to be predicted. Nevertheless, thermodynamics can often help to decide which of various conceivable mechanisms are the more probable ones.

The activation energy of an endothermic step is necessarily at least as high as the molar reaction enthalpy, $\Delta H°$ (see Figure 2.2 in Section 2.2), and a step with very high activation energy is apt to be quite slow. Accordingly, a pathway with one or more highly endothermic steps is suspect if there are alternatives without these.

Thermochemical data for possible reaction intermediates may not be available. However, approximate average bond energies [19-21] can serve as a rough guide (see Table 7.5; the actual energies depend on the substituents). The skilled kineticist will first look for pathways in which no high-energy bond is broken or, if none such is possible, for those in which a new high-energy bond is formed as the other is broken.

Table 7.5. Approximate average bond energies of importance in organic chemistry.

bond	energy kJ mol^{-1}	bond	energy kJ mol^{-1}
C$-$C	285-430	C$-$N	260-300
C$=$C	320-740	H$-$H	436
C$-$O	210-280	O$=$O	498
C$-$H	360-450	O$-$H	360-500

Unfortunately, this procedure is not conclusive because an apparently plausible mechanism that avoids highly endothermic steps may well be blocked for other reasons. The hydrogen-iodide reaction

$$H_2 + I_2 \longleftrightarrow 2\,HI$$

provides a telling example (see Section 4.2). The single-step, bimolecular reaction mechanism is thermodynamically plausible and compatible with observed rate behavior, yet the true mechanism

$$I_2 \longleftrightarrow 2I \overset{H_2}{\longleftrightarrow} 2HI \qquad (4.18)$$

involves the fairly highly endothermic dissociation of I_2 (as well as a trimolecular forward step). This is because the single-step mechanism violates the Woodward-Hoffmann exclusion rules (see also Example 7.8 farther below).

Tolman's 16- or 18-electron rule [22,23]. Loosely related to thermodynamics is the 16- or 18-electron rule, suggested by Tolman. This rule is a very helpful guideline for postulating intermediates in reactions involving transition-metal complexes, particularly in homogeneous catalysis. The rule states that a great majority of such reactions proceed through intermediates with 16 or 18 valence electrons. This is because such species are energetically favored over those with 14, 20 or odd-numbered electrons.

Cobalt hydrocarbonyl-catalyzed olefin hydroformylation with network 6.9 may serve as an example. Cobalt, with atomic number 27, contributes nine valence electrons to its complexes (the other eighteen occupy the inner 1-*s*, 2-*s*, 2-*p*, 3-*s*, and 3-*p* orbitals); H, the alkyl group, and the acyl group contribute one each, CO contributes two (of its fourteen electrons, four are shared by C and O in the double bond, an additional four each complete the inner octets of C and O), and an olefin ligand contributes the two π-electrons of its double bond. The contributions and totals for some key participants are:

valence-electron contributions

	Co	H	CO	-C=C-	-alkyl	-acyl	total
$HCo(CO)_4$	9	1	8				18
$HCo(CO)_3$	9	1	6				16
$\rangle\!\cdots\!\overset{H}{C}o(CO)_3$	9	1	6	2			18
$\rangle\!-Co(CO)_3$	9		6		1		16
$\rangle\!-\underset{O}{C}-Co(CO)_3$	9		6			1	16

There are exceptions to Tolman's rule, however [24,25]. For example, if the ligands are very bulky, the 16-electron complex may be sterically hindered, making a 14-electron species the more stable one. The complex Pd[P(*tert*-Bu)$_3$]$_2$ is a case in point [26]. Also, a solvent such as benzene can act as electron donor and thereby stabilize a nominally 14-electron complex as a 16-electron solvate [27]. A few reactions appear to proceed through paramagnetic, 17- or 19-electron complexes as intermediates [28,29]. 20-electron species are believed to be formed as intermediates in some associative ligand substitution reactions [30,31]. All such species are much less stable than the corresponding 16- or 18-electron complexes.

Woodward-Hoffmann exclusion rules. The Woodward-Hoffmann exclusion rules are based on the principle of conservation of molecular-orbital symmetry upon reactions, put forth by Robert B. Woodward and Roald Hoffmann in their landmark book on this topic [32,33]. Symmetry of the wave function is conserved while the valence electrons involved in the breaking and forming of bonds rearrange themselves from their orbitals in the reactants to those in the product or products. If this symmetry constraint permits formation of the product or products in the ground state from the reactants in the ground state, the reaction is "allowed;" if it requires either to be in an excited state, the reaction is "forbidden." In keeping with this picture, the exclusions are not iron-clad: Like traffic laws, they may be broken, but not too often and only for a good reason. An energy input high enough to overcome the energy barrier of a "forbidden" reaction can make it go. Specifically, a photochemical reaction in which a reactant is elevated to an excited state is "allowed" if the same thermal reaction is "forbidden." The reverse is also true. Moreover, the rules apply to "concerted" reactions, that is, to single reaction steps, and do not preclude the possibility that a multistep alternative pathway might exist.

Though only the highest occupied ground-state orbitals need be considered, a proper application of the rules requires familiarity with molecular-orbital theory (e.g., see [34]), a topic well beyond the scope of this book. However, two simple examples may serve to show how the Woodward-Hoffmann rules, apart from providing a better understanding of molecular mechanisms, can rule out seemingly plausible pathways (or at least show them to be highly improbable) and lead to general predictions.

Example 7.7. Cyclo-addition reactions and the 4n + 2 rule. The predictive power of the Woodward-Hoffmann principle becomes apparent, for example, with the application to cyclo-addition reactions [32,35,36]. Consider first the dimerization of ethene to cyclobutane:

The four π-electrons of the two ethene double bonds must redistribute themselves to the σ-orbitals of the newly formed cyclobutane single bonds. As it turns out in this case, the requirement of symmetry conservation does not allow two ground-state ethene molecules to combine in a concerted reaction to form a ground-state cyclobutane molecule: The reaction is "forbidden." The only allowed interactions are destabilizing.

In contrast, for the addition of ethene to butadiene to form cyclohexene

the result is the opposite. Symmetry conservation allows the stabilizing interactions, but not the destabilizing ones: The reaction is "allowed," with only a minor energy barrier to be overcome.

A more general rule emerges from these considerations: Concerted cyclo-addition reactions are "forbidden" if involving 4n π-electrons, as in ethene dimeri-zation (n is any integer), and are "allowed" if involving 4n + 2, as in cyclo-hexene formation [32,35,36].

Example 7.8. Hydrogen-halide reactions. The hydrogen-iodide reaction is known to proceed through the fairly highly endothermic dissociation of I_2 and a subsequent trimolecular step (see above and Section 4.2) rather than in a single, bimolecular step that avoids these normally unfavorable conditions. However, the single-step mech-anisms violates the orbital symmetry requirement [37]. The situation regarding the molecular orbitals of the valence electrons is similar to that in ethene dimerization (see Example 7.7 above): Symmetry conservation does not allow ground-state products to be formed from ground-state reactants. The reaction would have to proceed through an exited state, and the estimated height of the energy barrier is greater than for iodine dissociation.

An analogous argument applies to the hydrogen-bromide and hydrogen-chloride reactions and explains why these also proceed through dissociation of the halogen molecule (here followed by a chain reaction) rather than in a seemingly energetically favored single step.

7.5. Auxiliary techniques

In addition to the simplification of mathematics and the criteria and guidelines described so far, many other techniques can be brought to bear in network elucidation. Excellent literature on these is available, and only a brief overview will be given here. For quick orientation the reader is referred to a comprehensive review with copious references [38].

Determination of isomer distribution. A must in the elucidation of complex networks is to keep track of the isomer distribution of reactants and products wherever isomers are involved. Often, a reactant can add in different ways to another, giving rise to different product isomers. Also, a reactant might isomerize before reacting, with the same result. Today's analytical techniques, foremost among them gas chromatography, are sufficiently advanced to distinguish between isomers. In many instances, the results provide excellent evidence for or against postulated pathways.

A case in point is hydroformylation of mono-olefins, where the carbon atom of CO adds to either of the two double-bonded carbon atoms (see network 7.39 and Example 5.3 in Section 5.3). The discussion in Section 5.3 has shown how the basic structure of the network can be inferred from the dependence of the product isomer distribution on which reactant isomer is originally charged.

Isotope techniques. The labeling ("tagging") of reactants with distinguishable radioactive or stable isotopes is a powerful tool for the elucidation of mechanisms. Isotopes often used are deuterium, carbon-13, carbon-14, oxygen-17, and oxygen-18. In essence, selected atoms in a reactant are replaced by their isotopes, and the product or products are analyzed for the positions which these isotopes have taken up. Suitable methods of analysis for positions of stable isotopes are, among others, nuclear magnetic resonance and gas chromatography with mass-spectroscopic detection.

The classical example of such a study is that of ester hydrolysis by Polányi and Szabó [39]. The authors used $^{18}OH^-$ for the reaction and found that the majority of the heavy oxygen turned up in the acid anion rather than the alcohol:

$$RC\!\!\begin{array}{c} OR' \\ O \end{array} \quad + \quad {}^{18}OH^- \quad \longrightarrow \quad RC\!\!\begin{array}{c} {}^{18}O^- \\ O \end{array} \quad + \quad R'OH$$

This allowed them to conclude that the bond broken is that between oxygen and carbonyl carbon rather than that between oxygen and alkyl carbon.

An example already encountered in this section is the use of DCN in hydrocyanation of olefinic compounds, confirming expectations as to the addition of D and CN to the double-bond carbon atoms (see previous section).

The result of isotopic tagging can be obscured by fast self-exchange of the tag and the regular isotope between positions on the molecule or between different molecules, say, between reactant and solvent. Such "scrambling," if it occurs, makes the technique useless.

Another use of isotopically labeled reactants is for study of *kinetic isotope effects* [40,41]. The difference in zero-point energies between isotopes results in a difference in bond energies and thus in a difference in activation energies and reaction rates. The largest difference is that between hydrogen and deuterium. The effect can be of help especially in the identification of a rate-controlling step.

However, the interpretation of results is not straightforward as zero-point energies in the activated complex also play a role.

Synthesis of intermediates. An excellent technique for confirming or refuting a postulated pathway is to synthesize intermediates and use them as starting materials. Often, a key intermediate that is reactive enough to remain at trace level under reaction conditions is stable at very low temperatures (e.g., that of liquid nitrogen) and can be synthesized. If the reaction starting with the postulated intermediate yields the same products in the same ratios, this can be taken as evidence in favor of the presumed pathway. For example, the essential features of the Heck-Breslow mechanism of hydroformylation (see Example 6.2 in Section 6.3) with cobalt hydro-carbonyl catalysts have been verified in this way by synthesis and use of the alkyl- and acyl-cobalt species [42].

The main problem with this approach is that the intermediate must be brought to reaction conditions almost instantaneously. A transient of any length of time, say, to reach the elevated temperature and pressure of the reaction under study, is apt to falsify the results seriously. Equipment of the type described in Figure 3.3 in Section 3.1.1 or some other injection mechanism is needed.

Spectrophotometry. The theory of spectra is far advanced. In many cases, compounds can be unambiguously identified by their ultraviolet, visible, or infrared spectra (e.g., see Smith's book [43]). As an example, the double bond of a CO ligand in a complex has a strong characteristic infrared vibration frequency whose exact value depends on the electronic properties of the coordinating metal; these, in turn, are affected by the other substituents. In homogeneous catalysis by transition-metal complexes in particular, foremost among them hydrogenation, hydroformylation, and hydrocyanation, spectra have contributed much to the identification of reaction intermediates and thus of pathways.

It is essential that the spectra be taken under reaction conditions, usually involving elevated temperatures and pressures. Especially in homogeneous catalysis by metal complexes, ligand exchange and oxidation-reduction reactions are usually so fast that complex rearrangements keeps pace with equilibrium shifts as a sample is depressured and cooled. Until high pressure-high temperature cells were developed, information from spectra often was entirely misleading.

An amusing incident highlights this important point. In the 1960s, Shell Oil's patent position on phosphine-substituted cobalt hydrocarbonyls as hydroformylation catalysts was practically air-tight. A competitor, however, obtained a patent on a catalyst of this type, differing only by replacement of a CO ligand by a crotyl group, on the strength of the evidence that, at ambient conditions, its IR spectrum differed from those of the Shell-patented catalysts. Shell chemists thereupon took IR spectra of the competitor's catalyst under reaction conditions and found it to convert within seconds to what is present in the systems patented by Shell [44]. A costly lawsuit was avoided by the simple expedient of sending a preprint of the respective journal publication to the competitor to put him on notice that his patent would not stand up in court if he ever dared to use it.

Nuclear magnetic resonance. Spin-state and saturation labeling can be used to investigate relatively fast reactions [38,45]. In essence, the results show whether a reaction is fast or slow relative to the characteristic NMR frequency. For example, spin-state labeling can detect whether or not a fast exchange of ligands such as organic phosphines with ^{31}P nuclei occurs between different coordinative sites or between the complex and the solution [46]. Spin saturation transfer experiments have been used to clarify e.g. the mechanism of exchange between a methyl group and a methylene hydride:

$$HOs_3(CO)_{10}CH_3 \longleftrightarrow H_2Os_3(CO)_{10}CH_2$$

Saturation of the methyl group in the reactant leads to a decrease in the intensity of only one of the two hydride signals of the product, showing that the mechanism is hydride abstraction from methyl and that only one of the hydride sites is involved to a significant degree [47].

Electron spin resonance. Because of its extremely high sensitivity, electron spin resonance has been used in mechanistic studies of reactions of free radicals [48,49]. Direct observation of free radicals is possible. Often, however, radical traps such as nitroso compounds are used instead to catch the radical in form of a stable species [50]. Care must be taken to exclude other possible sources of free radicals [49]. Also, it may not be possible to analyze a sample under reaction conditions.

This brief survey does not include the many strictly analytical methods that can be used to for quantitative determination of concentrations of participants.

Summary

Networks can be effectively elucidated in either of two ways:

(1)   establishment of empirical rate equations that correctly reflect all available rate data, followed by deduction of a plausible network that can produce such rate equations; or

(2)   compilation of all plausible networks, followed by experiments to eliminate those that prove incompatible with observation.

Whether the first or second approach is better suited depends largely on how much quantitative kinetic data are already at hand. In either case, before the postulated network is accepted, it should have established a good track record of correctly predicting behavior under conditions not previously studied. Correct counterintuitive predictions are the most convincing.

The two most general features of a reaction are the apparent kinetic orders with respect to the participants (reactants, products, intermediates, catalysts, and silent partners) and the ranks of the intermediates and products. Reaction orders may vary with conversion, so accurate values are not sought. Ranks, established by Delplots, provide an indication of the sequence in which the respective species are formed, and are useful primarily in the study of reactions with many participants and about whose networks little is known to start with.

The conventional procedure of fitting a rate equation to experimental data is to use a power law reflecting the observed reaction orders. However, while fractional reaction orders may provide an acceptable fit, they cannot be produced by reasonable mechanisms. A better way is to fit the data to "one-plus" rate equations, that is, equations containing concentrations with integer exponents only, but with denominators composed of two or more additive terms of which the first is a "one." Such equations behave much like power laws with fractional exponents but, in contrast to these, can arise from reasonable mechanisms and therefore are more likely to hold over wide ranges of conditions. As an exception, rate equations with constant exponents of one half or integer (positive or negative) multiples of one half can result from chain reactions and reactions initiated by dissociation, and are acceptable if such a mechanism is probable or conceivable.

The general formula for the rate in simple pathways, derived in Section 6.3, can be used for deducing a large number of rules that relate observable kinetic behavior, such as reaction orders, to properties the network may have or definitely cannot have. (Catalytic reactions require qualifications; see Section 8.6.) These rules greatly facilitate network elucidation: Pathways or networks that include a feature producing behavior contrary to observation can be ruled out by whole groups rather than one at a time.

If a pathway or network turns out to be non-simple, a good strategy is to try to break it up into piecewise simple portions that can be studied independently. Whether and how this can be done depends on the reaction at hand. The job is easiest if the portions are irreversible, so that none of them feeds back into a preceding one, and if the non-trace intermediates can be synthesized.

Criteria and guidelines useful in network elucidation and supplementing the rules derived in this chapter include considerations of steric effects, molecularities of postulated reaction steps, and thermodynamic constraints as well as Tolman's 16- or 18-electron rule for reactions involving transition-metal complexes and the Woodward-Hoffmann exclusion rules based on the principle of conservation of molecular orbital symmetry. Auxiliary techniques that can be brought to bear include, among others, determinations of isomer distribution, isotope techniques, and spectrophotometry.

Examples include pyrolysis of an alkylbenzene; homogeneous aldehyde hydrogenation; olefin hydroformylation to alcohol with paraffin by-product formation, aldehyde condensation to heavy ends, and olefin isomerization; cyclo-addition reactions; and hydrogen-halide reactions.

References

1.   V. I. Dimitrov, *Kin. Katal. Lett.*, **7** (1977) 81.

2.   M. T. Klein and P. S. Virk, *I&EC Fundamentals*, **22** (1983) 35.

3.   N. A. Bhore, M. T. Klein, and K. B. Bischoff, *I&EC Research*, **29** (1990) 313.

4.   P. S. Myers and K. M. Watson, *Nat. Petr. News Tech. Soc.*, **38** (1946) 388.

5.   P. E. Savage and M. T. Klein, *I&EC Research*, **26** (1987) 488.

6.   I. Langmuir, *J. Am. Chem. Soc.*, **38** (1916) 2221; **40** (1918) 1361; see also any text on physical chemistry or reaction engineering.

7.   F. G. Helfferich and P. E. Savage, *Reaction kinetics for the practical engineer*, Course #195, AIChE Educational Services, New York, 7th ed. 1999, Section 7.3.

8.   J. L. Van Winkle, unpublished, 1977.

9.   F. G. Helfferich, *J.Phys. Chem.*, **93** (1989) 6676.

10. L. Marko, *Proc. Chem. Soc.*, **1962**, 67.

11. J. L. Van Winkle, personal communication, 1974.

12. *Landolt-Börnstein, New Series, Radical reaction rates in liquids*, H. Fischer, ed., Springer, Berlin, Part II Vol. 13 (5 subvolumes), 1983-1985, ISBN 0387126074, 0387132414, 0387117253, 0387121978, 0387136762; Vol. 18 (5 subvolumes), 1994-1997, 3540560548, 3540560556, 3540560564, 3540603573, 3540560572.

13. D. L. Baulch, C. J. Cobos, R. A. Cox, C. Esser, P. Frank, T. Just, J. A. Kerr, M. J. Pilling, J. Troe, R. W. Walker, and J. Warnatz, *J. Phys. Chem. Ref. Data*, **21** (1992) 411.

14. D. L. Baulch, C. J. Cobos, R. A. Cox, P. Frank, G. Hayman, T. Just, J. A. Kerr, T. Murrells, M. J. Pilling, J. Troe, R. W. Walker, and J. Warnatz, *J. Phys. Chem. Ref. Data*, **23** (1994) 847.

15. W. G. Mallard, F. Westley, J. T. Herron, R. F. Hampson, and D. H. Frizzell, *NIST chemical kinetics data base: Windows Version 2Q98*, National Institute of Standards and Technology, Gaithersburg, MD, 1998.

16. E. T. Denisov and T. Denisova, *Handbook of antioxidants: bond dissociation energies, rate constants, activation energies, and enthalpies of reactions*, CRC Press, Boca Raton, 2nd ed., 2000, ISBN 0849390044, Chapter 2.

17. J. D. Druliner, *Organometallics*, **3** (1984) 205.

18. I. J. Goldfarb and M. Orchin, *Adv. Catal.*, **9** (1957) 609.

19. T. L. Cottrell, *The strengths of chemical bonds*, Butterworths, London, 2nd ed., 1958.

20. N. N. Semenov, *Some problems in chemical kinetics and reactivity* (English translation), Vol. I, Princeton Univerity Press, 2nd ed., 1958.

21. *Handbook of chemistry and physics*, D. R. Lide, ed.-in-chief, CRC Press, Boca Raton, 80th ed., 1999-2000, ISBN 0849304806, Chapter 9, pp. 51-74.

22. C. A. Tolman, *Chem. Soc. Rev.*, **1** (1972) 337.

23. J. P. Collman, L. S. Hegedus, J. R. Norton, and R. G. Finke, *Principles and applications of organotransition metal chemistry*, University Science Books, Mill Valley, CA, 2nd ed., 1987, ISBN 0935702512, p. 33.

24. S.-Y. Chu and R. Hoffmann, *J. Phys. Chem.*, **86** (1982) 1289.

25. Collman *et.al.* (ref. 21), Section 2.3.

26. S. Otsuka, T. Yoshida, M. Matsumoto, and K. Nakatsu, *J. Am. Chem. Soc.*, **98** (1976) 5850.

27. M. H. J. M. de Croon, P. F. M. T. van Nisselrooij, H. J. A. M. Kuipers, and J. W. E. Coenen, *J. Mol. Catal.*, **4** (1978) 325.

28. J. Halpern, in *Fundamental research in homogeneous catalysis*, M. Tsutsui, ed., Plenum, New York, Vol. 3, 1978, ISBN 0306401991, p. 25.

29. A. E. Stiegman and D. R. Tyler, *Comments Inorg. Chem.*, **5** (1986) 215.

30. G. Yagupsky, C. K. Brown, and G. Wilkinson, *J. Chem. Soc.*, **A 1970**, 1392.

31. Collman *et al.* (ref. 22), Section 4.5.c.

32. R. B. Woodward and R. Hoffmann, *The conservation of orbital symmetry*, VCH Publishers, Weinheim, 1970, 5th printing 1989, ISBN 0895731096.

33. R. B. Woodward and R. Hoffmann, *Angew. Chem. Internat. Ed. Engl.*, **8** (1969). 781.

34. R. McWeeny, *Coulson's Valence*, Oxford University Press, 1979 (3rd ed. of C. A. Coulson, *Valence*), ISBN 0198551452.

35. R. Hoffmann and R. W. Woodward, *J. Am. Chem. Soc.*, **87** (1965) 2046.

36. A. Pross, *Theoretical and physical principles of organic reactivity*, Wiley, New York, 1995, ISBN 0471555991, Section 10.4.1.

37. R. Hoffmann, *J. Chem. Phys.*, **49** (1968) 3739.

38. C. A. Tolman and J. W. Faller, in *Homogeneous catalysis with metal phosphine complexes*, L. H. Pignolet, ed. Plenum, New York, 1983, ISBN 030641211X, Chapter 2.

39. M. Polányi and A. L. Szabó, *Trans. Farad. Soc.*, **30** (1934) 508.

40. L. Melander and W. H. Saunders, Jr., *Reaction rates of isotopic molecules*, Wiley, New York, 1980, ISBN 0471043966.

41. K. B. Wiberg, *Physical organic chemistry*, Wiley, New York, 1964, pp. 351-363.

42. R. F. Heck and D. S. Breslow, *J. Am. Chem. Soc.*, **83** (1961) 4023.

43. B. C. Smith, *Infrared spectral interpretation: a systematic approach*, CRC Press, Boca Raton, 1998, ISBN 0849324637.

44. W. W. Spooncer, A. C. Jones, and L. H. Slaugh, *J. Organomet. Chen.*, **18** (1969) 327.

45. L. M. Jackman and F. A. Cotton, eds., *Dynamic nuclear magnetic resonance spectroscopy*, Academic Press, New York, 1975, ISBN 0123788501.

46. J. W. Faller, *Adv. Organomet. Chem.*, **16** (1977) 211.

47. R. B. Calvert and J. R. Shapley, *J. Am. Chem. Soc.*, **100** (1978) 7726.

48. J. K. Kochi, *Accts. Chem. Res.*, **7** (1974) 351.

49. T. L. Hall, M. F. Lappert, and W. P. Lednor, *J. Chem. Soc., Dalton Trans.,* **1980**, 1448.
50. S. Terabe and R. Konaka, *J. Chem. Soc., Perkin Trans. II*, **1972**, 2136.

Chapter 8

Homogeneous Catalysis

Homogeneous catalysis is one of the most interesting fields of chemistry, especially for its mechanisms and kinetics. Unencumbered by the complications of non-uniform and easily contaminated surfaces and of adsorption equilibria and rates that make heterogeneous catalysis more difficult to interpret, homogeneous catalysis provides excellent opportunities for the study of the molecular causes of reactivity, of what makes reactions go. Homogeneous catalysis, by enzymes, can produce both very high rates and very high selectivities. Life itself could not have evolved without it. Moreover, homogeneous catalysis can be translated into heterogeneous catalysis by anchoring the catalytic species onto a solid polymeric support or, if it is ionic, having it held by electrostatic forces in a gel with fixed groups of opposite charge sign. Lastly, a thorough understanding of the molecular effects in homogeneous catalysis can serve as a guide in the development of new or more efficient metal and metal oxide catalysts and their supports.

From the point of view of kinetics and mechanisms, homogeneous catalysis can be classified into single-species and complex catalysis. In the former, a single molecule or ion combines with a reactant molecule to initiate the reaction and eventually reappears in its original form. An example is catalysis by hydrogen ions. In complex catalysis, the catalyst exists in several different forms, thus constituting a chemical system of its own that interacts with the reactants. Acid-base catalysis and catalysis by metal-organic complexes are the most prominent members of this category.

A further complication arises if a significant fraction of the total catalyst material may be present in the form of reaction intermediates rather than as the free catalyst. If the catalyst is highly active, a minute amount suffices to produce a high reaction rate, and even a trace-level intermediate may then contain a large fraction or possibly most of the catalyst material. Such behavior is typical for enzyme catalysis, but not confined to it. In such cases, the concentration of free catalyst may vary with conversion, may not be known, and may be very difficult to measure. Rather, what is known is the total amount of catalyst material added, and rate equations in terms of the latter are therefore needed. Such systems will be discussed in the later sections of this chapter.

Regardless of these complications, the Bodenstein approximation of quasi-stationary states of trace-level intermediates remains the principal tool for reduction of mathematical complexity. In fact, it is more often applied to catalytic than to thermal reactions. This is because, usually, the amount of catalyst needed to produce the desired reaction rate is small, and the catalyst-containing intermediates thus are apt to remain at very low concentrations, as the validity of the approximation requires. In trace-level catalysis, of course, this requirement is automatically met. It will be taken for granted in this chapter.

Moreover, this chapter remains restricted to "simple" reactions (no steps with two or more molecules of intermediates as reactants). Non-simple reactions are relatively rare in homogeneous catalysis. They can be handled as outlined in Section 6.5. A more detailed discussion is beyond the scope of this book.

8.1. Single-species catalysis

Many homogeneous reactions are catalyzed by a single chemical species, usually by solvated protons or metal ions. A typical example is the hydrolysis of sucrose in aqueous solution:

$$\text{sucrose} + H_2O + H^+ \longrightarrow \text{glucose} + \text{fructose} + H^+$$

alleged to be the first reaction whose kinetics were studied [1], and with its simple and easily observed rate behavior one of the staples of the undergraduate chemical laboratory. Many other hydrolysis, hydration, dehydration, and esterification reactions are of this type. Certain metal ions such as copper and zinc in aqueous solution can function in a similar fashion. Like the proton, they act as Lewis acids, initiating the reaction by inducing a positive charge on the reactant [2,3]. A typical example is decarboxylation of organic acids.

To arrive at rate equations of catalysis by a single species that is present almost exclusively as free catalyst, the formalism developed for noncatalytic reactions can be used with the catalyst appearing as both a reactant and a product. In fact, many of the examples used in earlier chapters to illustrate the deduction and application of rate equations were from catalytic reactions of this type. Thus, all the rules derived in Chapter 7 for network elucidation remain valid in such cases.

Typically, this kind of catalysis involves a single free catalyst (cat) which converts a reactant A to a product P in a single cycle with arbitrary number k of members, and possibly with co-reactants and co-products (not shown):

$$A \rightleftharpoons X_1 \rightleftharpoons \ldots \rightleftharpoons X_{k-1} \rightleftharpoons P \tag{8.1}$$

The rate equation, obtained with eqns 6.4 to 6.6, is

$$r_{\mathrm{P}} = \frac{\left[\displaystyle\prod_{i=0}^{k-1} \lambda_{i,i+1} - \prod_{i=0}^{k-1} \lambda_{i+1,i} \right] C_{\mathrm{cat}}}{D_{00}} \tag{8.2}$$

where both indices 0 and k refer to the free catalyst, and D_{00} is given by eqn 6.6. The concentrations of A and P do not appear explicitly, but are co-factors in the coefficients λ_{01} and $\lambda_{k,k-1}$, respectively.

In general, as the appearance of C_{cat} as a factor in the numerator of eqn 8.2 shows:

Forward and reverse rates in single-species bulk catalysis
are first order in free catalyst.

This is in keeping with the rule that a reaction is first order in a reactant that participates with one molecule in only the first step of a simple pathway (see Rule 7.13 in Section 7.3.1). However, even in single-species, single-cycle, bulk catalysis there are exceptions to this rule, as will be shown later in this section.

Example 8.1. Acetal hydrolysis [4,5]. The catalytic cycle of acetal hydrolysis in aqueous solution, catalyzed by dilute acid

$$\begin{array}{c} \mathrm{R'} \\ \mathrm{R''} \end{array}\!\! \begin{array}{c} \mathrm{OR} \\ \mathrm{C} \\ \mathrm{OR} \end{array} + H_2O + H^+ \longrightarrow \begin{array}{c} \mathrm{R'} \\ \mathrm{R''} \end{array}\!\!\mathrm{C}{=}\mathrm{O} + 2\,\mathrm{ROH} + H^+ \tag{8.3}$$

is believed to be

$$\tag{8.4}$$

(catalytic cycle diagram)

The second step is commonly assumed to be irreversible and rate-controlling, so that the first is in quasi-equilibrium:

$$-r_{acetal} = k_{12} C_1$$

$$C_1 \cong K_{01} C_{acetal} C_{H^\cdot}$$

(indices 1, 2, ... refer to intermediates X_1, X_2, ..., and 0 to cat; K_{01} is the equilibrium constant of the first step). Accordingly, the rate in terms of bulk participants is

$$-r_{acetal} \cong k C_{acetal} C_{H^\cdot} \qquad (k \equiv K_{01} k_{12}) \qquad (8.5)$$

and is first order in acetal and hydrogen ion, in agreement with experimental observation.

The assumption that the second step is rate-controlling follows tradition, but is needlessly restrictive. Granted that the intermediates remain at trace level, as they almost certainly do, the observed kinetic behavior results even if the first step is not close to equilibrium. For the cycle 8.4 with irreversible second step, as assumed, the general formula 6.4 to 6.6 gives

$$-r_{acetal} = k C_{acetal} C_{H^\cdot} \qquad \left[k \equiv \frac{k_{01} k_{12}}{k_{10} + k_{12}} \right] \qquad (8.6)$$

This rate equation is of the same algebraic form as eqn 8.5, but with different physical significance of the coefficient k. This is another example of the fact that an observed rate behavior can usually be explained in several different ways.

A distinction between the two rival explanations may be possible. An Arrhenius plot of $\ln(K_{01} k_{12})$ should give a reasonably straight line, whereas a plot of $\ln[k_{01} k_{12} /(k_{10} + k_{12})]$ may well be curved. This is because the first plot is for the logarithm of a *product* of two constants giving straight-line logarithmic plots, whereas the second is for a logarithm of an expression involving a *sum* of such quantities (see Section 12.1.4). Accordingly, a distinctly curved Arrhenius plot would be evidence against quasi-equilibrium in the first step. However, straight-line behavior would still allow both interpretations because the plot of $\ln[k_{01} k_{12} /(k_{10} + k_{12})]$ would be straight if the activation energies of k_{10} and k_{12} were equal.

In acid-catalyzed reactions, the distinction between single-species and complex catalysis is not always clear-cut. The actual catalyst is the solvated proton, H_3O^+ in aqueous solution, and H_2O (or a molecule of the nonaqueous solvent) may thus appear as a co-product in the first step and as a co-reactant in the step reconstituting the original solvated proton, possibly also in other additional steps, e.g., if the overall reaction is hydrolysis or hydration. Moreover, the acid added as catalyst may not be completely dissociated, and its dissociation equilibrium then affects the concentration of the solvated proton. At high concentrations this is true even for fairly strong acids such as sulfuric, particularly in solvents less polar than water. Such cases are better described as acid-base catalysis (see Section 8.2.1).

At high acid concentrations, hydrolysis or hydration reactions may no longer be first order in hydrogen ion as in eqns 8.5 and 8.6. Consider a hypothetical hydration reaction with catalytic cycle

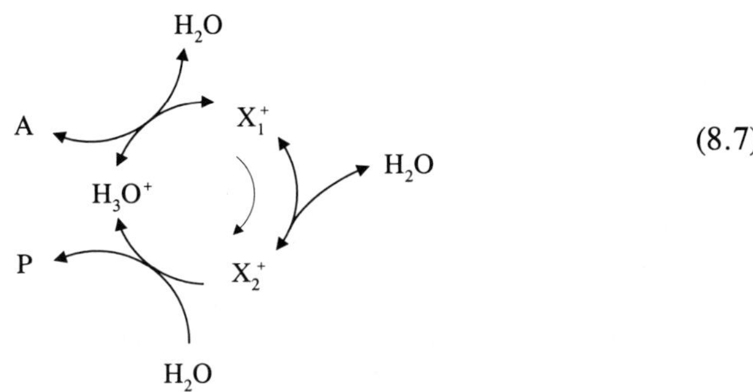

$$(8.7)$$

with protonated intermediates X_1^+ and X_2^+ at trace level and irreversible third step. The general formula 6.4 to 6.6 (with index 0 for H_3O^+) gives for this case

$$r_P \;=\; \frac{k_{01}k_{12}k_{20}\,C_{H_3O^+}\!\cdot C_A\,C_{H_2O}^2}{(k_{12}+k_{10})k_{20}\,C_{H_2O}^2 + k_{10}k_{21}\,C_{H_2O}}$$

or, in one-plus form after cancellations,

$$r_P \;=\; \frac{k_a\,C_{H_3O^+}\!\cdot C_A\,C_{H_2O}}{1 + k_b\,C_{H_2O}} \qquad\qquad (8.8)$$

$$\text{where}\qquad k_a \equiv \frac{k_{01}k_{12}k_{20}}{k_{10}k_{21}} \qquad\text{and}\qquad k_b \equiv \frac{(k_{12}+k_{10})k_{20}}{k_{10}k_{21}}$$

The rate is first order in A and H_3O^+ and of order between zero and one in H_2O (becoming first order in H_2O if the first two steps are at quasi-equilibrium). At very high acid concentration, more water is bound in the form of hydration shells of the ions if more acid is added. This reduces the concentration of *free* water (C_{H_2O} in eqn 8.8). Such a reduction of availability of water as reactant has a retarding effect and may even overcompensate the increase in catalytic hydrogen ion concentration. If it does, the rate as a function of acid concentration goes through a maximum [6].

The effect just described produces an apparent reaction order in H_3O^+ that is less than one and may even become negative at very high acid concentrations. The opposite behavior, an apparent order higher than one, is also found in a large

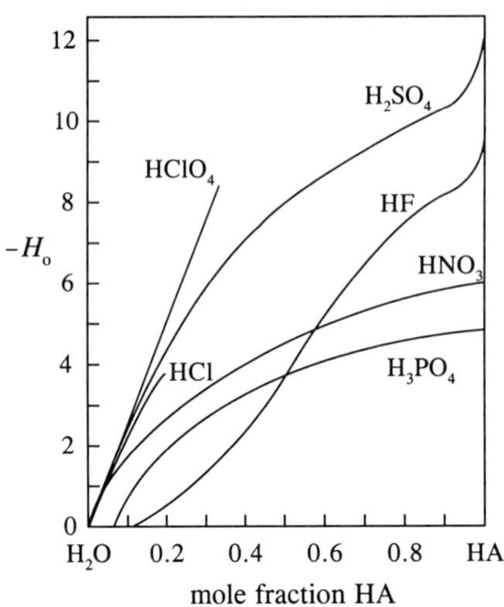

Figure 8.1. Hammett acidity functions for various strong acids (from Olah [8]).

number of instances. At very high concentrations of a strong acid, species other than H_3O^+ may function as additional proton donors. A semi-empirical procedure to account for this effect is to correlate the rate to the Hammett acidity function H_0 instead of the pH of the solution [7-10]. In brief review and without discussion of the theoretical basis, H_0 is given by

$$H_0 = pK_a - \log(C_{BH^+}/C_B)$$

and characterizes the tendency of the solution to protonate a neutral basic indicator B. Hammett acidity functions as a function of acid concentration are shown in Figure 8.1 for various strong acids (see also a detailed treatment by Connors [11]).

8.2. Complex catalysis

The term *complex catalysis* indicates that the catalyst exists in several different forms that interact with one another and play different roles. Usually, inter-conversion of the catalyst species is by dissociation-association or ligand exchange, reactions that are orders of magnitude faster than the catalyzed formation or cleavage of bonds within molecules. If so, the assumption that the catalyst species remain very close to equilibrium is justified.

8.2.1. Acid-base catalysis

The most common and most thoroughly studied type of homogeneous catalysis is acid-base catalysis. It includes hydrolysis, alcoholysis, esterification, and con-densation reactions among many others. It is characterized by the fact that the equilibrium between base and conjugate acid, or between acid and conjugate base, is coupled with the actual catalytic cycle.

Several examples in previous sections fall under the heading of acid-base catalysis. Nitration of aromatics with catalyst acid HB and its conjugate base B^- is one of these (see reaction 4.6 in Section 4.1). Also, acid-catalyzed hydrolysis and hydration reactions such as 8.3 and 8.7 can be viewed as belonging to this category because the original catalyst actually is H_3O^+ rather than H^+ and is reconstituted in a step that involves its conjugate base, H_2O, as co-reactant.

A classical example of acid-base catalysis is base-catalyzed aldol condensation of aldehydes [12-14]. It may serve here to illustrate typical facets of reactions of this kind. The catalytic cycle is

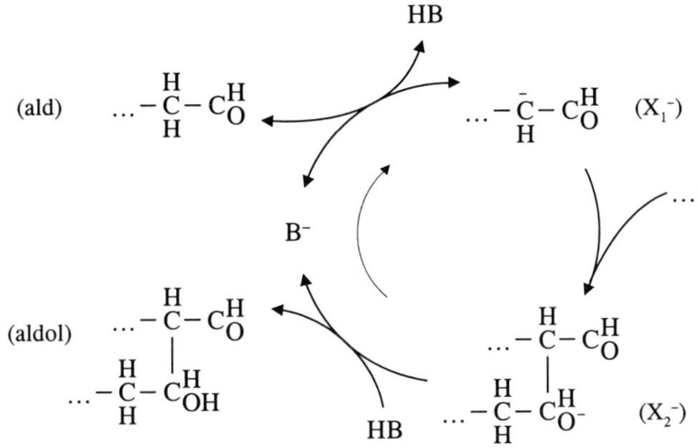

$$(8.9)$$

where ald is aldehyde, B^- and HB are the catalyst base and its conjugate acid, respectively, and the intermediates are at trace level. The second step, $X_1^- + $ ald $\longrightarrow X_2^-$, is commonly assumed to be irreversible.

In such a pathway, aldehyde is a reactant in both the first and second steps. Accordingly, this is a typical case of competing steps, discussed in Section 5.5. With X_1^- at trace level, the general formula for simple pathways (eqns 6.4 to 6.6.) gives

$$r_{\text{aldol}} \cong \frac{k_{01} k_{12} C_{\text{ald}}^2 C_{B^-}}{k_{12} C_{\text{ald}} + k_{10} C_{\text{HB}}} \tag{8.10}$$

with two possible limiting cases (see Section 5.5):

Case I	*Case II*
first step rate-controlling	second step rate-controlling
$k_{10} C_{\text{HB}} \ll k_{12} C_{\text{ald}}$	$k_{10} C_{\text{HB}} \gg k_{12} C_{\text{ald}}$
$r_{\text{aldol}} \cong k_{01} C_{\text{ald}} C_{B^-}$	$r_{\text{aldol}} \cong (k_{01} k_{12} / k_{10}) C_{\text{ald}}^2 (C_{B^-} / C_{\text{HB}})$
rate first order in aldehyde and first order in B^-	rate second order in aldehyde and first order in OH^-
general base catalysis	*specific base catalysis*

(Note that C_{B^-}/C_{HB} is proportional to C_{OH^-} because of the dissociation equilibrium $HB + OH^- \longleftrightarrow B^- + H_2O$ with $K_{\text{HB}} = C_{B^-} C_{H_2O} / C_{\text{HB}} C_{OH^-} = $ const. and $C_{H_2O} \cong$ const.)

If the first step is rate-controlling (Case I), the rate is first order in the respective base, B^-, *not* in hydroxide ion. If several bases are present, their contributions to the rate are additive. This is called *general base catalysis*. In contrast, if the second step is rate-controlling (Case II), the rate is first order in hydroxide ion, regardless of what base or bases are used as catalyst. This is called *specific base catalysis*. Intriguingly, mathematics produces first order in OH^- even though that ion is not a reactant in any of the kinetic steps.

While the two limiting cases of general and specific base catalysis are encountered reasonably often, other types of behavior are possible. Obviously, the two terms in the denominator of the rate equation 8.10 may be comparable. The reaction then is of varying order between one and two in aldehyde and between zero and one in OH^- [13].

A more interesting deviation from textbook behavior is found in some organic solvents. The rate may be strictly second order in aldehyde, but not first order in either OH^- or B^-. The catalytic pathway 8.9 cannot be questioned even in such cases. Rather, the results can be quantitatively explained if the second step, the addition of aldehyde to the carbanion X_1^-, is made reversible. Instead of eqn 8.10, the rate equation then is

$$r_{\text{aldol}} \cong \frac{k_{01}k_{12}k_{20}\, C_{\text{ald}}^2 C_{B^-} C_{\text{HB}}}{k_{12}k_{20}\, C_{\text{ald}} C_{\text{HB}} + k_{10}k_{20}\, C_{\text{HB}}^2 + k_{10}k_{21}\, C_{\text{HB}}} \tag{8.11}$$

If the rate is second order in aldehyde, the first term of the denominator must be negligible (this implies quasi-equilibrium in the first step). After cancellations and rearrangement to one-plus form, eqn 8.11 then becomes

$$r_{\text{aldol}} \cong \frac{k_a\, C_{\text{ald}}^2 C_{B^-}}{1 + k_b\, C_{\text{HB}}} \tag{8.12}$$

where

$$k_a \equiv \frac{k_{01}k_{12}k_{20}}{k_{10}k_{21}} \qquad \text{and} \qquad k_b \equiv k_{20}/k_{21}$$

Example 8.2. Aldol condensation of n-butanal in alcohol solvent [15]. Results obtained for aldol condensation of *n*-butanal in *n*-octanol are shown in Table 8.1. The reaction is second order in aldehyde. Its apparent second-order rate coefficient, $k_{\text{app}} = r_{\text{alc}}/C_{\text{ald}}^2$, is tabulated as a function of buffer composition. Although the reaction is second order in aldehyde, its rate is not proportional to the B^--to-HB ratio and so is not first order in OH^-. Equation 8.10 thus cannot represent the results.

Equation 8.12, more general for reactions second order in aldehyde, can be rearranged:

$$\frac{C_{\text{ald}}^2 C_{B^-}}{r_{\text{aldol}}} = \frac{1}{k_a} + \frac{k_b}{k_a} C_{\text{HB}}$$

Accordingly, if eqn 8.12 is applicable, a plot of $C_{ald}^2 C_{B^-}/r_{aldol}$ versus C_{HB} should give a straight line with slope k_b/k_a and intercept $1/k_a$. Such a plot, shown in Figure 8.2, is found to be linear with only some random scatter. The values of the coefficients calculated from slope and intercept or by linear regression are

$$k_a = 0.257 \text{ M}^{-2}\text{min}^{-1}$$

$$k_b = 10.9 \text{ M}^{-1}$$

Equation 8.12 with these values of the coefficients provides a reasonably good fit to the observed rates.

Table 8.1. Observed second-order rate coefficients for aldol condensation of *n*-butanal (11%wt in *n*-octanol, potassium-octoate buffer, 171°C) [16].

buffer composition [M]		k_{app}
B^-	HB	$M^{-1}min^{-1}$
0.099	0.011	0.0237
0.099	0.033	.0173
0.099	0.068	.0139
0.099	0.090	.0119
0.030	0.030	.0065
0.059	0.059	.0093
0.171	0.171	.0157

8.2.2. Catalysis by metal complexes

Many highly interesting and important reactions of organic chemistry are catalyzed by complexes of transition-metal ions in solution. These include hydrogenation, hydrocyanation, hydroformylation, and some oxidation and polymerization reactions, many of them practiced in industry on a large scale [G1,G3,G5,G10]. The most useful metal ions for this purpose are those of Group VIII, in particular cobalt, nickel, palladium, platinum, and rhodium. The most common ligands are carbon monoxide, organic phosphorous compounds, chloride, and the nitroso and cyano groups.

A common feature of catalysts of this type is that they exist as different complexes in or close to equilibrium with one another, some of which are able to accept reactants such as olefins, aldehydes, etc., as ligands. A knowledge of the catalyst equilibria is essential for an understanding of catalyst reactivity, selectivity, and kinetics. Moreover, a proper model of the reaction must include the effect of catalyst equilibria.

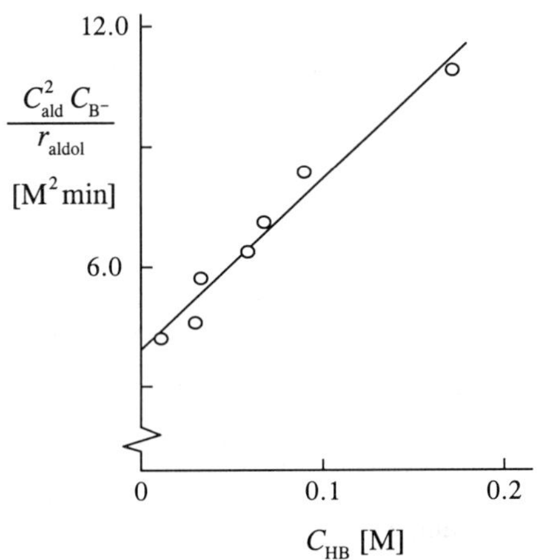

Figure 8.2. Straight-line plot of data from Table 8.1 with eqn 8.12 [15].

Example 8.3.   Phosphine-substituted cobalt hydrocarbonyls as hydroformylation catalysts.   Extensively studied catalyst systems with complex equilibria include phosphine-substituted hydrocarbonyls of cobalt, $HCo(CO)_3Ph$, where Ph stands for a tertiary organic phosphine.   They are modifications of the original oxo catalyst, $HCo(CO)_4$.   Like the latter, they catalyze the oxo or hydroformylation reaction of olefins to aldehydes one carbon number higher:

$$olefin + H_2 + CO \longrightarrow aldehyde$$

(see Example 6.2 in Section 6.3).   Unlike the oxo catalyst, they also have hydrogenation activity, so that the aldehyde produced is immediately converted to alcohol. This is an advantage because, usually, alcohol rather than aldehyde is the desired product.   The phosphine stabilizes the hydrocarbonyl.   This reduces the reaction rate under comparable conditions, but makes it possible to carry out the reaction at much lower pressure and to separate the products from the catalyst by short residence-time vacuum distillation.   Moreover, the phosphine-substituted hydrocarbonyls give a much higher selectivity to the usually desired straight-chain over branched-chain aldehydes. However, their hydrogenation activity also results in formation of paraffin as by-product [17-19].

The catalyst forms spontaneously from a cobalt salt or dicobalt octacarbonyl under reaction conditions (synthesis gas pressure and elevated temperature) in the presence of the phosphine and exists in a ligand-exchange equilibrium with the hydro-tetracarbonyl, $HCo(CO)_4$ [20,21].   Even only a minute amount of the much more reactive $HCo(CO)_4$ contributes significantly to the rate and adversely affects selectivity, apart from being the main cause of catalyst loss through cobalt deposition. Both hydrocarbonyls are acids, $HCo(CO)_4$ a much stronger one than $HCo(CO)_3Ph$ [22].   As the stronger acid, the undesired $HCo(CO)_4$ can be preferentially neutralized by addition of the right amount of base.   Its anion, $Co(CO)_4^-$, is stable under hydroformylation conditions, but is catalytically inactive, and so is $Co(CO)_3Ph^-$. With base added to maintain selectivity and catalyst stability, the key cobalt species under reaction conditions thus are the two hydrocarbonyls and their anions, linked by their ligand-exchange and neutralization equilibria as shown in the scheme 8.13 on the facing page.

In this catalyst system, addition of phosphine or decrease in partial pressure of CO (or of total pressure at constant synthesis-gas ratio) shifts the phosphine-carbon monoxide ligand-exchange equilibrium toward the phosphine-substituted species (right to left in scheme 8.13).   A decrease in phosphine concentration or an increase in CO pressure shifts it in the opposite direction (left to right).   Addition of base shifts equilibrium downward, from the acids to the anions.   In the presence of phosphine in an at least moderate excess, thermodynamics strongly favors $HCo(CO)_3Ph$ over $HCo(CO)_4$, but $Co(CO)_4^-$ remains very strongly favored over $Co(CO)_3Ph^-$ under all conditions of practical interest.

The equilibria 8.13 of the cobalt-containing species explain a puzzling observation: The addition of base turns the effects of CO pressure (or total pressure at constant synthesis-gas ratio) and phosphine concentration on rate into their opposites! In the *absence* of base, an increase in CO pressure or decrease in phosphine

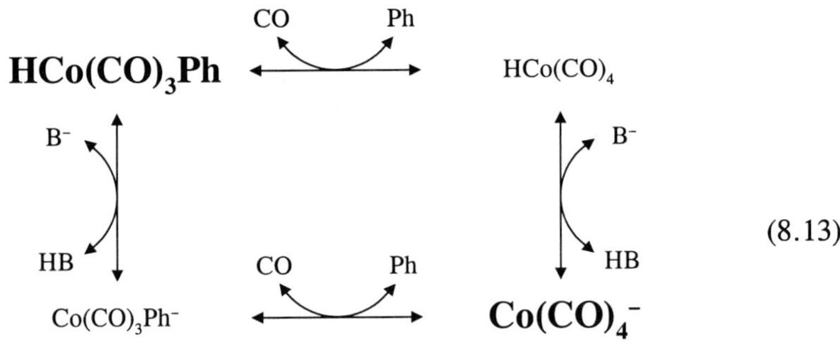

(predominant cobalt-containing species in larger font)

concentration shifts cobalt from $HCo(CO)_3Ph$ to $HCo(CO)_4$, to a more active catalyst, and so *increases* the rate. In contrast, in the *presence* of base, the shift is primarily from $HCo(CO)_3Ph$ to $Co(CO)_4^-$, from a catalytically active to an inactive species, and so *decreases* the rate. The decrease in rate with increase in pressure (in the presence of base) in a reaction consuming gas is counterintuitive all by itself.* Lastly, in the absence of base, the shift toward $HCo(CO)_4$ caused by an increase in pressure or a decrease in phosphine concentration results in poorer selectivity and higher catalyst losses.

If the amounts of the catalyst ingredients (cobalt, phosphine, and base) are increased by the same factor without change in pressure and H_2-to-CO ratio, the increase in phosphine concentration uncompensated by an increase in CO pressure shifts ligand exchange equilibrium toward $HCo(CO)_3Ph$. Without base, this shift is at the expense of the more active $HCo(CO)_4$, producing an apparent reaction order lower than one in "catalyst." In the presence of base, the shift is mostly at the expense of the inactive $Co(CO)_4^-$, resulting in an apparent order higher than one.

Granted quasi-equilibrium of ligand exchanges and neutralizations and the fact that $Co(CO)_3Ph^-$ is catalytically inactive and accounts for only an insignificant fraction of total cobalt, the system 8.13 can be modeled with two of the three equilibrium constants linking the concentrations of $HCo(CO)_3Ph$, $HCo(CO)_4$, and $Co(CO)_4^-$. This will be shown in more detail in Example 8.11 in Section 8.8.2.

The species shown in the system 8.13 are those present under reaction conditions. At lower temperatures and pressures, a wealth of different complexes are present, including dicobalt octacarbonyl and cationic complexes of cobalt that differ depending on what cobalt compound was initially charged [17,23-25].

Equilibria between different complexes are the rule in catalyst systems of this kind. However, the co-existence of two or more different complexes that catalyze the same reaction, but have different activities and selectivities, as do $HCo(CO)_3Ph$

---

* There is no violation of the Le Châtelier principle, which applies to equilibria, but not to rates. Equilibrium conversion is indeed higher at higher pressure, but is close to 100% anyway even at lower pressure.

and $HCo(CO)_4$ in the example above, is less common (another example of such behavior are phosphine-substituted rhodium hydrocarbonyl catalysts for hydroformylation [26]). In such cases, the rate is given by additive rate equations, one for each active species (see eqn 8.96 in Example 8.11).

In some important metal-organic catalyst systems, a significant fraction of the catalyst metal may be bound in the form of reaction intermediates rather than as free catalyst. Reaction mathematics under such conditions will be discussed in the next two sections of this chapter.

Apart from these two complications, the formalism developed in Chapter 6 for noncatalytic reactions remains applicable as far as rate equations are concerned, but must be combined with modeling of the catalyst equilibria: Where rate equations of transition-metal catalysis appeared in examples in earlier chapters, the quantity C_{cat} is the concentration of the free catalyst and may have to be supplied by a catalyst equilibrium model.

8.3. Classical models of enzyme kinetics

As has already been pointed out, any rate equation containing the concentration of the free catalyst is of little practical use if that concentration is unknown, is difficult or impossible to measure, and may vary with conversion, as is the case if a significant fraction of the total catalyst material is present in the form of intermediates of the reaction. This is often true in catalysis by enzymes or other trace-level catalysts. To be sure, the equations in terms of free-catalyst concentration remain correct. However, unless practically all the catalyst material is present as free catalyst, they no longer reflect the actual reaction orders. This is because the concentrations of the participants affect the rate not only directly as expressed explicitly in those equations, but also indirectly and implicitly through their effect on the free-catalyst concentration: As the reactant concentration decreases, so do those of the intermediates; in turn, this produces an increase in the free-catalyst concentration that boost the rate and, thereby, decreases the apparent reaction order. To reflect this facet correctly, what is needed are rate equations in terms of the *total* amount of catalyst material, a quantity that is constant and known.

The Bodenstein approximation is not invalidated by this complication: As long as the total amount of catalyst is extremely small compared with those of the reactants, as is largely true in enzyme kinetics, the catalytic cycle attains quasi-stationary conditions. (Exceptions are fast biological reactions of substances themselves at very low concentrations, a topic beyond the scope of this book.)

The earliest quantitative theory of enzyme kinetics dates back to 1913, when Michaelis and Menten [27] succeeded in explaining a key feature of enzyme reactions with a very simple model. As an introduction and to establish the relationship between trace-level and bulk-species catalysis, this classical work and its subsequent refinements will now be reviewed.

8.3.1. Michaelis-Menten kinetics.

The catalytic cycle considered by Michaelis and Menten [27,28], the simplest to exhibit the most characteristic feature of enzyme catalysis, is

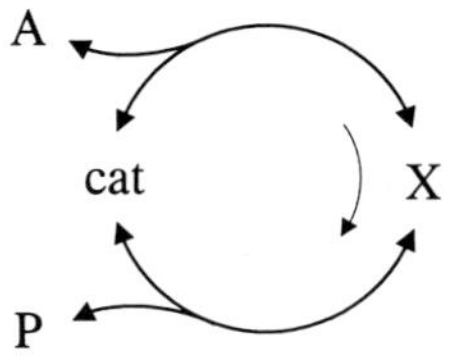

$$\text{(8.14)}$$

with quasi-equilibrium in the step A + cat $\longleftrightarrow$ X:

$$C_X/C_A C_{\text{cat}} \cong K_{AX} = \text{const.} \qquad (8.15)$$

The catalyst is distributed over free catalyst (cat) and the intermediate (X), so that

$$C_{\text{cat}} + C_X = C_{\Sigma\text{cat}} \qquad (8.16)$$

where $C_{\Sigma\text{cat}}$ is the total amount of catalyst per unit volume.

The initial rate of product formation is that of the second step:

$$r_P^o = k_{XP} C_X \qquad (8.17)$$

(reverse reaction still insignificant). Expressing C_X in eqn 8.17 in terms of $C_{\Sigma\text{cat}}$ by means of eqns 8.15 and 8.16 one obtains:

$$r_P^o \cong \frac{K_{AX} k_{XP} C_A C_{\Sigma\text{cat}}}{1 + K_{AX} C_A} \qquad (8.18)$$

An empirical equation of this form had been suggested as early as 1903 by Henri [29]. Ten years later, Michaelis and Menten provided its theoretical basis [27].

According to eqn 8.18, the initial rate is first order in total catalyst material and of order between zero and one in the reactant, A. Note that Rule 7.13 for simple thermal pathways, which states that a reaction is first order in a reactant that participates in the first step, is *not* obeyed (it will be modified in Section 8.6).

Limiting cases of eqn 8.18 are

$$r_P^o \cong K_{AX} k_{XP} C_A C_{\Sigma\text{cat}} \qquad \text{if} \quad K_{AX} C_A \ll 1 \qquad \text{(first order in A)}$$

$$r_P^o \cong k_{XP} C_{\Sigma\text{cat}} \qquad \text{if} \quad K_{AX} C_A \gg 1 \qquad \text{(zero order in A)}$$

The first extreme, at low reactant concentration, corresponds to a first-step equilibrium very strongly in favor of the free catalyst, so that $C_{\Sigma cat} \cong C_{cat}$; the second, at high reactant concentration, to an equilibrium very strongly in favor of the intermediate, X, so that $C_{\Sigma cat} \cong C_X$. Between the two extremes, the rate varies nonlinearly with the reactant concentration, approaching a limit at very high concentrations (see Figure 8.3).

In a Michaelis-Menten cycle, the initial rate is first order in the reactant at low reactant concentration and approaches a limiting value when that concentration is increased.

Such behavior is generally referred to as *Michaelis-Menten kinetics* or *saturation kinetics*. As will be seen, it is a rather common feature of trace-level catalysis, not restricted to cycles as simple as 8.14.

How saturation kinetics comes about is easy to see. If the reactant concentration is very low, most of the catalyst material is present as the free catalyst, and the behavior is the same as in bulk-species catalysis. At very high reactant concentrations, however, most of the catalyst material gets converted to the intermediate X. Regardless of how much farther the reactant concentration is raised, the rate, given by eqn 8.17, can at most approach the limit $(r_P)_{max} = k_{XP}C_{\Sigma cat}$ because C_X (catalyst bound as intermediate) cannot possibly exceed $C_{\Sigma cat}$ (total catalyst material).

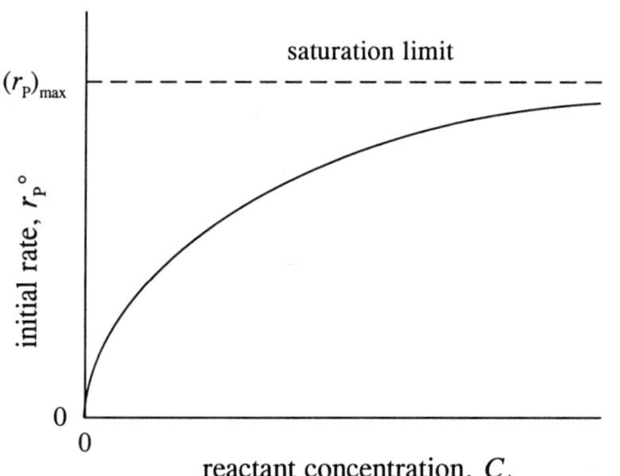

Figure 8.3. Rate as function of reactant concentration in reaction with Michaelis-Menten saturation kinetics (schematic).

Equation 8.18 for Michaelis-Menten kinetics has been shown here in a form consistent with the formalism used elsewhere in this book. Usually, however, the equation is written in a different but equivalent form [27,28]:

$$r_P^{\,o} = \frac{(r_P)_{max}\, C_A}{K_{XA} + C_A} \qquad (8.19)$$

where $(r_P)_{max} = k_{XP} C_{\Sigma cat}$ is the maximum possible rate. The so-called Michaelis constant K_{XA} is the dissociation constant of $X \longleftrightarrow A + cat$ and is the reciprocal of K_{AX} in eqn 8.18.

8.3.2. Briggs-Haldane kinetics.

As first pointed out by Briggs and Haldane [30], the assumption of quasi-equilibrium in the first step is inconveniently restrictive. They relaxed that postulate by replacing the quasi-equilibrium condition 8.15 with the Bodenstein approximation for the trace intermediate X:

$$r_X = k_{AX} C_{cat} C_A - C_X (k_{XA} + k_{XP}) \cong 0 \tag{8.20}$$

so that

$$C_X \cong \frac{k_{AX} C_{cat} C_A}{k_{XA} + k_{XP}}$$

With the catalyst balance 8.16 and rate equation 8.17 this gives

$$r_P^o \cong \frac{k_{AX} k_{XP} C_A C_{\Sigma cat}}{k_{XA} + k_{XP} + k_{AX} C_A} \tag{8.21}$$

or, in one-plus form,

$$r_P^o \cong \frac{k_a C_A C_{\Sigma cat}}{1 + k_b C_A} \tag{8.22}$$

where $\qquad k_a \equiv \dfrac{k_{AX} k_{XP}}{k_{XA} + k_{XP}} \qquad$ and $\qquad k_b \equiv \dfrac{k_{AX}}{k_{XA} + k_{XP}}$

The one-plus rate equation 8.22 is of the same algebraic form as the Michaelis-Menten equation 8.18, only the physical significance of the coefficients is different [instead of the constant K_{AX}, the expression $k_{AX} /(k_{XA} + k_{XP})$ now appears]. Accordingly, the behavior is the same as for Michaelis-Menten kinetics, and that name is often used for Briggs-Haldane kinetics as well.

8.3.3. Reversible cycles

As a rule, enzyme reactions are reversible. To account for the reverse reaction at significant conversion and so remove the restriction to initial rates, the reverse step $P + cat \longrightarrow X$ in the cycle 8.14 must be included. Both the rate equation 8.17 and the Bodenstein approximation 8.20 then contain an additional term to account for this reverse step. The resulting rate equation is

$$r_P \cong \frac{(k_{AX} k_{XP} C_A - k_{XA} k_{PX} C_P) C_{\Sigma cat}}{k_{XA} + k_{XP} + k_{AX} C_A + k_{PX} C_P} \tag{8.23}$$

or, in one-plus form

$$r_{\mathrm{P}} \cong \frac{(k_a C_{\mathrm{A}} - k_c C_{\mathrm{P}})\, C_{\Sigma\mathrm{cat}}}{1 + k_b C_{\mathrm{A}} + k_d C_{\mathrm{P}}} \tag{8.24}$$

where

$$k_a \equiv \frac{k_{\mathrm{AX}} k_{\mathrm{XP}}}{k_{\mathrm{XA}} + k_{\mathrm{XP}}}, \qquad k_c \equiv \frac{k_{\mathrm{XA}} k_{\mathrm{PX}}}{k_{\mathrm{XA}} + k_{\mathrm{XP}}}, \qquad k_b \equiv \frac{k_{\mathrm{AX}}}{k_{\mathrm{XA}} + k_{\mathrm{XP}}}, \qquad k_d \equiv \frac{k_{\mathrm{PX}}}{k_{\mathrm{XA}} + k_{\mathrm{XP}}}$$

Equation 8.23 is the most general rate equation for a trace-level catalyst cycle A $\longleftrightarrow$ P with one intermediate. It reduces to the Briggs-Haldane equation 8.21 if $k_{\mathrm{PX}} \to 0$ or $C_{\mathrm{P}} = 0$, that is, if the second step is irreversible or only the initial rate is considered. It reduces further to the Michaelis-Menten equation 8.18 if, in addition, $k_{\mathrm{XP}} \ll k_{\mathrm{XA}}$, that is, if the first step is in quasi-equilibrium.

A comparison of eqn 8.23 for arbitrary distribution of catalyst material with the formula 8.2 developed for bulk-species catalysts is instructive. Application of the latter to the cycle 8.14 yields

$$r_{\mathrm{P}} \cong \frac{(k_{\mathrm{AX}} k_{\mathrm{XP}} C_{\mathrm{A}} - k_{\mathrm{XA}} k_{\mathrm{PX}} C_{\mathrm{P}})\, C_{\mathrm{cat}}}{k_{\mathrm{XA}} + k_{\mathrm{XP}}} \tag{8.25}$$

Equation 8.23 derived in the present section differs by having the total instead of the free catalyst concentration in the numerator, as well as two additional terms in the denominator. The equivalence of the two equations can be shown as follows: Replacement of C_{cat} in eqn 8.25 with use of eqn 8.16 and subsequent replacement of C_{X} with the Bodenstein approximation (eqn 8.20 with additional term $k_{\mathrm{PX}} C_{\mathrm{cat}} C_{\mathrm{P}}$) yields eqn 8.23. Equation 8.25 is simpler and therefore preferable if one can be sure that all but an insignificant fraction of catalyst is in free form. Equation 8.23 is not subject to this restriction, an advantage one must pay for with the inconvenience of having to handle two more denominator terms.

8.3.4. Common features and plots

A common feature of all single-cycle kinetics discussed so far is a one-plus rate behavior with reaction order between zero and one with respect to the reactant, A (and for a possible reverse rate, with respect to the product, P). The Michaelis-Menten and Briggs-Haldane rate equations 8.18 and 8.22 have the same algebraic form, and so has the initial rate in the reversible cycle, that is, eqn 8.24 with terms involving C_{P} still being insignificant. This common one-plus form can be re-arranged:

$$\frac{C_{\Sigma\mathrm{cat}}}{r_{\mathrm{P}}^{\,\mathrm{o}}} = \frac{k_b}{k_a} + \frac{1}{k_a C_{\mathrm{A}}} \tag{8.26}$$

Accordingly, a plot of $C_{\Sigma\text{cat}}/r_P^o$ versus $1/C_A$ gives a straight line if kinetics is of this type. If so, the coefficients can be calculated from the intercept k_b/k_a and slope $1/k_a$. The often used *Lineweaver-Burk plot* of $1/r_P^o$ versus $1/C_A$ [31] achieves the same goal, but gives different straight lines for different catalyst concentrations $C_{\Sigma\text{cat}}$. Other straight-line plots are those of C_A/r_P^o versus C_A (*Hanes plot* [32]) and r_P^o versus r_P^o/C_A (*Eadie-Hofstee plot* [33,34]) (see also Segel's book [28]).

Reactions orders between zero and one in accordance with one-plus rate equations are very common in enzyme catalysis, even if the cycle is more complex and involves additional reactants or products. The plots just described thus are more broadly applicable. On the other hand, straight lines in such plots are only evidence of saturation kinetics, not an indication that the catalyst cycle has only one intermediate.

Example 8.4. Phosphate transfer catalyzed by hexokinase. The enzyme hexokinase catalyzes transfer of phosphate to C_6 sugars. Table 8.2 shows results reported for the reaction

$$\text{Mg-ATP + glucose} \quad \longleftrightarrow \quad \text{Mg-ADP + glucose-P} \qquad (8.27)$$

where Mg-ATP is magnesium adenosine-5'-triphosphate, Mg-ADP is magnesium adenosine-5'-diphosphate, and glucose-P is glucose 6-phosphate. The table lists values of $C_{\Sigma\text{cat}}/r_P^o$ obtained under various conditions (r_P^o = initial rate).

A cursory inspection of the results reveals that the reaction orders with respect to both Mg-ATP and glucose are positive, but less than one. For the initial rate this suggests an equation of the form

$$r_P^o \;=\; \frac{k_a C_A C_B C_{\Sigma\text{cat}}}{1 + k_b C_A + k_c C_B}$$

so that:

$$\frac{C_{\Sigma\text{cat}}}{r_P^o} \;=\; \frac{1}{k_a C_A C_B} + \frac{k_b}{k_a C_B} + \frac{k_c}{k_a C_A}$$

(A and B are Mg-ATP and glucose). Plots of $C_{\Sigma\text{cat}}/r_P^o$ versus $1/C_{\text{Mg-ATP}}$ and $1/C_{\text{glucose}}$ are shown in Figure 8.4 (next page). Both plots give reason-ably good straight lines for groups of data for which the concentration of the other reactant is the same, as the hypothetical rate equation demands. The equation thus appears to be acceptable.

Table 8.2. Initial rates of phosphate transfer reaction 8.27 catalyzed by hexokinase at 25° C [35].

$C_{\text{Mg-ATP}}$ mM	C_{glucose} mM	$C_{\Sigma\text{cat}}/r_P^o$ s
0.473	0.10	$7.48*10^{-3}$
	.20	4.54
	.50	2.90
	1.00	2.55
0.920	0.10	6.10
	.20	3.90
	.50	2.51
	1.00	2.21
1.94	0.10	4.87
	.20	3.17
	.50	2.17
	1.00	1.83
4.00	0.10	4.44
	.20	2.93
	.50	2.03
	1.00	1.72

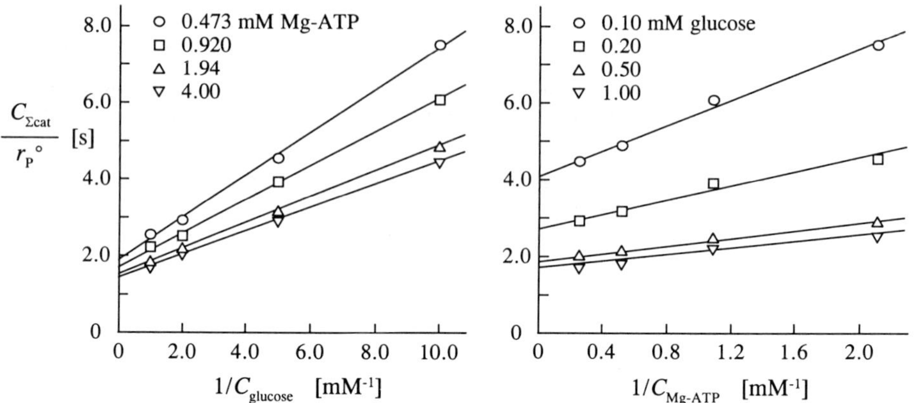

Figure 8.4. Plots of $C_{\Sigma cat}/r_P^{\,\circ}$ versus reciprocal reactant concentrations in phosphate transfer reaction (data from Table 8.2). Left: plots versus $1/C_{glucose}$ at different $C_{Mg\text{-}ATP}$; right plots versus $1/C_{Mg\text{-}ATP}$ at different $C_{glucose}$.

A type of mechanism compatible with the form of the hypothetical rate equation is the cycle

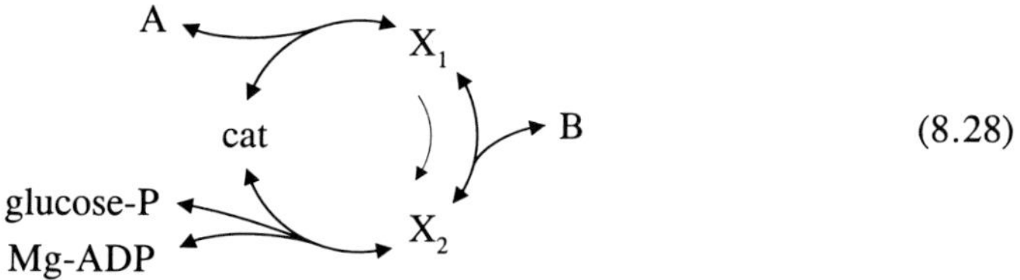

$$(8.28)$$

with very fast decomposition of X_2 into products and catalyst, probably in more than one step. The equation for the initial rate in a cycle of this type is the Briggs-Haldane equation 8.21 with appropriately changed indices and replacement of k_{XP} (i.e., k_{12}) by $k_{12}C_B$ as the second denominator term:

$$r_P^{\,\circ} \;\cong\; \frac{k_{01}k_{12}C_{\Sigma cat}C_A C_B}{k_{10} + k_{12}C_B + k_{01}C_A} \qquad (8.29)$$

(index 0 = catalyst). Whether glucose or Mg-ATP is first to enter the catalyst cycle (i.e., is reactant A) cannot be decided without additional information.

The actual system is more complex than shown here and involves dissociation of the Mg-ATP and Mg-ADP complexes and inhibition by free ATP. However, the cycle 8.28 (with glucose entering first) appears to be essentially correct [35].

8.4. General formula for single catalytic cycles: Christiansen mathematics

For catalytic cycles with more than three or four members, the long-hand derivation of rate equations gets out of hand. However, a general formula comparable to that given in Section 6.3 for noncatalytic simple pathways was established as early as 1931 by Christiansen [36-38].*

As at the start of this chapter, we consider the single catalytic cycle

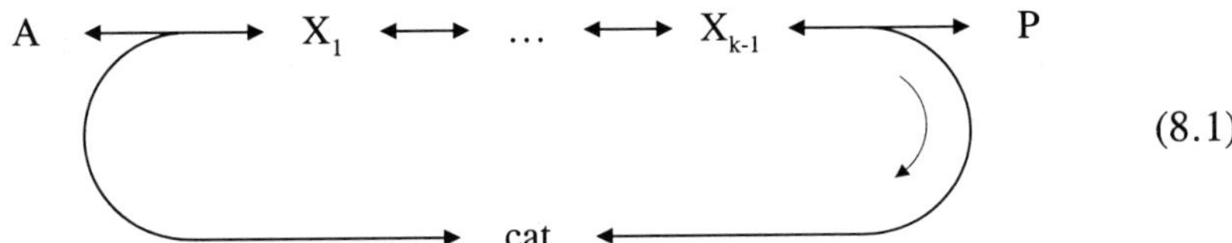

$$(8.1)$$

with any number k of members and possibly with co-reactants and co-products (not shown), but this time seeking a rate equation in terms of total catalyst material and admitting that the latter may be distributed in any arbitrary way over free catalyst and intermediates. The rate equation based on Christiansen's mathematics is

$$r_{\mathrm{P}} = \frac{\left[\prod_{i=0}^{k-1} \lambda_{i,i+1} - \prod_{i=0}^{k-1} \lambda_{i+1,i} \right] C_{\Sigma\mathrm{cat}}}{\mathfrak{C}} \qquad (8.30)$$

where indices 0 and k refer to the free catalyst, and where the denominator $\mathfrak{C}$ is the sum of all elements in the "Christiansen matrix," shown here for a four-membered cycle (k = 4)

$$\begin{bmatrix} \lambda_{12}\lambda_{23}\lambda_{30} & \lambda_{10}\lambda_{23}\lambda_{30} & \lambda_{10}\lambda_{21}\lambda_{30} & \lambda_{10}\lambda_{21}\lambda_{32} \\ \lambda_{23}\lambda_{30}\lambda_{01} & \lambda_{21}\lambda_{30}\lambda_{01} & \lambda_{21}\lambda_{32}\lambda_{01} & \lambda_{21}\lambda_{32}\lambda_{03} \\ \lambda_{30}\lambda_{01}\lambda_{12} & \lambda_{32}\lambda_{01}\lambda_{12} & \lambda_{32}\lambda_{03}\lambda_{12} & \lambda_{32}\lambda_{03}\lambda_{10} \\ \lambda_{01}\lambda_{12}\lambda_{23} & \lambda_{03}\lambda_{12}\lambda_{23} & \lambda_{03}\lambda_{10}\lambda_{23} & \lambda_{03}\lambda_{10}\lambda_{21} \end{bmatrix} \qquad (8.31)$$

The numerator of the Christiansen rate equation 8.30 is the same as that of eqn 8.2 used earlier for cycles with bulk catalyst, except that the total catalyst concentration, $C_{\Sigma\mathrm{cat}}$, takes the place of the concentration of the free catalyst, C_{cat}. However, the

---

* Christiansen provided all necessary mathematics but omitted to write an explicit equation for the reaction rate. Perhaps for this reason—also no doubt because much of his work was published in Danish or German—he has not received as much credit as he deserved. More often cited later authors, most prominent among them King and Altman [39], apparently unaware of his work, largely reinvented his approach and elaborated upon it.

denominator contains additional terms. The terms of the first row of the matrix 8.31 are seen to be those of the denominator D_{00} of the equation for bulk-catalyst cycles. The second row is generated from the first by increase of all index numbers by 1 and replacement of any resulting k by 0. Each successive row is obtained from the preceding one by this same recipe, and so is the first row from the last.

A rigorous derivation of Christiansen's rate equation is laborious. However, it is easy to see how this formula comes about, as the following argument will show.

If all but an insignificant fraction of the catalyst material were present as free catalyst, $C_{\Sigma cat}$ would practically equal C_{cat}, so that the Christiansen numerator would equal that in eqn 8.2 for bulk catalyst cycles. The Christiansen denominator must then also equal the denominator in eqn 8.2. It does so if all terms except those of the first row are insignificant, i.e., if $\mathfrak{C} = D_{00}$.

To generalize from here: Nothing distinguishes the free catalyst mathematically from the other cycle members, and the place where index numbering along the cycle "starts" and "ends" can be chosen arbitrarily. Thus, if practically all catalyst material were present as, say, the first intermediate, we could start numbering at the latter. Since the predominant cycle member then is that with index 0, the rate equation must be the same as it was with the free catalyst as the predominant member and labeled 0. But to restore the original indexing, which we intend to keep, we have to increase all index numbers by 1. The denominator now matches the second row of the Christiansen matrix 8.31, i.e., $\mathfrak{C} = D_{11}$ (see definition 6.6, with numbering clockwise around the cycle to and past zero). By the same token, each clockwise shift of the predominant cycle member by one position increases all index numbers by 1 and makes the next matrix row farther down the only one whose terms can be significant. In other words, if one cycle member contains practically all catalyst material, only one matrix row is significant. In this light, the complete Christiansen equation 8.30 can be recognized as the general formula that reduces as required to each of the special cases with practically all catalyst contained in one member of the cycle.

This argument also leads directly to one of Christiansen's key conclusions:

In the Christiansen matrix 8.31, the sum of the elements of each row is proportional to the concentration of one of the members of the catalyst cycle:

Sum of row 1	proportional to	free catalyst
row 2		intermediate X_1
row 3		intermediate X_2
	...	
row j		intermediate X_{j-1}
	...	
last row		last intermediate

More specifically:

$$\frac{C_j}{C_{\Sigma cat}} = \frac{D_{jj}}{\mathbb{C}} = \frac{\text{sum of row } j+1}{\text{sum of all elements}} \qquad (j \geq 0) \qquad (8.32)$$

($j = 0$ refers to free catalyst).

The general catalyst cycle 8.1 in terms of λ coefficients allows for any number of reactants and products and any nature of the intermediates X_j, subject only to the condition that no step involves more than one molecule of intermediates as reactant. It thus is applicable to a great number of different types of systems. Only one example will be shown here.

Example 8.5. Ping-pong transfer reactions. Some enzymatic transfer reactions proceed by so-called *ping-pong mechanisms* [40,41]. In these, the conversion of a reactant to a product leaves the enzyme in a different form. The modified enzyme then converts a second reactant to another product while itself being restored to its original form. Enzymatic transaminase reactions interconverting amino and keto acids provide a typical example [40]:

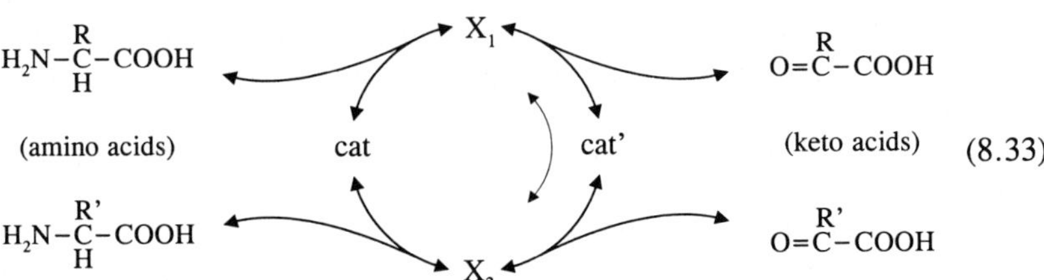

Equation 8.30 applies with $k = 4$, index 2 referring to cat' (the amino derivative of the enzyme), and with the concentrations of the amino and keto acids appearing as cofactors in the respective λ coefficients.

8.5. Reduction of complexity

Rigorous rate equations for multistep catalytic reactions in terms of total amount of catalyst material are enormously cumbersome. Just the reduction to the level complexity of the Christiansen formula calls for the Bodenstein approximation of quasi-stationary behavior of the intermediates, requiring these to remain at trace concentrations, and that formula still entails a lot more algebra than does the general rate equation for noncatalytic simple pathways: For a reaction with three intermediates, the Christiansen denominator contains sixteen terms instead of four; for a reaction with six intermediates, forty-nine instead of seven! Although the mathematics is simple and easy to program for modeling purposes and, usually, some terms can be consolidated, the equations are too unwieldy for network elucidation if the cycle has more than three or four members and several co-reactants and co-products. Here, even more than in the case of noncatalytic pathways, further reduction of complexity is imperative, so much so that the practitioner will apply less stringent criteria as to closeness of his approximations. This is true for what is allowed to pass as "trace level" for the Bodenstein approximation as much as for the demands of applicability of additional simplifications.

The three principal tools of reduction of complexity, discussed in Chapter 4, are the approximations of a rate-controlling step, of quasi-equilibrium steps, and of quasi-stationary behavior of intermediates. The Christiansen formula has already invoked the last of these three. The other two can be used for additional simplification. A further, new and very powerful tool is the concept of *relative abundance of catalyst-containing species*. Moreover, much can sometimes be gained if one or several steps can be taken as irreversible. To summarize:

Tools for Reduction of Complexity
of Christiansen Formula:

relative abundance of catalyst-containing species (*macs* and *lacs*)
rate-controlling step
quasi-equilibrium steps
irreversible steps

8.5.1. *Relative abundance of catalyst-containing species.**

If the catalyst is present almost completely in the form of one member of the cycle, be it as the free catalyst or an intermediate X_j, that species is called the "most abundant catalyst-containing species," or *macs* for short. If a *macs* exists, one row of the matrix 8.31 dominates all others: the first row if the *macs* is the free catalyst, the $(j+1)$'th row if the *macs* is X_j. The Christiansen rate equation 8.30 is thereby reduced to the lower degree of complexity of those of bulk-catalytic and noncatalytic simple pathways, with only k instead of k^2 terms in the denominator (see eqns 8.2 and 6.4 to 6.6).

A comparison of eqn 8.30 and matrix 8.31 with eqns 6.4 to 6.6 leads to a very important rule. If, say, X_j is the *macs*, then eqn 8.30 reduces to a form whose denominator consists only of the elements of row $j+1$ of the matrix 8.31; more-

---

* The concept of a predominant cycle member was first introduced in heterogeneous catalysis by Boudart [42], who coined the term "most abundant surface intermediate," abbreviated *masi* (see Section 8.10).

over, $C_{\Sigma\text{cat}}$ approximately equals C_j. Comparison with the general rate equation for noncatalytic reactions, eqns 6.4 to 6.6, then shows the reduced form of eqn 8.30 to be the same as for a noncatalytic pathway $X_j \longleftrightarrow X_{j+1} \longleftrightarrow \ldots \longleftrightarrow X_j$:

- *The rate equation of a catalytic cycle with a* macs *is the same as that of an imaginary simple pathway of same steps that "starts" and "ends" with the* macs.

In other words, to obtain the rate equation for a catalytic cycle with a *macs*, the cycle can be "snipped" at the latter to give a linear pathway. The rate equation of that imaginary pathway approximates that of the cycle (granted the validity of the assumption that practically all of the total catalyst material is present as the *macs*). This simple rule allows the rate equation of any catalytic reaction with a *macs* to be written down as quickly and easily as those for linear simple pathways. It will be used extensively in the next section (see also Figure 8.5 in that section).

As an example, consider a four-member cycle with a co-reactant B in the second step:

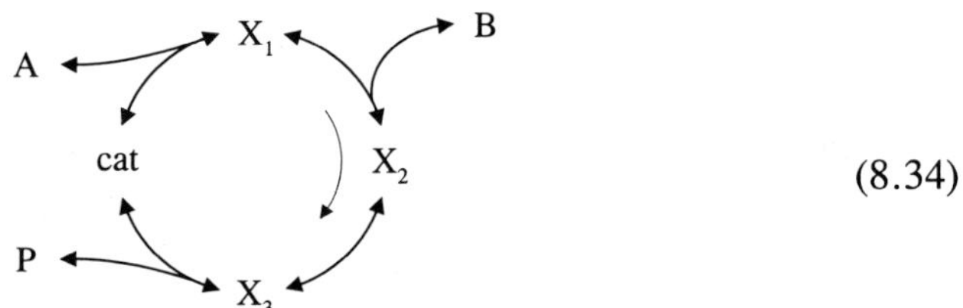

$$(8.34)$$

If X_2 is the *macs,* the Christiansen matrix retains only its third row:

$$
\begin{bmatrix}
0 & 0 & 0 & 0 \\
0 & 0 & 0 & 0 \\
\lambda_{30}\lambda_{01}\lambda_{12} & \lambda_{32}\lambda_{01}\lambda_{12} & \lambda_{32}\lambda_{03}\lambda_{12} & \lambda_{32}\lambda_{03}\lambda_{10} \\
0 & 0 & 0 & 0
\end{bmatrix}
\qquad (8.35)
$$

The denominator of the rate equation—the sum of the four elements of the third row—is seen to be the same as for an imaginary simple pathway with same steps as in the cycle 8.34 and starting and ending with X_2, the *macs*. After replacement of the λ coefficients and collection of terms, the rate equation is

$$
r_P = \frac{(k_{01}k_{12}k_{23}k_{30}C_A C_B - k_{10}k_{21}k_{32}k_{03}C_P)C_{\Sigma\text{cat}}}{(k_{30} + k_{32})k_{01}k_{12}C_A C_B + k_{32}k_{03}k_{12}C_B C_P + k_{32}k_{03}k_{10}C_P} \qquad (8.36)
$$

(C_A is co-factor in λ_{01}, C_B in λ_{12}, and C_P in λ_{03}).

Unless the *macs* is the free catalyst, the denominator of the rate equation differs from that in eqn 8.2. This affects the reaction orders, as will be discussed in the next section.

If one intermediate contains no more than an insignificant fraction of the total catalyst material, it will be called the "least abundant catalyst-containing species," or *lacs* for short. The matrix row corresponding to that intermediate is negligible. If, say, X_1 in the cycle 8.34 is a *lacs*, the matrix 8.31 reduces to

$$\begin{bmatrix} \lambda_{12}\lambda_{23}\lambda_{30} & \lambda_{10}\lambda_{23}\lambda_{30} & \lambda_{10}\lambda_{21}\lambda_{30} & \lambda_{10}\lambda_{21}\lambda_{32} \\ 0 & 0 & 0 & 0 \\ \lambda_{30}\lambda_{01}\lambda_{12} & \lambda_{32}\lambda_{01}\lambda_{12} & \lambda_{32}\lambda_{03}\lambda_{12} & \lambda_{32}\lambda_{03}\lambda_{10} \\ \lambda_{01}\lambda_{12}\lambda_{23} & \lambda_{03}\lambda_{12}\lambda_{23} & \lambda_{03}\lambda_{10}\lambda_{23} & \lambda_{03}\lambda_{10}\lambda_{21} \end{bmatrix} \qquad (8.37)$$

and the denominator of the rate equation loses four of its original sixteen terms.

More than one intermediate may qualify as a *lacs* (linguistic purity would demand the abbreviation then to be read as "low-abundance catalyst-containing species"). If so, more than one matrix row becomes negligible. Of course, if there is a *macs*, all other members of the catalytic cycle are automatically in the *lacs* category.

8.5.2. Rate-controlling step

Whenever the forward and reverse λ coefficients of a step are much smaller than all others, the denominator terms involving those coefficients become negligible (see Section 4.1). Say, the step $X_j \longleftrightarrow X_{j+1}$ (co-reactants or -products not shown) is rate-controlling. The terms involving $\lambda_{j,j+1}$ and $\lambda_{j+1,j}$ will be then small compared with those which, instead, involve $\lambda_{j,j-1}$ and $\lambda_{j+1,j+2}$, respectively. This makes all terms but one disappear in each row of the Christiansen matrix. In the general case of a k-membered cycle, this reduces the number of terms from k^2 to k.

For example, assume the step $X_1 + B \longleftrightarrow X_2$ in the cycle 8.34 to be rate-controlling. Inspection of the matrix 8.31 shows that it will be reduced to

$$\begin{bmatrix} 0 & \lambda_{10}\lambda_{23}\lambda_{30} & 0 & 0 \\ \lambda_{23}\lambda_{30}\lambda_{01} & 0 & 0 & 0 \\ 0 & 0 & 0 & \lambda_{32}\lambda_{03}\lambda_{10} \\ 0 & 0 & \lambda_{03}\lambda_{10}\lambda_{23} & 0 \end{bmatrix} \qquad (8.38)$$

After replacement of the λ coefficients and collection of terms, the resulting rate equation is

$$r_P \;\cong\; \frac{(k_{01}k_{12}k_{23}k_{30}C_A C_B \;-\; k_{10}k_{21}k_{32}k_{03}C_P)\,C_{\Sigma\text{cat}}}{k_{23}k_{30}(k_{10} + k_{01}C_A) \;+\; k_{03}k_{10}C_P(k_{23} + k_{32})} \qquad (8.39)$$

This equation contains the rate coefficients of all steps including those at quasi-equilibrium. Instead, the rate can be expressed in terms of the more easily accessible equilibrium constants of those steps:

$$k_{j,j+1}/k_{j+1,j} = K_{j,j+1} \qquad \text{or} \qquad k_{j+1,j}/k_{j,j+1} = K_{j+1,j}$$

with the respective substitutions the rate equation becomes

$$r_{\mathrm{P}} \;\cong\; \frac{(K_{01}k_{12}C_{\mathrm{A}}C_{\mathrm{B}} - k_{21}K_{32}K_{03}C_{\mathrm{P}})\,C_{\Sigma\mathrm{cat}}}{1 + K_{01}C_{\mathrm{A}} + (1 + K_{32})K_{03}C_{\mathrm{P}}} \qquad (8.40)$$

8.5.3. *Quasi-equilibrium steps*

If the forward and reverse λ coefficients of a step are much larger than all others, that step is at quasi-equilibrium (see Section 4.2). The participants in that step then are present at all times in concentrations related to one another by the thermodynamic equilibrium condition, and so can be lumped into one pseudo-component. Since the number of denominator terms in the rate equation equals the square of the number of the cycle members, this reduces the amount of algebra considerably.

With quasi-equilibrium in the step $X_j \longleftrightarrow X_{j+1}$ and its participants lumped into X_L, the general cycle is

(lumped species shown in box).

The numerical values of the rate coefficients of formation of the lumped species from X_{j-1} and X_{j+2} are not affected by the lumping, that is, they are the same as those of formation of X_j from X_{j-1} and of X_{j+1} from X_{j+2}, respectively, in the actual cycle:

$$\lambda_{j-1,L} = \lambda_{j-1,j} \qquad \text{and} \qquad \lambda_{j+2,L} = \lambda_{j+2,j+1} \qquad (8.41)$$

This is because for both the actual and the reduced cycle the respective rate equations have the concentrations of X_{j-1} and X_{j+2}, respectively, as factors. However, in the rate equations for decay of the lumped species to X_{j-1} and X_{j+2}, the concentration of the lumped species appears instead of that of X_j or X_{j+1}. Therefore, to obtain the coefficients of the reduced cycle, those of the actual cycle must be multiplied by the fractions C_j/C_L and C_{j+1}/C_L, respectively, where $C_L \equiv C_j + C_{j+1}$ is the concentration of the lumped species:

$$\lambda_{L,j-1} = (C_j/C_L)\lambda_{j,j-1}, \qquad\qquad \lambda_{L,j+2} = (C_{j+1}/C_L)\lambda_{j+1,j+2} \qquad (8.42)$$

The concentration ratios in these equations can be replaced with use of the equilibrium condition for the fast step:

$$C_{j+1}/C_j\,\mathfrak{R}_{j,j+1} = K_{j,j+1}$$

where $\mathfrak{R}_{j,j+1}$ is the product of the concentrations of any co-reactants of the step divided by that of the concentrations of any co-products. With this condition, the fractions in eqn 8.42 become

$$\frac{C_j}{C_L} = \frac{1}{1 + K_{j,j+1}\mathfrak{R}_{j,j+1}}, \qquad\qquad \frac{C_{j+1}}{C_L} = \frac{K_{j,j+1}\mathfrak{R}_{j,j+1}}{1 + K_{j,j+1}\mathfrak{R}_{j,j+1}} \qquad (8.43)$$

Several steps in a cycle may be fast enough for quasi-equilibrium. If two or more such steps are successive, all the cycle members involved can be lumped into a single pseudo-component.

The derivation of the rate equation in final form is best illustrated with a specific example:

Example 8.6. Rate for a hypothetical cycle with quasi-equilibrium steps. Assume that the second and third steps in the four-membered cycle 8.34 are in quasi-equilibrium and the fourth step is irreversible ($k_{03} = 0$):

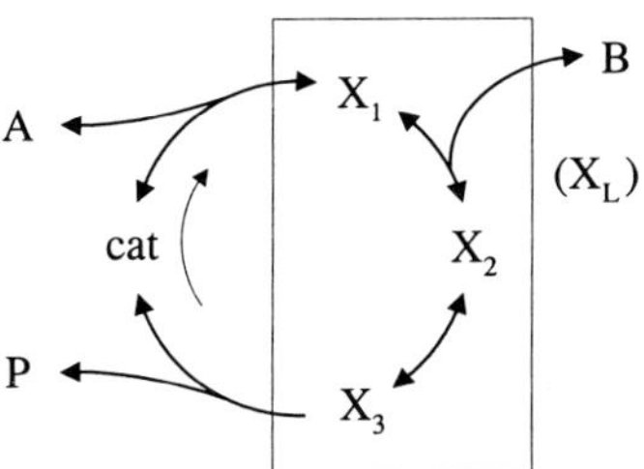

With lumping of X_1, X_2, and X_3 into a single pseudo-species X_L (shown in box), the reduced cycle now has only two members: cat and X. The Briggs-Haldane equation 8.21 applies, without restriction to initial rate and with index X replaced by L:

$$r_P \cong \frac{k_{AL}k_{LP}C_A C_{\Sigma cat}}{k_{LA} + k_{LP} + k_{AL}C_A} \qquad (8.44)$$

(indices A and P instead of 0 are retained for distinction between the two steps of the reduced cycle). With the equilibrium condition $C_1 : C_2 : C_3 = 1 : K_{12}C_B : K_{13}C_B$, the concentration fractions corresponding to eqns 8.43 are

$$\frac{C_1}{C_L} = \frac{1}{1 + (K_{12} + K_{13})C_B}, \qquad \frac{C_3}{C_L} = \frac{K_{13}C_B}{1 + (K_{12} + K_{13})C_B}$$

so that

$$k_{LA} = \frac{k_{10}}{1 + (K_{12} + K_{13})C_B}, \qquad k_{LP} = \frac{k_{30}K_{13}C_B}{1 + (K_{12} + K_{13})C_B}$$

(see eqns 8.42). With these substitutions, eqn 8.44 becomes

$$r_P = \frac{k_{01}K_{13}k_{30}C_A C_B C_{\Sigma cat}}{k_{10} + K_{13}k_{30}C_B + k_{01}C_A[1 + (K_{12} + K_{13})C_B]}$$

Accordingly, the reaction is of order between zero and one in A and B and first order in total catalyst material, showing typical saturation kinetics.

It may surprise that the rate decreases with increasing K_{12}, a "forward" equilibrium constant. The reason is that an increase in K_{12} (at constant K_{13}) increases the concentration of X_2 at the expense of X_1 and X_3, making the lumped species more sluggish to react.

8.5.4. Irreversible steps.

Whenever a step in the catalytic cycle is practically irreversible, the respective reverse rate coefficient $\lambda_{j+1,j}$ and all terms involving it become negligible. This includes the negative term in the numerator.

Assume the second step in the four-membered cycle 8.34 is irreversible, so that $\lambda_{21} \cong 0$. The Christiansen matrix then reduces to

$$\begin{bmatrix} \lambda_{12}\lambda_{23}\lambda_{30} & \lambda_{10}\lambda_{23}\lambda_{30} & 0 & 0 \\ \lambda_{23}\lambda_{30}\lambda_{01} & 0 & 0 & 0 \\ \lambda_{30}\lambda_{01}\lambda_{12} & \lambda_{32}\lambda_{01}\lambda_{12} & \lambda_{32}\lambda_{03}\lambda_{12} & \lambda_{32}\lambda_{03}\lambda_{10} \\ \lambda_{01}\lambda_{12}\lambda_{23} & \lambda_{03}\lambda_{12}\lambda_{23} & \lambda_{03}\lambda_{10}\lambda_{23} & 0 \end{bmatrix} \qquad (8.45)$$

The number of denominator terms has decreased from sixteen to ten. In general, for a k-membered cycle the reduction is from k^2 to $k(k+1)/2$. If the irreversible step is $X_j \longrightarrow X_{j+1}$, the (j+2)th row remains intact, and each successive row loses one more element.

More than one step may be irreversible. The approximation can then be applied to each of them, with further simplification of the rate equation.

As the reduced matrix 8.45 shows, steps following the irreversible second one in the cycle do affect the rate. This is another example for the fact that the rules for noncatalytic simple pathways do not apply unless the free catalyst is the *macs*. A more general set of rules will be given in the next section.

8.5.5. *Combinations of approximations*

The approximations in this section can be combined in many different ways. Significant further simplification may result. For example, for a cycle 8.34 with X_1 as the *macs* and irreversible step $X_1 + B \longrightarrow X_2$, the only remaining row of the Christiansen matrix is the second, and that row has but one single element, $\lambda_{23}\lambda_{30}\lambda_{01}$ (see reduced matrix 8.45). In this case, the rate equation reduces to

$$ r_P \;\cong\; \frac{\lambda_{01}\lambda_{12}\lambda_{23}\lambda_{30}\,C_{\Sigma\mathrm{cat}}}{\lambda_{23}\lambda_{30}\lambda_{01}} \;=\; \lambda_{12}\,C_{\Sigma\mathrm{cat}} \;=\; k_{12}\,C_B\,C_{\Sigma\mathrm{cat}} \tag{8.46} $$

A rather simple rate equation is also obtained from a combination of the approximations of a *macs* and a rate-controlling step. The existence of a *macs* leaves only one matrix row, and that of a rate-controlling step ensures that the row has only one contributing term. In many such cases, the rate is controlled by the step consuming the *macs*. This is because slow consumption of a cycle member favors its attainment of a relatively high concentration. Again returning to the example of the cycle 8.34 with X_1 as the *macs*, if the step $X_1 + B \longleftrightarrow X_2$ is rate-controlling, the only surviving element of the matrix is $\lambda_{23}\lambda_{30}\lambda_{01}$ (see reduced matrix 8.38), and the rate equation becomes

$$ r_P \;\cong\; \frac{(\lambda_{01}\lambda_{12}\lambda_{23}\lambda_{30} - \lambda_{10}\lambda_{21}\lambda_{32}\lambda_{03})\,C_{\Sigma\mathrm{cat}}}{\lambda_{23}\lambda_{30}\lambda_{01}} \;=\; \left[\lambda_{12} - \frac{\lambda_{10}\lambda_{21}\lambda_{32}\lambda_{03}}{\lambda_{01}\lambda_{23}\lambda_{30}}\right] C_{\Sigma\mathrm{cat}} $$

$$ =\; \left[k_{12}\,C_B - \frac{k_{21}\,C_P}{K_{01}K_{23}K_{30}\,C_A}\right] C_{\Sigma\mathrm{cat}} \tag{8.47} $$

Comparison with eqn 8.46 shows that the forward rate is the same as in the previous example, in which the step consuming the *macs* was irreversible.

Many other combinations of the approximations can be used if warranted.

*Example 8.7. Adiponitrile synthesis via hydrocyanation.** Hydrocyanation of 4-pentenenitrile (4-PN) to adiponitrile (ADN) in the presence of an arylborane modifier:

$$ \text{\small$\diagup\!\diagdown\,$CN} \quad + \quad \text{HCN} \quad \longrightarrow \quad \text{NC}\diagdown\!\diagup\!\diagdown\text{CN} \tag{8.48} $$

is an important step in du Pont's current *Nylon*-6 process. The network 8.49 was established by Tolman and co-workers [43–45]. The pathway of the preceding hydrocyanation of butadiene to 3-pentenenitrile is similar. By-product formation, not included in the network 8.49, will be examined later (see Example 8.12 in Section 8.8.3).

---

* The kinetics analysis in this example is based in part on unpublished work by A. A. Gupte (1992).

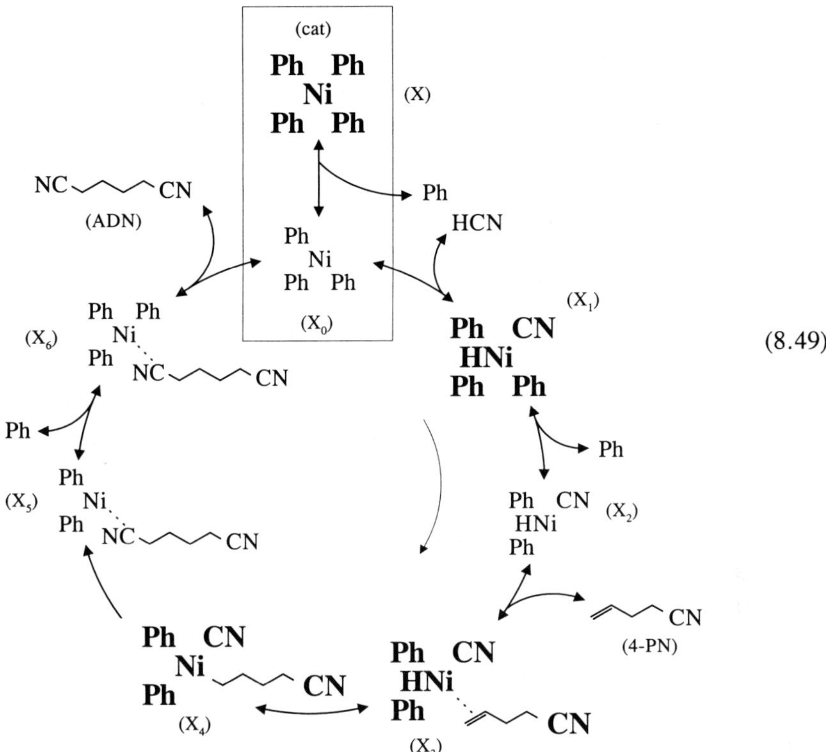

(8.49)

NiPh$_4$ is a complex of zero-valent nickel with tertiary organic phosphine or phosphite ligands, Ph. Under reaction conditions, the ligand-deficient species NiPh$_3$ (X$_0$), HNiPh$_2$CN (X$_2$), and 4PN-NiPh$_2$ (X$_5$) and apparently also 4PN-NiPh$_3$ (X$_6$) contain only insignificant fractions of total nickel, as evident from IR, UV, and NMR spectra [44,46]. The reaction is irreversible, in all likelihood owing to irreversible carbon-carbon bond formation in the step forming X$_5$. Ligand loss NiPh$_4 \longrightarrow$ NiPh$_3$ + Ph is at quasi-equilibrium, so that these two nickel complexes can be lumped into a pseudo-single species, X (shown in box). In the network 8.49, the principal nickel-containing species are shown in larger font, and the notation X$_i$ is retained for all intermediates although this requires a more liberal interpretation of "trace level." The reaction is not inhibited by ADN, so that λ_{X6}, which contains the ADN concentration as co-factor, may not appear in the denominator of the rate equation. Therefore, either λ_{X6} is zero or, more likely, X$_6$ is a *lacs* so that the entire last Christiansen matrix row, the only one containing λ_{X6}, becomes negligible.

Application of the Christiansen formula to this cycle gives the rate equation

$$r_{ADN} \cong \lambda_{X1}\lambda_{12}\lambda_{23}\lambda_{34}\lambda_{45}\lambda_{56}\lambda_{6X}C_{\Sigma cat}/\mathfrak{C} \qquad (8.50)$$

($\mathfrak{C}$ = sum of all elements of Christiansen matrix). With irreversible carbon-carbon coupling ($\lambda_{54} = 0$) and X$_2$, X$_5$, and X$_6$ as *lacs*, the Christiansen matrix reduces to

$$\begin{bmatrix}
\lambda_{12}\lambda_{23}\lambda_{34}\lambda_{45}\lambda_{56}\lambda_{6X} & \lambda_{1X}\lambda_{23}\lambda_{34}\lambda_{45}\lambda_{56}\lambda_{6X} & \lambda_{1X}\lambda_{21}\lambda_{34}\lambda_{45}\lambda_{56}\lambda_{6X} & \lambda_{1X}\lambda_{21}\lambda_{32}\lambda_{45}\lambda_{56}\lambda_{6X} & \lambda_{1X}\lambda_{21}\lambda_{32}\lambda_{43}\lambda_{56}\lambda_{6X} & 0 & 0 \\
\lambda_{23}\lambda_{34}\lambda_{45}\lambda_{56}\lambda_{6X}\lambda_{X1} & \lambda_{21}\lambda_{34}\lambda_{45}\lambda_{56}\lambda_{6X}\lambda_{X1} & \lambda_{21}\lambda_{32}\lambda_{45}\lambda_{56}\lambda_{6X}\lambda_{X1} & \lambda_{21}\lambda_{32}\lambda_{43}\lambda_{56}\lambda_{6X}\lambda_{X1} & 0 & 0 & 0 \\
0 & 0 & 0 & 0 & 0 & 0 & 0 \\
\lambda_{45}\lambda_{56}\lambda_{6X}\lambda_{X1}\lambda_{12}\lambda_{23} & \lambda_{43}\lambda_{56}\lambda_{6X}\lambda_{X1}\lambda_{12}\lambda_{23} & 0 & 0 & 0 & 0 & 0 \\
\lambda_{56}\lambda_{6X}\lambda_{X1}\lambda_{12}\lambda_{23}\lambda_{34} & 0 & 0 & 0 & 0 & 0 & 0 \\
0 & 0 & 0 & 0 & 0 & 0 & 0 \\
0 & 0 & 0 & 0 & 0 & 0 & 0
\end{bmatrix}$$

With eqns 8.42 and 8.43 for coefficients involving lumped reactants and with $C_0 \ll C_{cat}$ and the quasi-equilibrium condition $C_0 C_{Ph}/C_{cat} \cong K_{cat}$, one finds λ_{X1} to be

$$\lambda_{X1} \cong \lambda_{01} C_0/C_{cat} = \lambda_{01} K_{cat}/C_{Ph}$$

while $\lambda_{1X} = \lambda_{10}$ and $\lambda_{6X} = \lambda_{60}$. With these replacements, cancellation of λ_{56} and λ_{60}, and the substitutions

$$\lambda_{01} = k_{01} C_{HCN}, \quad \lambda_{12} = k_{12}, \quad \lambda_{23} = k_{23} C_{4PN}, \quad \lambda_{34} = k_{34}, \quad \lambda_{45} = k_{45},$$

$$\lambda_{10} = k_{10}, \quad \lambda_{21} = k_{21} C_{Ph}, \quad \lambda_{32} = k_{32}, \quad \lambda_{43} + k_{43}$$

the rate equation becomes

$$r_{ADN} \cong K_{cat} k_{01} k_{12} k_{23} k_{34} k_{45} C_{HCN} C_{4PN} C_{\Sigma cat}/\mathbb{C}^* \tag{8.51}$$

where $\mathbb{C}^* \equiv \mathbb{C} C_{Ph}$ is given by

$$\begin{aligned}
\mathbb{C}^* &= (k_{12}k_{23}k_{34}k_{45} + k_{10}k_{23}k_{34}k_{45})C_{4PN}C_{Ph} + (k_{10}k_{21}k_{34}k_{45} + k_{10}k_{21}k_{32}k_{45} + k_{10}k_{21}k_{32}k_{43})C_{Ph}^2 \\
&\quad + k_{23}k_{34}k_{45}k_{01}K_{cat}C_{4PN}C_{HCN} + (k_{21}k_{34}k_{45}k_{01} + k_{21}k_{32}k_{45}k_{01} + k_{21}k_{32}k_{43}k_{01})K_{cat}C_{HCN}C_{Ph} \\
&\quad + (k_{45}k_{01}k_{12}k_{23} + k_{45}k_{01}k_{12}k_{23})K_{cat}C_{HCN}C_{4PN} \\
&\quad + k_{01}K_{cat}k_{12}k_{23}k_{34}C_{HCN}C_{4PN}
\end{aligned} \tag{8.52}$$

The rate equation 8.51, written with collective coefficients, then becomes:

$$r_{ADN} \cong \frac{k_a C_{HCN} C_{4PN} C_{\Sigma cat}}{k_b C_{4PN} C_{Ph} + k_c C_{Ph}^2 + k_d C_{HCN} C_{4PN} + k_e C_{HCN} C_{Ph}} \tag{8.53}$$

[If X_6 is not a *lacs*, but λ_{06} is zero, the last matrix row contributes one single term $k_{01}k_{12}k_{23}k_{34}k_{45}K_{cat}C_{HCN}C_{4PN}$ with the same concentration co-factors as that in the fifth row, so that eqn 8.53 remains valid with only a different significance of k_d.] The first two steps of the cycle are likely to be at quasi-equilibrium. If so, the first denominator term in eqn 8.53 is negligible. The rate equation in one-plus form then has only three phenomenological coefficients. In any event, the reaction orders are: plus one for nickel, between zero and plus one for HCN and 4-pentenenitrile, and between zero and minus two for the organic phosphine.

Note that the ratio of the first two to all four denominator terms in eqn 8.53 equals the fraction of nickel present as $NiPh_4$ (plus a negligible amount as $NiPh_3$). A knowledge of that fraction—e.g., from spectra—can facilitate the calculation of coefficient values from rate data.

The traditional way of handling this kind of hydrocyanation networks has been to postulate rate control by carbon-carbon coupling ($X_4 \longrightarrow X_5$) and quasi-equilibrium in all other steps. On this basis a simpler rate equation has been proposed [44]:

$$r_{ADN} = \frac{k_a C_{HCN} C_{4PN} C_{cat}}{C_{Ph}^2} \tag{8.54}$$

However, this equation is in terms of the concentration of $NiPh_4$ rather than total nickel, and so does not reflect the actual reaction orders. For orders as eqn 8.54 suggests, the denominator in eqn 8.51 would have to consist exclusively of terms with only C_{Ph}^2 as co-factor. As eqn 8.52 shows, only three terms meet this condition, and all of them stem from the first matrix row. Accordingly, $NiPh_4$ would have to be the *macs*, contrary to spectroscopic and NMR evidence. Moreover, the other two terms from the first matrix row would also have to be negligible, requiring the first and second step of the cycle to be at quasi-equilibrium. With cat taken to be total nickel, eqn 8.54 can therefore be only a rough approximation, except for catalyst systems in which $NiPh_4$ is indeed the *macs* and the two first steps are at quasi-equilibrium.

8.6. Relationships between network properties and kinetic behavior

As has already become apparent, some of the rules governing the relationships between network properties and kinetic behavior, derived in Section 7.3.1 for noncatalytic simple pathways, can no longer be relied upon in catalysis, except if the free catalyst is the *macs*. A re-examination therefore is called for.

Rules of general validity. The following rules apply without qualifications:

- *A pathway is irreversible if one or more of its steps are irreversible* (Rule 7.11).

This follows from a simple free-energy argument: To be practically irreversible, a step must involve an extremely large free-energy loss. Unless that loss is offset by a comparable free-energy gain (necessitating a practically irreversible reverse step and so being academic), the overall reaction must involve such a loss and be irreversible. This argument is independent of any catalyst effects.

- *A positive reaction order with respect to a product requires a step in which the product acts as reactant* (Rule 7.20, referring to forward reaction).

Whether or not the reaction is catalytic, the concentration of the respective product must appear as a co-factor in the numerator of the equation for the forward rate. This happens only if that participant acts as reactant in a forward step.

- *If the forward and reverse rate coefficients of a step are much smaller than all others, the other steps are in quasi-equilibrium* (Rule 7.21).

The general principle that a step much slower than all others gives the latter enough time to attain and maintain quasi-equilibrium applies to catalytic reactions as well. Lastly, Rule 7.24 for step consolidation must be qualified:

A step with co-reactant entry can be consolidated with a subsequent step with co-product exit or a rearrangement step, and a rearrangement step can be consolidated with a subsequent step of co-product exit or another rearrangement step, *but only if the intermediate is a lacs.*

(8.55)

Reaction orders in cycles with macs [47]. The rules for reaction orders in simple pathways do not apply to catalytic cycles with arbitrary distribution of catalyst material over the cycle members. In the general case, any cycle with more than two or three members gives rise to a Christiansen matrix with a profusion of terms, many of which are apt to involve different combinations of reactant and product concentrations as co-factors. This makes it impossible to formulate general rules for such cycles. All that can be said is that a distribution of catalyst material over several species makes for fractional and varying reaction orders as concentrations tend to appear in the numerator and some but not all terms of the denominator.

None the less, a set of rules can be given for catalyst systems with a *macs*. As shown in the previous section, the rate equation for a catalytic reaction with a *macs* is the same as that for an imaginary (linear) simple pathway that "starts" and "ends" with the *macs*. Think of the catalytic cycle as being "cut" at the *macs* to give a linear pathway with the *macs* at both end (see Figure 8.5). That imaginary equivalent pathway has the same rate equation as the actual catalytic cycle. With this principle, the rules for reaction orders deduced in Section 7.3.1 can be reformulated for catalytic cycles with a *macs* (as in the earlier section, the rules are for the forward rate if the reaction is reversible).

The first such rule is:

Steps following an irreversible step, up to and including the step forming the *macs*, have no effect on the rate. The reaction orders are zero with respect to any species that participate in only those steps.

(8.56)

(see Rules 7.12 and 7.19). In the equivalent imaginary linear pathway, those and only those steps are preceded by the irreversible one. For example, if X_2 is the *macs* and the step $X_2 \longrightarrow X_3$ is irreversible, all other steps of the pathway are preceded by it (see Figure 8.5). Only that step then affects the forward rate. Among the steps that do not are cat $+ A \longleftrightarrow X_1$ and $X_1 \longleftrightarrow X_2$, and the orders with respect to any species that participate in them (and not in the irreversible one) are zero.

The forward step consuming the *macs* corresponds to the first step in the equivalent imaginary linear pathway. This correspondence immediately yields the following rules regarding reaction orders from the rules 7.13 to 7.16 in Section 7.3.1:

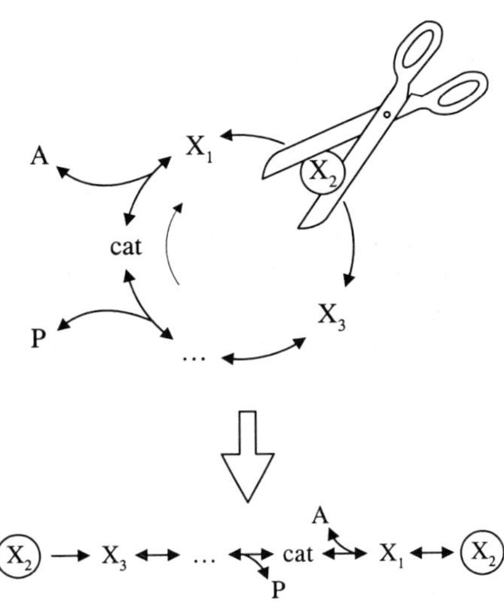

Figure 8.5. Catalytic cycle with intermediate X_2 as *macs* (shown encircled), cut to give equivalent linear pathway starting and ending with *macs* X_2.

A reaction is first order with respect to any reactant that participates (with one molecule) in only the forward step consuming the *macs*.

A reaction is of order between zero and plus one with respect to any reactant that participates (with one molecule) in a forward step other than that consuming the *macs*.

A reaction is second order with respect to any reactant that participates with two molecules in only the forward step consuming the *macs*.

etc.

(8.57)

Rules 7.17 and 7.18 regarding necessary conditions for negative reaction orders must also be reformulated. They are best expressed with reference to the equivalent linear pathway:

> For a reaction order to be negative, the respective participant (reactant, product, or silent partner) must be a product in a reversible step. In the equivalent linear pathway, this step may not be the last nor be preceded by an irreversible one.
>
> For a reactant or silent partner, the step sequence of the equivalent linear pathway must be such that the step in which the species in question is a product must precede that or those in which it is a reactant.

(8.58)

Similarly, Rule 7.21 for the sequence of co-reactant entries becomes

> The later a reactant enters the equivalent linear pathway, the lower is its reaction order.

(8.59)

Example 8.8. Reaction orders for a hypothetical four-membered catalytic cycle with different most abundant catalyst-containing species. Consider the catalytic cycle

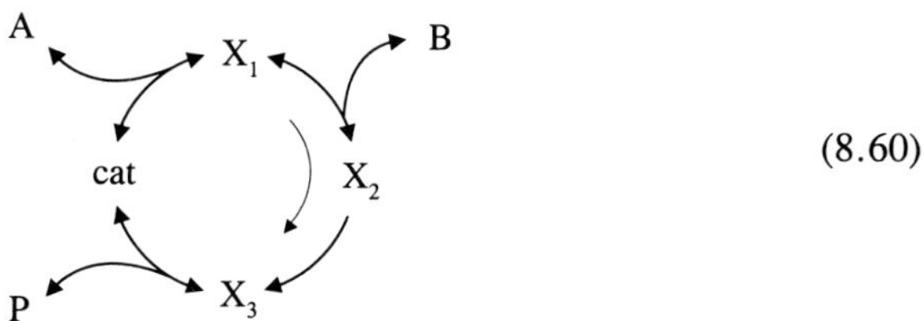

(8.60)

with irreversible step $X_2 \longrightarrow X_3$. The forward rate is

$$r_P = k_{01}k_{12}k_{23}k_{30}\,C_A\,C_B\,C_{\Sigma cat}/\mathfrak{C} \tag{8.61}$$

where $\mathfrak{C}$ is the sum of the elements of the matrix

$$
\begin{bmatrix}
k_{12}k_{23}k_{30}C_B & k_{10}k_{23}k_{30} & k_{10}k_{21}k_{30} & 0 \\
k_{23}k_{30}k_{01}C_A & k_{21}k_{30}k_{01}C_A & 0 & 0 \\
k_{30}k_{01}k_{12}C_A C_B & 0 & 0 & 0 \\
k_{01}k_{12}k_{23}C_A C_B & k_{03}k_{12}k_{23}C_B C_P & k_{03}k_{10}k_{23}C_P & k_{03}k_{10}k_{21}C_P
\end{bmatrix}
$$

(Elements involving the zero reverse coefficient k_{32} are zero.) The concentrations of the participants other than the catalyst—reactants A and B and product P—appear in some, but not all matrix elements. Thus, in the general case of arbitrary distribution of the catalyst material over the members of the cycle, the forward rate is of order between zero and plus one in A and B, and between zero and minus one in P. However, if a *macs* exist, only one matrix row contributes, and some or all orders may be at one of their limits. Table 8.3 summarizes the algebraic forms of the rate equations and the reaction orders for the different possible *macs*.

Table 8.3. Algebraic forms of rate equations and reactions orders for catalytic cycle 8.60 with different *macs* (collective coefficients k_a etc. are defined differently in the different equations).

macs	rate	reaction orders
cat	$$r_P = \dfrac{k_a C_A C_B C_{\Sigma cat}}{k_b + k_c C_B}$$	A: $+1$ B: between 0 and $+1$ P: 0
X_1	$$r_P = k_a C_B C_{\Sigma cat}$$	A: 0 B: $+1$ P: 0
X_2	$$r_P = k_a C_{\Sigma cat}$$	A: 0 B: 0 P: 0
X_3	$$r_P = \dfrac{k_a C_A C_B C_{\Sigma cat}}{k_b C_A C_B + k_c C_B C_P + k_d C_P}$$	A: between 0 and $+1$ B: between 0 and $+1$ P: between 0 and -1

As a comparison shows, the reaction orders derived above with the Christiansen matrix could equally well have been predicted with the rules set forth earlier in this section. The example confirms the usefulness of the rules. It also demonstrates how profoundly kinetics can be affected by the distribution of catalyst material.

8.7. Cycles with external reactions

In many reactions of homogeneous catalysis, one or several linear pathways are connected to the actual catalytic cycle. The most common example is catalysis by a ligand-deficient complex, initiated by reversible ligand loss. Also in this category are certain types of inhibition, activation, and poisoning.

In general terms, such networks are of the type

$$ S \; \text{———} \; X_j \; \bigcirc \tag{8.62} $$

(no details of the catalytic cycle and external pathway shown). The cycle member X_j to which the external pathway is connected may be the actual catalyst or an intermediate, and the external pathway may consist of one or several steps. Also, several external pathways may be connected to the cycle at the same or different intermediates; the extensions to such cases will be obvious.

For the cycle itself, the general Christiansen formula 8.30 can be used:

$$ r_P \;=\; \frac{\left[\displaystyle\prod_{i=0}^{k-1} \lambda_{i,i+1} \;-\; \prod_{i=0}^{k-1} \lambda_{i+1,i} \right] C_{\Sigma^\circ}}{\mathfrak{C}} \tag{8.63} $$

where Σ° is the total amount of catalyst material in the cycle, and $\mathfrak{C}$ is the sum of all elements of the Christiansen matrix for the cycle. However, that equation is in terms of only the catalyst material in the cycle, to the exclusion of that in the external pathway and its end member, S, and must be expanded as will be shown now with practical examples of increasing complexity.

8.7.1. Ligand-deficient catalysts.

Catalysis by metal-organic complexes often involves as a first step the loss of a ligand from the catalyst, freeing a coordinative site that can bind a reactant:

$$ \text{cat} \;\longleftrightarrow\; X_0 \;+\; L \quad \bigcirc \tag{8.64} $$

The external pathway cat $\longleftrightarrow$ X_0 + L involves no species that are formed or consumed by the overall reaction. As only a reversible ligand loss it is also apt to be at least as fast as the cycle, if not very much faster. If so, it will be at quasi-equilibrium if X_0 is quasi-stationary, and the rate becomes

$$r_P \cong \frac{\left[\prod_{i=0}^{k-1}\lambda_{i,i+1} - \prod_{i=0}^{k-1}\lambda_{i+1,i}\right]C_{\Sigma\text{cat}}}{\mathbb{S} + D_{00}C_L/K_{\text{cat}}} \qquad (8.65)$$

where D_{00} is the sum of the first-row elements of the Christiansen matrix for the cycle, and $K_{\text{cat}} = C_0 C_L/C_{\text{cat}}$ is the equilibrium constant of ligand loss. The external reaction is seen to add a term $D_{00}C_L/K_{\text{cat}}$ to the denominator. This new term is proportional to the amount of catalyst metal present as the coordinatively saturated complex, cat, and reflects the fact that the external reaction retards the rate by shifting metal out of the producing cycle into that inactive form. The ligand acts as a "silent partner" and inhibitor with a reaction order between zero and minus one.

Equation 8.65 is an alternative to lumping of the saturated and ligand-deficient species, as was done in the hydrocyanation Example 8.7 in the preceding section, and gives identical results with slightly less algebraic handling.

Derivation of eqn 8.65. At quasi-equilibrium of the catalyst:

$$C_0 C_L/C_{\text{cat}} \cong K_{\text{cat}}$$

so that

$$C_{\text{cat}} \cong (C_L/K_{\text{cat}})C_0$$

The balance of total catalyst metal is

$$C_{\Sigma\text{cat}} = C_{\Sigma^\circ} + C_{\text{cat}} \cong C_{\Sigma^\circ} + (C_L/K_{\text{cat}})C_0 \qquad (8.66)$$

Of the metal in the cycle, Σ°, the fraction present as X_0 is given by the ratio of the sum of the first-row elements to the sum of all elements of the Christiansen matrix of the cycle (see eqn 8.32), so that

$$C_0 = (D_{00}/\mathbb{S})C_{\Sigma^\circ}$$

Substitution into eqn 8.66 gives

$$C_{\Sigma\text{cat}} \cong C_{\Sigma^\circ}\left(1 + (D_{00}C_L/\mathbb{S}K_{\text{cat}})\right)$$

Using this to replace C_{Σ° in eqn 8.63 one obtains eqn 8.65.

The external pathway may consist of more than one step, possibly involving additional co-reactants or co-products. If so, each species S_i along the pathway adds a denominator term that is proportional to its concentration. The new terms are of the form $D_{00}/\mathfrak{R}'_{i0}K'_{i0}$, where K'_{i0} is the equilibrium constant of the partial reaction $S_i + ... \longleftrightarrow X_0 + ...$, and $\mathfrak{R}'_{i0}$ is the ratio of the product of the co-reactant concentrations to that of the co-product concentrations of this reaction (primes indicate that the quantities refer to the external pathway). With m species along the external pathway:

$$r_\mathrm{P} \;\cong\; \frac{\left[\prod_{i=0}^{k-1}\lambda_{i,i+1} \;-\; \prod_{i=0}^{k-1}\lambda_{i+1,i}\right] C_{\Sigma\mathrm{cat}}}{\mathfrak{C} \;+\; D_{00}\Big/ \sum_{i=1}^{m}(\mathfrak{R}'_{i0}\,K'_{i0})} \tag{8.67}$$

(Note that at least one $\mathfrak{R}'_{i0}$ will contain C_L as a factor). For example, with an external pathway

$$S_m \longleftrightarrow \cdots \longleftrightarrow S_2 \rightleftharpoons S_1 \underset{L}{\overset{M}{\rightleftharpoons}} X_0 \tag{8.68}$$

the additional denominator term proportional to the concentration of S_2 would be $D_{00}(C_\mathrm{L}/C_\mathrm{M}K'_{20})$.

Example 8.9. Olefin hydrogenation with Wilkinson's catalyst. Wilkinson's catalyst is a dihydrido-chloro-phosphino complex of rhodium, $H_2RhClPh_3$, where Ph is an organic phosphine such as triphenyl phosphine [48-52]. The dominant mechanism of olefin hydrogenation with this catalyst, established chiefly by Halpern [53-55] in detailed studies that included measurements of equilibria in the absence of reactants and of reaction rates of isolated participants, backed by independent NMR studies [56] and *ab initio* molecular orbital calculations [57], is shown as 8.69 on the facing page (without minor parallel pathways and side reactions).

The overall reaction is irreversible and its rate is independent of the concentration of the hydrogenated product. Accordingly, the reverse step X_0 + paraffin $\longrightarrow$ X_3 must be kinetically insignificant, so that all terms with λ_{03} can be disregarded. Furthermore, no double-bond migration in the olefin occurs, nor does exchange of H for D on the olefin in experiments with D_2 [58]. This indicates that metal-carbon bond formation ($X_2 \longrightarrow X_3$) is also irreversible, i.e., terms with λ_{32} are insignificant. Without further assumptions other than quasi-stationary behavior and validity of Henry's law, eqn 8.67 applied to the network 8.69 gives:

$$r_\mathrm{par} \;\cong\; \frac{k_{01}k_{12}k_{23}k_{30}\,C_\mathrm{ole}\,p_{H_2}\,C_{\Sigma\mathrm{cat}}}{\mathfrak{C} \;+\; D_{00}\!\left(C_\mathrm{Ph}/K_\mathrm{cat} \;+\; C_\mathrm{Ph}p_{H_2}/K_\mathrm{M}\right)} \tag{8.70}$$

where $\mathfrak{C}$ is the sum of all elements of the matrix

$$\begin{bmatrix} k_{12}k_{23}k_{30}C_\mathrm{ole} & k_{10}k_{23}k_{30} & k_{10}k_{21}k_{30} & 0 \\ k_{23}k_{30}k_{01}p_{H_2} & k_{21}k_{30}k_{01}p_{H_2} & 0 & 0 \\ k_{30}k_{01}k_{12}p_{H_2}C_\mathrm{ole} & 0 & 0 & 0 \\ k_{01}k_{12}k_{23}p_{H_2}C_\mathrm{ole} & 0 & 0 & 0 \end{bmatrix}$$

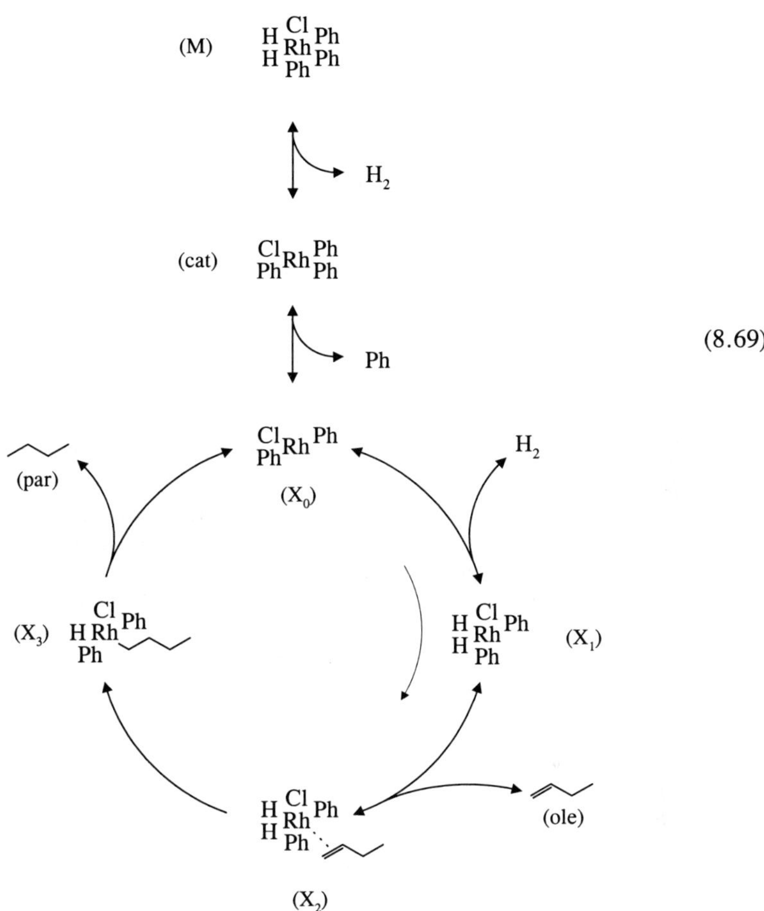

$$(8.69)$$

Also, D_{00} is the sum of the elements of the first matrix row, and K_{cat} and K_M are the equilibrium constants of cat $\longleftrightarrow$ X_0 + Ph and M $\longleftrightarrow$ X_0 + H_2 + Ph, respectively. Written in terms of phenomenological coefficients, the rate equation is

$$r_{par} \cong \frac{k_a C_{ole} p_{H_2} C_{\Sigma cat}}{1 + k_b C_{ole} + k_c p_{H_2} + k_d C_{ole} p_{H_2} + k_e C_{Ph} + k_f C_{ole} C_{Ph} + k_g p_{H_2} C_{Ph} + k_h C_{ole} p_{H_2} C_{Ph}} \qquad (8.71)$$

The reaction orders are: plus one in rhodium, between zero and plus one in olefin and hydrogen, and between zero and minus one in phosphine.

Equation 8.70 is the most general rate equation for the network 8.69. Simplifications may be justified. At high hydrogen pressure, rhodium may exist practically exclusively in the form of its hydrides; if so, the first matrix row (for ClRhPh$_2$) and the term $D_{00}C_{Ph}/K_{cat}$ (for ClRhPh$_3$) become insignificant, and the rate equation 8.70 or 8.71 loses all denominator terms not involving H$_2$ and thus becomes zero order in hydrogen. Moreover, if there is reason to assume that the π-allyl complex (X$_2$) and the rhodium alkyl species (X$_3$) also contain only an insignificant fraction of the rhodium, the last two matrix rows can be omitted, and eqn 8.71 loses its k_d term as well.

Originally, on the basis of batch studies without hydrogen atmosphere, sole rate control by metal-carbon bond formation (X$_2 \longrightarrow$ X$_3$) and rhodium distribution over H$_2$RhClPh$_3$ (M) and H$_2$RhClPh$_2$ole (X$_1$) was assumed. This gives a simpler rate equation of the form [54]:

$$r_{par} \;=\; \frac{k_a C_{ole} C_{\Sigma cat}}{k_b C_{ole} + k_c C_{Ph}} \tag{8.72}$$

However, the usually observed dependence on hydrogen pressure remains unaccounted for.

In more recent work on hydrogenation of butadiene polymers and copolymers, the attempt was made to explain the dependence on hydrogen pressure with sole rate control by olefin addition to H$_2$RhClPh$_2$ (X$_1$) and quasi-equilibrium rhodium distribution over the complexes with and without hydrogen [59] instead of kinetic significance of the step X$_0$ + H$_2 \longrightarrow$ X$_1$. This gives a rate equation for double-bond disappearance of the form

$$-r_{\underline{=}} \;=\; \frac{k_a C_{ole} p_{H_2} C_{\Sigma cat}}{1 + k_b p_{H_2} + k_c p_{H_2} C_{Ph}} \tag{8.73}$$

Olefin addition is normally fast, but could conceivably be slow if the olefin is a polymer. However, a fast metal-carbon bond formation is an unlikely assumption. The observed, unusual, close-to-first order dependence on olefin (in this case, degree of unsaturation of the polymer) is accounted for by eqn 8.73, but can be explained in a less contrived fashion with the more general equation 8.70: If X$_2$ and X$_3$ are *lacs* and the first step of the cycle is at quasi-equilibrium ($k_{12}C_{H_2} \ll k_{10}$), all denominator terms containing the olefin concentration become negligible and the reaction orders with respect to H$_2$ and phosphine remain unchanged, even with X$_0$ being a *lacs* in accordance with spectrophotometric evidence. More importantly, a fit of the constants in eqn 8.73 to the observed rates with the assumption of sole rate control by olefin addition would require an equilibrium rhodium distribution favoring the unsaturated and notoriously labile complex ClRhPh$_2$ even at fairly high hydrogen pressures and phosphine concentrations.   This is at odds with spectrophotometric evidence, molecular-orbital calculations, and the observation that hydrogen uptake by ClRhPh$_3$ is fast and stoichiometric.

It has been said that all members of the catalytic cycle remain at concentrations too low to be detected [60]. However, if all were *lacs*, $\mathfrak{C}$ in the denominator of eqn 9.70 would have to be negligible compared with the other two terms, resulting in a rate that is of order minus one in phosphine. That is contrary to most observations.

8.7.2. Inhibition, activation, decay, and poisoning.

Inhibition. As seen in the previous subsection, the presence of excess ligand reduces the reaction rate produced by a ligand-deficient catalyst because it promotes syphoning off catalyst material from the active cycle into the inactive external pathway. This can be viewed as a special case of inhibition: Although a necessary ingredient of the catalyst system, the ligand depresses the rate.* More generally, any substance that reduces the rate through removal of catalyst material from the cycle by reaction with one of the cycle members is called an inhibitor.

The inhibitor may react with the true catalyst X_0, i.e., the cycle member that initiates the reaction by binding the reactant A. This is called *competitive inhibition* because inhibitor and reactant compete directly for a binding site on the catalyst. Alternatively, the inhibitor may react with another cycle member X_j. This is called *noncompetitive inhibition*.

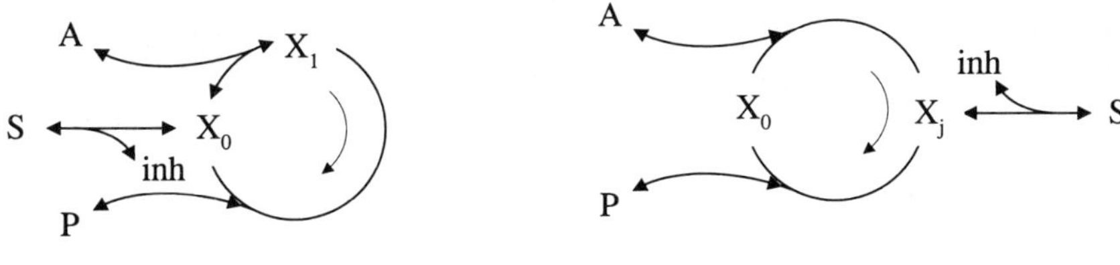

competitive inhibition   noncompetitive inhibition

For competitive inhibition with single-step external pathway, eqn 8.65 is directly applicable, except that the inhibitor concentration replaces that of the free ligand. For noncompetitive inhibition by reaction with X_j, D_{00} must be replaced by D_{jj}, the sum of the elements of row $j+1$ of the Christiansen matrix. Both cases are covered by:

$$r_P \cong \frac{\left[\prod_{i=0}^{k-1}\lambda_{i,i+1} - \prod_{i=0}^{k-1}\lambda_{i+1,i}\right] C_{\Sigma\mathrm{cat}}}{\mathfrak{C} + D_{jj} K_{\mathrm{inh}} C_{\mathrm{inh}}} \qquad \text{(j may be zero)} \qquad (8.74)$$

---

* The conclusion "lowest ligand concentration, highest rate" would be unwarranted, however. At too low a ligand concentration, the catalyst loses more than one of its ligands and either becomes inactive (if another molecule can occupy the coordinative site) or decomposes.

Here, K_{inh} is the equilibrium constant of the inhibition reaction inh $+ X_j \longrightarrow$ S, and index j refers to the cycle member with which the inhibitor reacts (j = 0 for competitive inhibition, and j $\neq$ 0 for noncompetitive inhibition). The extension to external pathways consisting of more than one step is as in eqn 8.67.

In both competitive and noncompetitive inhibition, the reaction is of order between zero and minus one with respect to the inhibitor. However, there is a kinetic difference between competitive and noncompetitive inhibition. In the former, the action of the inhibitor can be effectively countered by an increase in reactant concentration; direct competition by the reactant for a catalyst site can "crowd out" the inhibitor. In noncompetitive inhibition, this is not the case; even a large excess of reactant does not impair the inhibitor's access to the cycle member X_j. [Mathematically, in competitive inhibition the new and retarding denominator terms have D_{00} as factor, the sum of the first matrix row and only row that lacks the coefficient λ_{01}, the only coefficient with C_A as co-factor. In contrast, in noncompetitive inhibition the terms have D_{jj} as factor and contain λ_{01} and thus C_A as co-factor; the result is that an increase in C_A, apart from a direct beneficial effect on the rate, also strengthens the adverse effect of the noncompetitive inhibitor.]

If the reaction involves two reactants that enter the cycle at different places, inhibition may be competitive with respect to one and noncompetitive with respect to the other. Also, there may be two or more inhibitors reacting with the same or different cycle members. The formulas given here are easily extended to such cases.

Various other forms of inhibition are possible. The simplest of these is *reactant inhibition*, in which an inhibitor competes with the catalyst for the reactant, reducing the latter's likelihood to enter the catalytic cycle. As an example, consider reactant inhibition in a three-membered cycle whose last step is irreversible:

$$
\begin{array}{l}
\text{S} \rightleftharpoons \text{A} \rightleftharpoons \text{X}_1 \\
\qquad \text{inh} \quad \text{cat} \\
\text{P} \longleftarrow \text{X}_2
\end{array}
\qquad\qquad (8.75)
$$

The quasi-equilibrium condition for the inhibition reaction is $K_{inh} \cong C_S / C_A C_{inh}$ and the material balance for A is $C_{\Sigma A} = C_A + C_S$. Combination of these conditions shows that the fraction of total A that is still free to react with the catalyst is

$$
C_A \cong C_{\Sigma A} / (1 + K_{inh} C_{inh})
$$

In effect, the reaction rate is retarded by a factor $1/(1 + K_{inh} C_{inh})$ and thus is of order between zero and minus one with respect to the inhibitor. For example, for a reaction with network 8.75 (three-membered cycle, last step irreversible), the rate is given by

$$r_{\mathrm{P}} \cong \frac{k_{01}k_{12}k_{20}C_{\Sigma \mathrm{A}}C_{\Sigma\mathrm{cat}}}{\mathfrak{C}(1 + K_{\mathrm{inh}}C_{\mathrm{inh}})} \qquad (8.76)$$

where $\mathfrak{C}$ is the sum of the elements of the Christiansen matrix

$$\begin{bmatrix} k_{12}k_{20} & k_{10}k_{20} & k_{10}k_{21} \\ k_{20}k_{01}C_{\mathrm{A}} & k_{21}k_{01}C_{\mathrm{A}} & 0 \\ k_{01}k_{12}C_{\mathrm{A}} & 0 & 0 \end{bmatrix}$$

A much more comprehensive and in-depth discussion of different types of inhibition can be found in Segel's book *Enzyme Catalysis* [61].

Activation. Activation is the opposite of inhibition: The reaction rate is increased rather than decreased by a silent partner. The catalyst may require activation to function at all, as in a network

$$\mathrm{cat} \longleftrightarrow \overset{\mathrm{act}}{\longrightarrow} \mathrm{X_0} \qquad (8.77)$$

in which the true catalyst is $\mathrm{X_0}$ and is produced by reaction of the catalyst species charged, cat, with the activator, act. This is called *essential activation.*

Application of the general equation 8.65 to the network 8.77 gives

$$r_{\mathrm{P}} \cong \frac{\left[\prod_{i=0}^{k-1}\lambda_{i,i+1} - \prod_{i=0}^{k-1}\lambda_{i+1,i}\right]C_{\Sigma\mathrm{cat}}}{\mathfrak{C} + D_{00}/K_{\mathrm{act}}C_{\mathrm{act}}} \qquad (8.78)$$

The rate is of order between zero and plus one with respect to the activator. The additional denominator term reflects the fact that some catalyst material is present as the not yet activated species, cat. This retarding term is infinitely large if no activator is present, and decreases with increasing activator concentration.

A much more detailed discussion of activation mechanisms is given in Segel's book [62].

Decay and poisoning. In all situations studied to this point, the external pathway was that of a reversible reaction fast enough to become quasi-stationary. Of course, the reaction may also be irreversible. If so, catalyst material that has left the cycle does not return and is permanently lost, and the reaction eventually comes to a stand-still unless the catalyst is replenished. If the loss of catalyst material is fast, the system obviously is of no practical interest. In contrast, slow irreversible loss is a problem the practical engineer often has to contend with.

Any member of the catalytic cycle, or more than one of the members, may be subject to the loss reaction. That reaction may be a rearrangement into an inert form, a breaking apart into inert fragments, or a reaction with another substance to yield an inert species. The first two cases are called *decay*; the last one, *poisoning*.

Inasmuch as catalyst loss in any practical situation is very slow compared with the catalytic reaction, the quasi-stationary behavior of the latter is not significantly impaired. Accordingly, it makes no difference in kind which of the cycle members is (or are) decaying or being poisoned: The principal result in any event is a slow drain of catalyst material from the producing cycle.

A practical approach to quantitative kinetics of catalyst systems subject to decay or poisoning is to model the catalytic reaction as though no loss occurred but, unless the loss is compensated by make-up, to make the total amount of catalyst material time-dependent in accordance with the loss reaction. Often, little is known about the exact cause and mechanism of loss, and an empirical loss rate is the best one can come up with. In the rare instances in which the exact loss reaction has been identified, a more detailed modeling is possible, as a specific example will illustrate.

Let us say, the loss reaction is a single step $X_j + poi \longrightarrow S$, where poi is the poisoning substance. The loss rate then is

$$-r_{\Sigma\,cat} \; = \; k_{poi}\, C_j\, C_{poi}$$

and equals $-dC_{\Sigma cat}/dt$ if there is no make-up. The loss rate involves the concentration of the affected cycle member, X_j. According to eqn 8.32 this concentration is

$$C_j \; = \; D_{jj}\, C_{\Sigma^\circ}\, /\mathfrak{C} \tag{8.79}$$

where D_{jj} is the sum of the elements of row $j+1$ of the Christiansen matrix for the cycle, $\mathfrak{C}$ is the sum of all elements of that matrix, and Σ° is the total catalyst material in the cycle.

Proper modeling requires knowledge of the affected cycle member and such an approach. This is because the distribution of total catalyst material over the cycle members varies with conversion. For instance, with saturation kinetics, most of the catalyst material may be in the form of a reaction intermediate as long as the reactant concentration is still high, but will then shift to the free catalyst as the reactant is used up.

8.8. Multiple cycles*

Catalyst systems with more than one cycle are not unusual. There is a wealth of possible configurations, and only some relatively simple and often encountered ones can be discussed here. The examples are selected for their suitability to illustrate the application of the tools developed in this chapter. With a good understanding of these tools, the reader should have no problems in deriving the mathematics for situations of his that are not covered here.

8.8.1. *Competing reactions* (cycles with common member)

In practice, one and the same catalyst often catalyzes more than one reaction of the same or different reactants. An example is hydrogenation of a mixture of unsaturated compounds. Such networks are of the general type

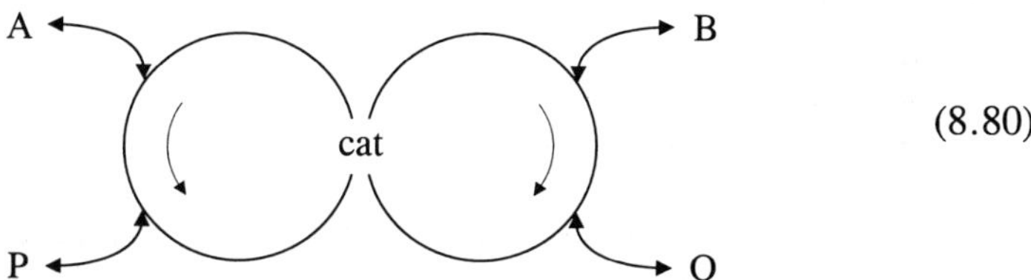

$$(8.80)$$

here shown for just two reactions of different reactants A and B to products P and Q, respectively (details of cycles omitted). If the free catalyst, cat, is the *macs*, the reactions proceed independently side by side. However, if significant amounts of catalyst material are present in the form of other cycle members, catalyst material in one cycle is not contributing to the reaction in the other. In effect, the two reactants then inhibit one another.

The general rate equation for the cycle converting A to P (with m members) can be written

$$
r_{\mathrm{P}} = \frac{\left[\prod_{i=0}^{m-1} \lambda_{i,i+1} - \prod_{i=0}^{m-1} \lambda_{i+1,i} \right] C_{\Sigma\mathrm{cat}}}{\mathfrak{C} + (D_{00}/D_{00}') \sum_{i=1}^{n} D_{ii}'}
\tag{8.81}
$$

Here, primes refer to the B $\longrightarrow$ Q cycle (with n members), $\mathfrak{C}$ is the sum of all elements of the Christiansen matrix for the A $\longrightarrow$ P cycle, and the D_{00} are the sums

---

* The mathematical derivations in this section are based largely on unpublished work by J.-M. Chern (1992).

of the first-row elements of the respective matrices. Note that the summation over the D_{ii}' does not include D_{00}' (sum of first-row elements of matrix of B $\longrightarrow$ Q cycle). The second denominator term reflects the inhibition by the other reactant; it becomes negligible if no member of the B $\longrightarrow$ Q cycle except X_0 contains a significant amount of catalyst material.

　　　　In a network such as 8.80, conversion of one reactant is inhibited by the other if, and only if, at least one intermediate in the cycle involving the former contains a significant amounts of catalyst material. Experimentally observed inhibition thus constitutes evidence that this is indeed the case.

Derivation of eqn 8.81. In an ordinary, single, k-membered cycle, eqn 8.32 can be used to replace $C_{\Sigma cat}$ in the general rate equation 8.30 to give:

$$r_{\mathrm{P}} \;\cong\; \Xi\, C_j / D_{jj} \qquad (j=0,\ldots,k) \tag{8.82}$$

where

$$\Xi \;\equiv\; \prod_{i=0}^{k-1} \lambda_{i,i+1} \;-\; \prod_{i=0}^{k-1} \lambda_{i+1,i} \tag{8.83}$$

After rearrangement, eqn 8.82 gives

$$C_j \;\cong\; r_{\mathrm{P}} D_{jj} / \Xi \qquad (j=0,\ldots,k) \tag{8.84}$$

The overall balance of catalyst material is

$$C_{\Sigma cat} \;=\; \sum_{i=0}^{m} C_i \;+\; \sum_{i=1}^{n} C_i' \tag{8.85}$$

(note the second summation does not include C_0). With eqns 8.84 and noting that the summations over the D_{jj} and D_{jj}' give $\mathfrak{C}$ and $\mathfrak{C}'$, respectively, one obtains

$$C_{\Sigma cat} \;\cong\; r_{\mathrm{P}} \mathfrak{C} / \Xi \;+\; r_{\mathrm{Q}} \mathfrak{C}' / \Xi' \tag{8.86}$$

　　　　Dividing eqn 8.82 for the A $\longrightarrow$ P cycle by that for the B $\longrightarrow$ Q cycle and solving for r_{Q}, recognizing that C_0 (concentration of cat) is the same in both cycles, one finds

$$r_{\mathrm{Q}} \;\cong\; r_{\mathrm{P}} D_{00} \Xi' / D_{00}' \Xi \tag{8.87}$$

This allows r_{Q} in eqn 8.86 to be replaced. Solving the resulting equation for r_{P} yields eqn 8.81.

　　　　The member which two cycles have in common need not be the free catalyst. Equation 8.81 then applies with the appropriate changes in indices. The example to follow illustrates such a case.

Example 8.10. Osmium-tetroxide catalyzed asymmetric dihydroxylation of olefins.
Substituted olefins such as styrene can be dihydroxylated to give vicinal diols:

$$\ldots -CH=CH_2 + RO + H_2O \longrightarrow \ldots -CHOH-CH_2OH + R$$

The reaction, usually carried out in aqueous acetone and with an amine oxide as the
oxidant RO, is catalyzed by osmium tetroxide, activated by an alkaloid ligand, L. To
explain the higher enantio-selectivity achieved at lower olefin concentrations, a network with two cycles having a common member has been proposed [63,64]:

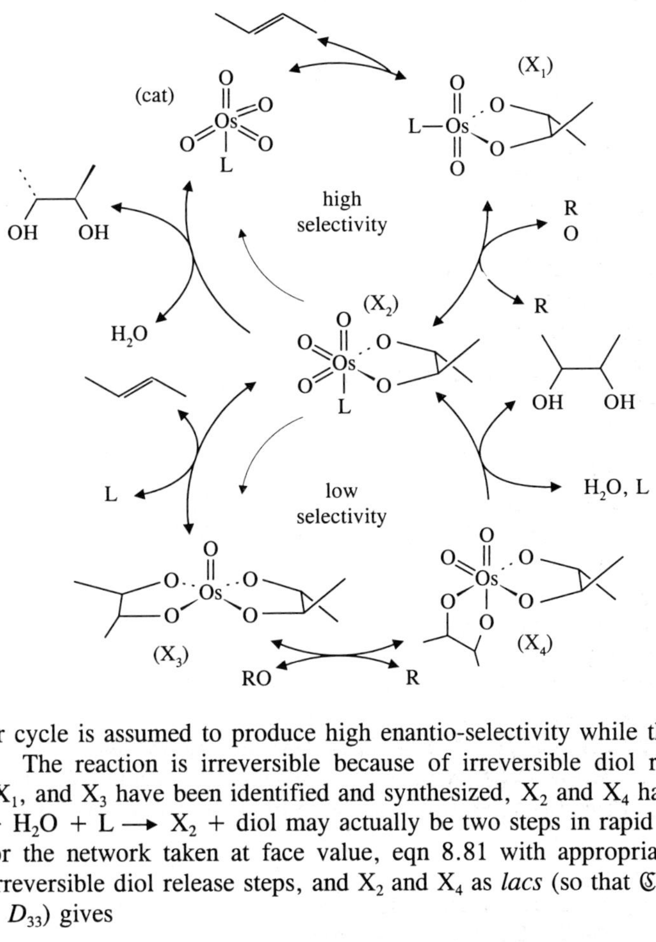

(8.88)

The upper cycle is assumed to produce high enantio-selectivity while the lower one
does not. The reaction is irreversible because of irreversible diol release. The
catalyst, X_1, and X_3 have been identified and synthesized, X_2 and X_4 have not. The
step $X_4 + H_2O + L \longrightarrow X_2 +$ diol may actually be two steps in rapid succession.

For the network taken at face value, eqn 8.81 with appropriately changed
indices, irreversible diol release steps, and X_2 and X_4 as *lacs* (so that $\mathfrak{C} = D_{00} + D_{11}$
and $\mathfrak{C}' = D_{33}$) gives

(8.89)

$$-r_{ole} \cong \left[\frac{k_{01}k_{12}k_{20}}{D_{00} + D_{11} + (D'_{22}/D_{22})D_{33}} + \frac{k_{34}k_{42}k_{23}C_L}{D_{33} + (D_{22}/D'_{22})(D_{00} + D_{11})} \right] C_{ole}C_{H_2O}C_{RO}C_{\Sigma cat}$$

with

$$D_{00} = k_{12}k_{20}C_{RO}C_{H_2O} + k_{10}k_{20}C_{H_2O} + k_{10}k_{21}C_R$$

$$D_{11} = k_{20}k_{01}C_{H_2O}C_{ole} + k_{21}k_{01}C_R C_{ole}$$

$$D_{22} = k_{01}k_{12}C_{ole}C_{RO}$$

$$D'_{22} = k_{34}k_{42}C_{RO}C_{H_2O}C_L + k_{32}k_{42}C_L^2 C_{H_2O} + k_{32}k_{43}C_L C_R$$

$$D_{33} = k_{42}k_{23}C_{H_2O}C_L C_{ole} + k_{43}k_{23}C_R C_{ole}$$

The reaction orders are: plus one for osmium; between zero and plus one for olefin, water, and oxidant; between plus one and minus one for the alkaloid ligand; and between zero and minus one for the spent oxidant (zero if oxidation is irreversible).

Because only the upper cycle gives high enantio-selectivity, the yield ratio of the two cycles is of interest. From eqn 8.87 with appropriate change in indices and after cancellations:

$$\frac{r_{diol}}{r'_{diol}} \cong \frac{\varXi D'_{22}}{\varXi' D_{22}} = \frac{k_{20}(k_{34}k_{42}C_{H_2O}C_{RO} + k_{32}k_{42}C_L C_{H_2O} + k_{32}k_{43}C_R)}{k_{34}k_{42}k_{23}C_{ole}C_{RO}} \tag{8.90}$$

The yield ratio is inversely proportional to the olefin concentration and increases less than proportionately with increasing concentrations of water, ligand, and spent oxidant and decreasing concentration of oxidant.

It has been said that the reaction is first order in olefin. Equation 8.89 shows that, if the network 8.88 is correct, this can be true only if X_1 and X_3 are also *lacs* (to make D_{11} and D_{33} negligible) and $r'_{diol} \ll r_{diol}$. That may well be the case if the olefin concentration is kept very low to achieve high enantio-selectivity.

8.8.2. *Dual- and multiple-form catalysts* (connected cycles)

A catalyst may exist in two or more forms with different catalytic activities. In the simplest systems of this type, two such forms interconvert in a quasi-equilibrium step. The conversion may or may not involve other species. It may, for example, be a ligand exchange. The two catalyst species may catalyze the same reaction or different ones. A network of this type, with ligand exchange and different reactions A $\longrightarrow$ P and B $\longrightarrow$ Q, is

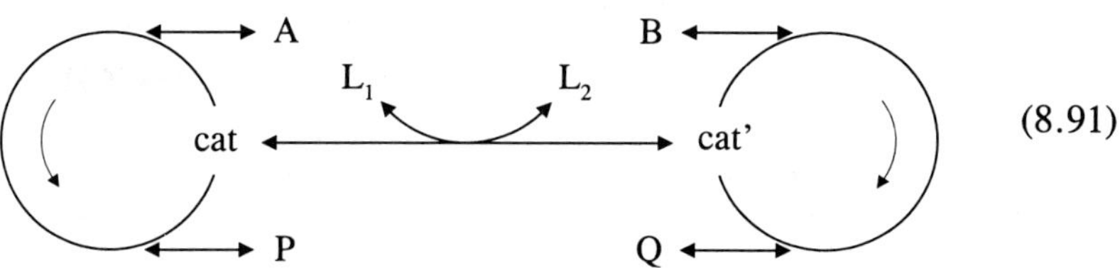

$$\tag{8.91}$$

The rate equation for the m-membered A $\longrightarrow$ P cycle, derived along the same lines as eqn 8.81, is

$$r_P \;\cong\; \frac{\left[\displaystyle\prod_{i=0}^{m-1}\lambda_{i,i+1} - \prod_{i=0}^{m-1}\lambda_{i+1,i}\right] C_{\Sigma\mathrm{cat}}}{\mathfrak{C} + K_{\mathrm{cat}}(C_{L1}/C_{L2})(D_{00}/D_{00}')\,\mathfrak{C}'} \tag{8.92}$$

where $\mathfrak{C}$ and $\mathfrak{C}'$ are the sums of the elements of the Christiansen matrices of the two cycles, and K_{cat} is the equilibrium constant of the ligand exchange reaction cat + L_1 $\longleftrightarrow$ cat' + L_2. The equation for the other cycle is, of course, analogous. The following example describes a slightly more complicated catalyst system of this type.

Example 8.11. Hydroformylation with phosphine-substituted cobalt hydrocarbonyl catalyst. The phosphine-substituted cobalt hydrocarbonyl catalyst used for hydroformylation of olefins has been described in Section 8.2 (see network 8.13). The principal reaction

$$\text{olefin} + H_2 + CO \longrightarrow \text{aldehyde}$$

follows the Heck-Breslow mechanism 6.9 shown in Section 6.3 (see also Section 7.3.2) and is catalyzed by both $HCo(CO)_3Ph$ and $HCo(CO)_4$, where Ph is a tertiary organic phosphine ligand. Stripped of complications—olefin isomerization, formation of isomeric products and paraffin, subsequent aldehyde hydrogenation, and ligand exchange of cycle members—but with the catalyst equilibria, the network is:

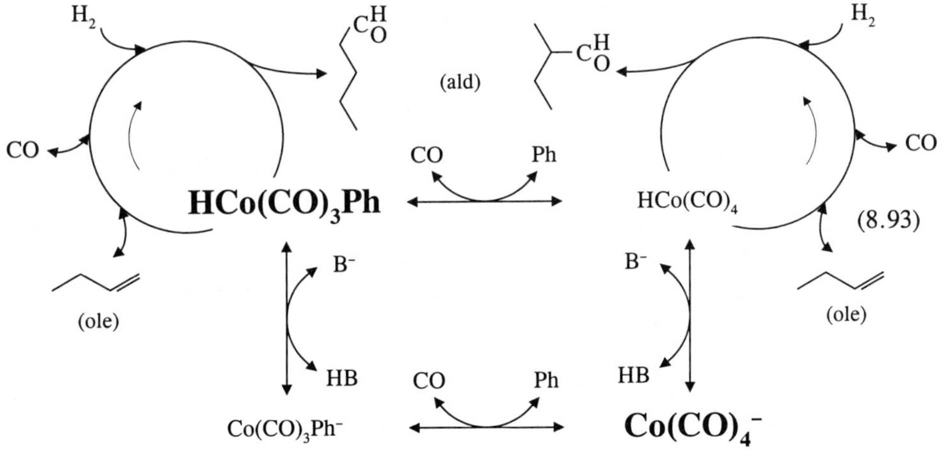

(main cobalt-containing species shown in larger font; B^- and HB are added base and its conjugate acid, respectively; both cycles produce both aldehyde isomers, but only the predominant product is shown; for details of the catalytic cycles, see network 6.9).

According to spectrophotometric evidence, all cobalt-containing species except $HCo(CO)_3Ph$ and $Co(CO)_4^-$ are *lacs* under typical reaction conditions. Although a *lacs*, $HCo(CO)_4$ nevertheless contributes to olefin conversion because its catalytic activity is several orders of magnitude higher than that of $HCo(CO)_3Ph$. Within its own cycle, $HCo(CO)_4$ is the *macs*. The rate is the sum of the rates produced by the two cycles.

For each cycle, the rate is given by a one-plus equation of the form

$$r_{ald} \cong \frac{k_a C_{ole} C_{cat}}{1 + k_b p_{CO}/p_{H_2}} \qquad (6.12)$$

(Martin equation, see Example 6.2 in Section 6.3).

The total catalyst balance, if including only the species containing significant fractions of cobalt, is

$$C_{\Sigma Co} \cong C_{HCo(CO)_3Ph} + C_{Co(CO)_4^-} \qquad (8.94)$$

With this and the equilibrium conditions for $HCo(CO)_3Ph + CO \longleftrightarrow Co(CO)_4^- + Ph$ and $HCo(CO)_3Ph + CO + B^- \longleftrightarrow HCo(CO)_4 + Ph + HB$

$$\frac{C_{HCo(CO)_4} C_{Ph}}{C_{HCo(CO)_3Ph} p_{CO}} = K_1 \qquad \text{and} \qquad \frac{C_{Co(CO)_4^-} C_{Ph} C_{HB}}{C_{HCo(CO)_3Ph} p_{CO} C_{B^-}} = K_2 \qquad (8.95)$$

respectively, one obtains

$$C_{HCo(CO)_3Ph} = C_{\Sigma Co} \frac{C_{HB} C_{Ph}}{C_{HB} C_{Ph} + K_2 p_{CO} C_{B^-}}$$

$$C_{HCo(CO)_4} = C_{\Sigma Co} \frac{K_1 C_{HB} C_{CO}}{C_{HB} C_{Ph} + K_2 p_{CO} C_{B^-}}$$

With these substitutions for $HCo(CO)_3Ph$ and $HCo(CO)_4$ and eqn 8.94, the overall rate as the sum of the rates of the two cycles is, after cancellations:

$$r_{ald} \cong \left[\frac{k_a C_{Ph}}{1 + k_b p_{CO}/p_{H_2}} + \frac{k_a' K_1 p_{CO}}{1 + k_b' p_{CO}/p_{H_2}} \right] \frac{C_{ole} C_{HB} C_{\Sigma Co}}{C_{HB} C_{Ph} + K_2 p_{CO} C_{B^-}} \qquad (8.96)$$

(primes refer to the catalyst cycle with $HCo(CO)_4$).

According to eqn 8.96, the reaction orders are: plus one for total cobalt; between zero and plus one for H_2 and HB; between zero and minus one for CO and B^-; and between minus one and plus one for phosphine. Although the reaction is first order in cobalt, it may be of higher or lower order than first in "catalyst" if the concentration ratios of the catalyst ingredients cobalt, phosphine, and base are held constant. This has been discussed in Example 8.3 in Section 8.2.

8.8.3. *Reactions with multiple products* (cycles with common pathway segments)

Many catalytic reactions form a range of products rather than only a single one. In most such cases, the pathway to a co-product branches off from that to the main product after the first or a few early steps. The network then consists of cycles that have a step or pathway segment in common. Typical examples are the formation of isomeric products in paraffin oxidation and olefin hydration, hydrohalogenation, hydroformylation, and hydrocyanation, as well as paraffin by-product formation in hydroformylation.

The simplest networks of this kind consist of two cycles with an initial, common pathway of k steps:

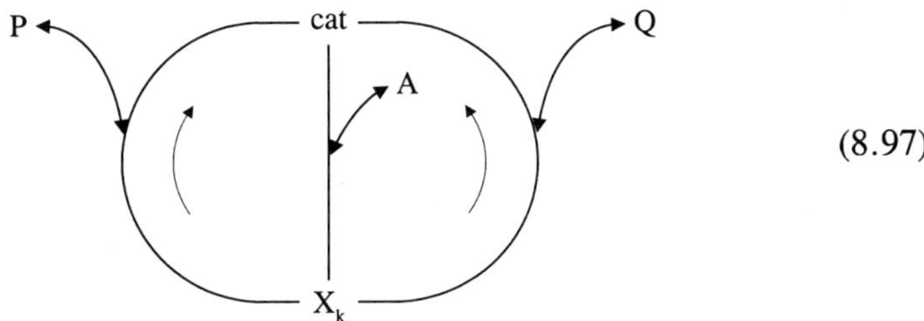

$$(8.97)$$

(details of pathway and cycles not shown). In general form, the rate equations are

$$
-r_A \;\cong\; \frac{\left[(\Lambda_{k0} + \Lambda'_{k0}) \prod_{i=0}^{k-1} \lambda_{i,i+1} \;-\; (\Lambda_{0k} + \Lambda'_{0k}) \prod_{i=0}^{k-1} \lambda_{-i+1,i} \right] C_{\Sigma\,\mathrm{cat}}}{\displaystyle \sum_{i=0}^{k} D_{ii} + \sum_{i=k+1}^{m-1} \left[\frac{\Lambda_{ki} D_{kk} + \Lambda_{0i} D_{00}}{\Lambda_{ik} + \Lambda_{i0}} \right] + \sum_{i=k+1}^{n-1} \left[\frac{\Lambda'_{ki} D_{kk} + \Lambda'_{0i} D_{00}}{\Lambda'_{ik} + \Lambda'_{i0}} \right]}
\tag{8.98}
$$

$$
r_P \;\cong\; \frac{(\Lambda_{k0} D_{kk} - \Lambda_{0k} D_{00}) C_{\Sigma\,\mathrm{cat}}}{\displaystyle \sum_{i=0}^{k} D_{ii} + \sum_{i=k+1}^{m-1} \left[\frac{\Lambda_{ki} D_{kk} + \Lambda_{0i} D_{00}}{\Lambda_{ik} + \Lambda_{i0}} \right] + \sum_{i=k+1}^{m-1} \left[\frac{\Lambda'_{ki} D_{kk} + \Lambda'_{0i} D_{00}}{\Lambda'_{ik} + \Lambda'_{i0}} \right]}
\tag{8.99}
$$

and an analogous equation for r_Q.* Here, the λ coefficients are those of the common pathway cat $+ A \longleftrightarrow \ldots \longleftrightarrow X_k$; the Λ and Λ' are the segment coeffi-

---

* It is not obvious that these equations meet the stoichiometry requirement $-r_A = r_P + r_Q$. That they do can be verified by replacement of the Λ and Λ' coefficients and comparison of the equations for $-r_A$ and $r_P + r_Q$ after resulting cancellations in the latter.

cients of the pathways $X_k \longleftrightarrow \ldots \longleftrightarrow$ cat + P (with m members) and $X_k \longleftrightarrow \ldots \longleftrightarrow$ cat + Q (with n members), respectively (see eqns 6.5 for segment coefficients); and the D_{ii} are the sums of the elements of the rows of the Christiansen matrix of the "collapsed" network 8.104 (single cycle with loop coefficients for the parallel pathways, see below). For example, for a two-step common pathway (k=2):

$$D_{00} = \lambda_{12}\mathscr{L}_{20} + \lambda_{10}\mathscr{L}_{20} + \lambda_{10}\lambda_{21}$$
$$D_{11} = \mathscr{L}_{20}\lambda_{01} + \lambda_{21}\lambda_{01} + \lambda_{21}\mathscr{L}_{02} \qquad (8.100)$$
$$D_{22} = \lambda_{01}\lambda_{12} + \mathscr{L}_{02}\lambda_{12} + \mathscr{L}_{02}\lambda_{10}$$

with loop coefficients

$$\mathscr{L}_{20} = \Lambda_{20} + \Lambda'_{20} , \qquad\qquad \mathscr{L}_{02} = \Lambda_{02} + \Lambda'_{02} \qquad (8.101)$$

In eqns 8.98 and 8.99, the denominator terms are proportional to the concentrations of the catalyst-containing participants: the D_{ii} of the first sum to the members of the common pathway, the terms of the second sum to the intermediates along $X_k \longleftrightarrow \ldots \longleftrightarrow$ cat + P, and those of the third sum to the intermediates along $X_k \longleftrightarrow \ldots \longleftrightarrow$ cat + Q. If any of these participants are *lacs*, their terms can be dropped.

Often, one product (say, P) is desired and the other is a by-product that must be removed and disposed of. If so, the P-to-Q yield ratio, Y_{PQ}, and the selectivity to P, S_P, are of interest (see Section 1.6 for yield, yield ratio, and selectivity). From eqns 8.98, 8.99, and the analogous equation for r_Q:

$$Y_{PQ} \equiv \frac{r_P}{r_Q} = \frac{\Lambda_{k0}D_{kk} - \Lambda_{0k}D_{00}}{\Lambda'_{k0}D_{kk} - \Lambda'_{0k}D_{00}} \qquad (8.102)$$

$$S_P \equiv \frac{r_P}{-r_A} = \frac{\Lambda_{k0}D_{kk} - \Lambda_{0k}D_{00}}{(\Lambda_{k0} + \Lambda'_{k0})\prod\limits_{i=0}^{k-1}\lambda_{i,i+1} - (\Lambda_{0k} + \Lambda'_{0k})\prod\limits_{i=0}^{k-1}\lambda_{i+1,i}} \qquad (8.103)$$

Derivation of eqns 8.98 and 8.99. The network 8.97 can be "collapsed" in a single cycle with the k+1 members of the common pathway and a pseudo-single step converting X_k to cat and P or Q:

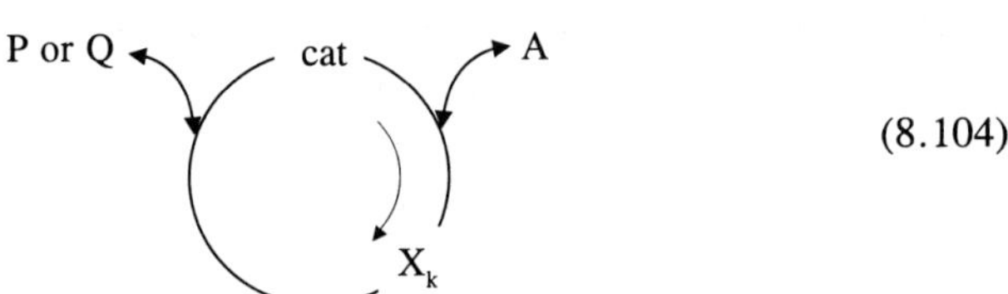

$$(8.104)$$

(see Section 6.4.1 for reduction of networks with loops, and eqn 6.15 for loop coefficients).

The concentrations of the network members are obtained as follows: Application of eqn 8.84 to the members of the common pathway gives

$$C_i \cong \frac{-r_A D_{ii}}{\mathcal{L}_{k0} \prod_{i=0}^{k-1} \lambda_{i,i+1} - \mathcal{L}_{0k} \prod_{i=0}^{k-1} \lambda_{i+1,i}} \qquad (0 \leq i \leq k) \qquad (8.105)$$

The concentrations of the members of the cycles can be found with a relationship expressing the concentration of an intermediate in a pathway as a function of those of the end members. Use of the Bodenstein approximation for any intermediate X_i in a pathway $X_j \leftrightarrow \cdots \leftrightarrow X_k$ (reduced to two pseudo-single steps) yields

$$C_i \cong \frac{\Lambda_{ji} C_j + \Lambda_{ki} C_k}{\Lambda_{ij} + \Lambda_{ik}} \qquad (j < i < k) \qquad (8.106)$$

Applied to the pathway $X_k \leftrightarrow \cdots \leftrightarrow \text{cat} + P$:

$$C_i \cong \frac{\Lambda_{ki} C_k + \Lambda_{0i} C_{cat}}{\Lambda_{ik} + \Lambda_{i0}} \qquad (k < i < m) \qquad (8.107)$$

With eqns 8.105 to replace C_k and C_{cat} this gives

$$C_i \cong \frac{-r_A (D_{kk} \Lambda_{ki} + D_{00} \Lambda_{0i})}{\left[\mathcal{L}_{k0} \prod_{i=0}^{k-1} \lambda_{i,i+1} - \mathcal{L}_{0k} \prod_{i=0}^{k-1} \lambda_{i+1,i} \right] (\Lambda_{ik} + \Lambda_{i0})} \qquad (k < i < m) \qquad (8.108)$$

The equation for the intermediates in the other pathway, $X_k \leftrightarrow \cdots \leftrightarrow \text{cat} + Q$, is analogous.

The rate $-r_A$ can now be obtained from the catalyst balance and the concentrations. The catalyst balance is

$$C_{\Sigma cat} = \sum_{i=0}^{k} C_i + \sum_{i=k+1}^{m-1} C_i + \sum_{i=k+1}^{n-1} C_i' \qquad (8.109)$$

Using eqns 8.105 to replace the concentrations under the first sum, eqns 8.108 and their equivalents for the other cycle to replace those of the second and third sums, respectively, then replacing the loop coefficients by means of eqns 8.101, and finally solving for $-r_A$ one obtains eqn 8.98.

The rate r_P is that of the pathway $X_k \leftrightarrow \cdots \leftrightarrow \text{cat} + P$:

$$r_P \cong \Lambda_{k0} C_k - \Lambda_{0k} C_{cat} \qquad (8.110)$$

With eqns 8.105 for C_k and C_{cat} and eqn 8.98 for $-r_A$ this gives eqn 8.99.

This derivation illustrates how, in general, rate equations for more complex networks with common pathway segments can be compiled. The essential steps are:

(1)   reduction of the cycles other than common pathways to pseudo-single steps with loop coefficients, to obtain a single-cycle collapsed network;

(2)   use of eqns 8.105 or their equivalents to relate the concentrations of the members of common segments, including the node members, to the overall rate through the collapsed network;

(3)   use of eqns 8.108 or their equivalents to relate the concentrations of the intermediates in loops to those of the node members and thereby to the overall rate;

(4)   use of the catalyst balance to obtain the overall rate from the so obtained concentrations of all catalyst-containing participants;

(5)   establishment of formation rates of individual products from equations equivalent to 8.108 for the respective segments, the concentrations of the segment end members, and the overall rate;

(6)   replacement of the loop, segment, and λ coefficients by the respective true rate coefficients and reactant and product concentrations.

Note that it may be possible to consolidate steps within segments in accordance with Rule 8.55. This can reduce the algebra considerably.

Example 8.12.   By-product formation in hydrocyanation of 4-pentenenitrile [44]. Hydrocyanation of 4-pentenenitrile (4-PN) to adiponitrile (ADN) was examined in some detail Example 8.7. The du Pont process maximizes the yield of adiponitrile, the desired product, by addition of a Lewis acid such as zinc chloride or tri-alkylborane, but nevertheless some 2-methyl-glutaronitrile (2-MGN) is formed as by-product:

$$
\begin{array}{c}
\text{HCN} + \\[4pt]
\text{(4-PN)}
\end{array}
\quad\longrightarrow\quad
\begin{array}{c}
\text{NC}\diagdown\diagup\diagdown\text{CN} \\
\text{(ADN)} \\[10pt]
\text{(2-MGN)}
\end{array}
\tag{8.111}
$$

To account for the by-product formation, the network 8.49 must be expanded. The branch is at the rearrangement $X_2 \longleftrightarrow X_3$ of a π-complex to a species with metal-carbon σ-bond:

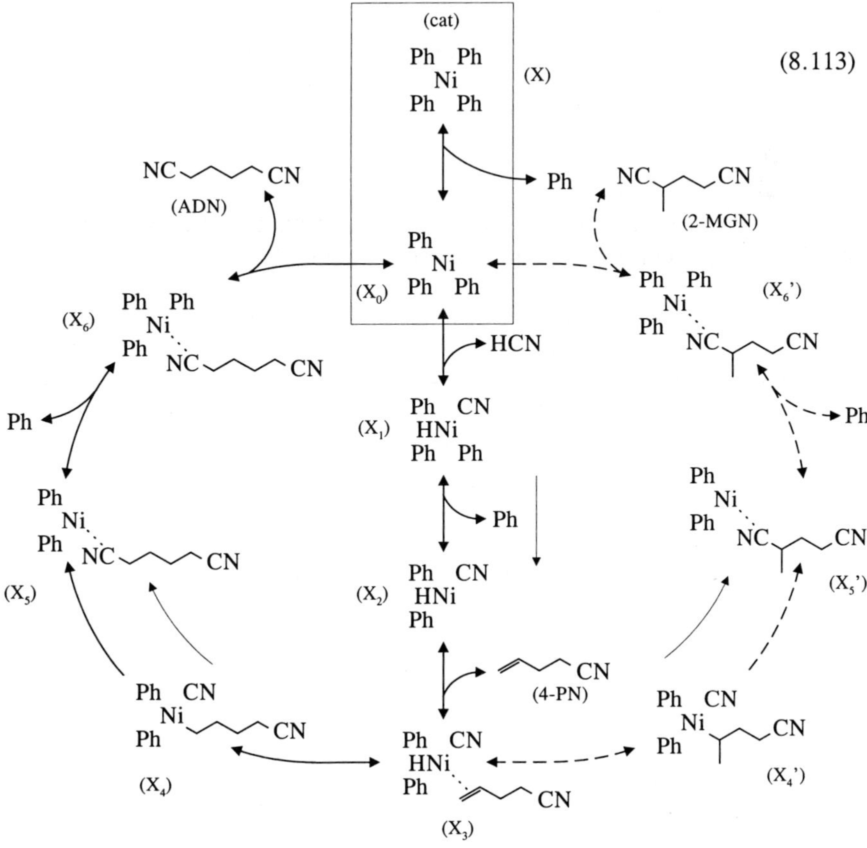

$$(8.112)$$

(Ph = organic phosphine). Accordingly, the expanded network that includes by-product formation is:

$$(8.113)$$

Equations 8.98, 8.99, 8.102, and 8.103 are directly applicable to the rates of 4-pentenenitrile consumption, adiponitrile formation, yield ratio, and selectivity to adiponitrile, with $k=3$ and lumping of cat and X_0 into X as in Example 8.7 (with index 0 in the equations replaced by X). The reaction is irreversible, so the negative terms in eqns 8.98 and 8.99 drop out. Species X_2, X_5, X_5', X_6, and X_6' are *lacs* and, therefore, D_{22} is zero (as are D_{55}, D_{55}', D_{66}, and D_{66}', not needed in eqns 8.102 and 8.103). With λ_{54} and $\lambda_{54}' = 0$, the loop and segment coefficients become:

$$\mathscr{L}_{3X} = \Lambda_{3X} + \Lambda_{3X}', \qquad\qquad \mathscr{L}_{X3} = 0,$$

$$\Lambda_{3X} = \frac{\lambda_{34}\lambda_{45}}{\lambda_{45} + \lambda_{43}}, \qquad\qquad \Lambda_{3X}' = \frac{\lambda_{34}'\lambda_{45}'}{\lambda_{45}' + \lambda_{43}'}$$

and the D_{ii} are

$$D_{XX} = (\lambda_{12}\lambda_{23} + \lambda_{1X}\lambda_{23} + \lambda_{1X}\lambda_{21})\mathscr{L}_{3X} + \lambda_{1X}\lambda_{21}\lambda_{32},$$

$$D_{11} = \lambda_{X1}(\lambda_{23} + \lambda_{21})\mathscr{L}_{3X} + \lambda_{X1}\lambda_{21}\lambda_{32},$$

$$D_{22} = 0, \qquad\qquad D_{33} = \lambda_{X1}\lambda_{12}\lambda_{32}$$

where $\mathscr{L}_{3X}$ is given above and the λ coefficients are the same as in Example 8.7, with the λ' analogous to the respective λ (see eqns 8.100 and 8.101).

With all these substitutions, explicit equations for 4-pentenenitrile consumption and adiponitrile formation are obtained from eqns 8.98 and 8.99. The equations are more lengthy, but of same algebraic form as eqn 8.53 for the network 8.49 without by-product formation.

The adiponitrile-to-2-methyl glutaronitrile yield ratio and the selectivity to adiponitrile are found to be

$$Y_{\text{ADN/MGN}} = \frac{\Lambda_{3X}}{\Lambda_{3X}'} = \frac{k_{34}k_{45}(k_{45}' + k_{43}')}{k_{34}'k_{45}'(k_{45} + k_{43})}$$

$$S_{\text{ADN}} = \frac{\Lambda_{3X}D_{33}}{(\Lambda_{3X} + \Lambda_{3X}')\lambda_{X1}\lambda_{12}\lambda_{23}} = \left[1 + \frac{k_{34}'k_{45}'(k_{45} + k_{43})}{k_{34}k_{45}(k_{45}' + k_{43}')}\right]^{-1}$$

(see eqns 8.102 and 8.103) and are seen to be concentration-independent. Not surprisingly, the best leverage is an increase of the $k_{23}:k_{23}'$ ratio for the steps at the branch, but increases in $k_{34}:k_{34}'$ and $k_{32}':k_{32}$ also help.

In this example, the rates in the networks with and without by-product formation were found to be of same algebraic form and the yield ratio and selectivity to be concentration-independent. This is due to the concentration independence of the segment coefficients of the two parallel pathways, Λ_{2X} and Λ_{2X}' and is not generally true even if the different pathways consist of strictly analogous steps.

The situation analyzed here in detail is a quite common one in metal-complex catalysis of additions to olefinic double bonds. As a rule, the olefin adds to the metal at a coordinative site, forming a π-complex. This is followed by a re-arrangement with formation of a σ-bond between the metal and either of the two double-bond carbon atoms (see reaction 8.112). In most cases, the rearrangement yields two isomeric species, depending on which of the two carbon atom forms the σ-bond. In such cases, insertion of another ligand (CN in hydrocyanation, CO in hydroformylation, etc.) between metal and carbon and release from the catalyst yields two isomeric final products. Moreover, if both π-complex formation and σ-bonding are reversible, the catalyst also promotes double-bond migration within the olefin (not accounted for in the network 8.113). The new olefin isomers may then give rise to further isomeric final products. A typical example of this is hydro-formylation of straight-chain olefins (see Example 5.3 in Section 5.3 and network 7.40 in Section 7.4).

8.9. Self-accelerating reactions (autocatalysis)

In most reactions, the rate decreases monotonically as conversion progresses. This is not universally true, however. In exceptional reactions, the rate increases with conversion, if possibly only within a certain range. Such self-acceleration is com-monly called *autocatalysis*. Although genuine catalysis might not be involved, it seems appropriate to discuss such behavior in the context of catalytic reactions.

8.9.1. Product-promoted reactions

The typical textbook example of autocatalysis is that of a hypothetical single-step reaction with stoichiometry A $\longleftrightarrow$ P, but mechanism

$$A + P \longleftrightarrow 2P \tag{8.114}$$

and rate

$$r_\mathrm{P} = k_{\mathrm{AP}}\, C_\mathrm{A}\, C_\mathrm{P} - k_{\mathrm{PA}}\, C_\mathrm{P}^2 \tag{8.115}$$

[65-70]. The rate increases with progressing conversion up to a maximum and then decays toward zero as equilibrium is approached.

Reaction 8.114 is genuine autocatalysis because the product P acts as catalyst. It is a strange reaction, though, because the product could never come into being if it were not present at start in at least a trace quantity (or could arise in some other fashion). The simple, single-step mechanism does not correspond to any known reaction. Its mathematical behavior reflects the most typical symptoms of auto-catalysis, but is of little quantitative value.

The most common example of genuine autocatalysis in real chemistry is acid-catalyzed ester hydrolysis

$$RCOOR' + H_2O \longrightarrow RCOOH + R'OH$$

This is a two- or three-step catalytic reaction whose kinetics depends on the type of ester [71-73]. The acid produced and hydrogen ion from its dissociations add to the amount of catalyst initially present, and there may or may not be a slow thermal reaction in parallel [66].

A different and more common type of self-acceleration results from promotion by an intermediate. Although habitually called autocatalysis, this is not genuine catalysis because the intermediate is consumed, whereas a catalyst is defined as a substance that accelerates the reaction, but re-emerges in its original state. The simplest example of this kind is the hypothetical reaction [74]

$$A \longrightarrow \text{product(s)} \qquad\qquad (8.116)$$

in which the intermediate, K, promotes further conversion of the reactant, A. (This can also be classified as competing steps, see Section 5.5.)

A real-life reaction involving promotion by an intermediate is autoxidation of hydrocarbons. Here, the principal reaction is

$$RH + O_2 \longrightarrow ROOH \longrightarrow \text{products}$$

Formation of the hydroperoxide is a chain reaction requiring free radicals as chain carriers. These are formed by an initiation mechanism, but additional ones may arise from decomposition of the hydroperoxide, and, if so, accelerate conversion. Hydrocarbon autoxidation will be examined in more detail in the next Chapter (see Example 9.3 in Section 9.6.2).

A situation of this general kind—a complex mechanism involving transient species that accelerate the main reaction—often is behind what goes as autocatalytic behavior. "Autocatalysis" as commonly understood, namely, reactions whose rates increase with progressing conversion at least within a certain range, is a broader term than its literal meaning implies. However, like many others in our technical vocabulary, this usage has become too firmly ingrained for change.

The variety of reactions in which a product or major intermediate boosts the rate, possibly while itself decomposing, is great, and each of them has its own mathematics and quirks. It seems impossible to develop universal equations that cover such behavior in a quantitative fashion, and no attempt to do so is made here.

8.9.2. *Reactant-inhibited reactions*

Rather than because of promotion by a product or intermediate, the rate may accelerate because a reactant that acts as inhibitor is consumed. For example, a small amount of inhibitor present initially may depress the rate of a chain reaction until used up (see Section 9.7). More interesting are reactions inhibited by one of the principal reactants (called *substrate-inhibited* in biochemistry parlance). An example is hydroformylation, in which CO is a reactant with negative reaction order (see Example 6.2 in Section 6.3). There is a subtle but important difference between product-promoted and reactant-inhibited reactions: The rate of a product-promoted reaction builds up to a maximum and then declines as the effect of reactant depletion overpowers that of promotion by the product. In contrast, the rate of a reactant-inhibited reaction may keep increasing, theoretically until the inhibiting reactant is completely consumed. If the negative apparent order arises from an additive denominator term in a one-plus rate equation, the rate cannot exceed a finite limit, but if the true order is negative and constant, the rate theoretically approaches infinity as the reactant is used up. This, of course, is physically impossible, and some other mechanism, event, or limitation kicks in. Possible mass-transfer implications of such behavior will be examined in Example 12.2 in Section 12.3. Not surprisingly, the chance of instability or other unusual reactor behavior is greater in reactant-inhibited than in product-promoted reactions.

8.10. Analogies to heterogeneous catalysis

Kinetics of heterogeneous catalysis has received much attention. Its mathematical theory is well advanced, largely thanks to extensive work of Boudart, Temkin, and others. On the other hand, heterogeneous catalysis has to deal not only with the same kind of difficulties homogeneous catalysis faces, but with the added complications of surface properties, adsorption/desorption equilibria and rates, and mass transfer to and from catalytic sites, phenomena whose effects often are more important than those of actual kinetics of the reaction on the surface.

In principle, the formulas and procedures for homogeneous catalysis can be applied to heterogeneous catalysis as well if extended to include adsorption and desorption. An unoccupied catalyst site in heterogeneous catalysis corresponds to a free-catalyst molecule in homogeneous catalysis; and reactant adsorption and product desorption correspond to addition of reactant to, and release of product by, the catalyst molecule, respectively. One possibility is to treat the reaction on the catalyst surface as homogeneous and combine it with adsorption and desorption kinetics or equilibria. A much more common approach is to view adsorption as a reversible, single-step "reaction" of the respective species with an unoccupied catalyst site, in order to obtain a formalism equivalent to that of homogeneous reactions (*Langmuir-Hinshelwood kinetics*). In both cases, numerical solutions are usually required unless further simplifications are introduced.

As this chapter has shown, rate equations of multistep homogeneous catalysis are still relatively simple if the catalyst-containing intermediates are at trace level, but the free catalyst is not. In heterogeneous catalysis this corresponds to an almost entirely unoccupied catalyst surface. Since adsorption is prerequisite for reaction, low surface coverage results in low rates and therefore is of practical interest only in exceptional situations. Heterogeneous catalysis cannot avoid dealing with substantially covered surface, and even in the best of cases these call for Christiansen mathematics with its profusion of denominator terms. Even Christiansen's formula becomes inapplicable if adsorbed species react with one another, thereby making the mechanism non-simple. A common, though crude procedure is to consider the surface reaction as occurring in a single step and appoint one step—surface reaction, or adsorption or desorption of one participant—as rate-controlling (*Hougen-Watson kinetics*).

Short of going that far, the tools that can be used for reduction of complexity are much the same in heterogeneous as in homogeneous catalysis (see Chapter 4 and Section 8.5). Specifically:

• The Bodenstein approximation of quasi-stationary behavior, required even by the Christiansen formula, is almost universally (if often tacitly) taken for granted in heterogeneous catalysis.

• The concept of rate control by a single step, with all other steps at quasi-equilibrium, is the norm in heterogeneous catalysis, as it is in almost all of the work on multistep homogeneous kinetics to date. Possible rate-controlling steps in heterogeneous catalysis include the attachment of a reactant to, or detachment of a product from, the catalyst surface rather than only chemical conversions.

• The idea of irreversible steps is also widely used in heterogeneous catalysis.

• The concept of a most abundant catalyst-containing species (*macs* in this book) was originally introduced by Boudart for heterogeneous catalysis under the name of *masi*, for "most abundant surface intermediate" [42].

Often, combinations of these approximations are invoked. The foremost example are the theorems formulated by Boudart [42], which have their counterparts in, or are corollaries to, rules stated in this book:

Theorem I: In a catalytic sequence of any number of irreversible steps, if the surface intermediate involved in the last step is the *masi*, there are only two kinetically significant steps, the first one and the last one.

In homogeneous catalysis this corresponds a special case of the combination of the concepts of a *macs* and an irreversible step (see Section 8.6).

Theorem II: In a catalytic sequence of steps, all steps that follow an irreversible step involving the *masi* as reactant are kinetically not significant.

In homogeneous catalysis this correspond to the rule that steps following an irreversible one, up to and including the step forming the *macs*, have no effect on the rate (Rule 8.56).

Theorem III: All equilibrated steps following a step involving the *masi* as a product can be combined in a single overall equilibrium that regulates the concentration of the *masi*. Vice versa, all equilibrated steps preceding a rate-controlling step involving the *masi* as reactant can be combined in a similar overall equilibrium.

This corresponds to the rules for rate-controlling steps in pathways with reversible steps (Sections 4.1.2 and 8.6): If the forward and reverse rate coefficients of a step in the pathway of a reversible reaction are much smaller than all others, the other steps are practically in equilibrium.

In essence, the procedures for reduction of complexity described in this book are applicable in principle to reactions on catalyst surfaces, provided the latter are uniform and no segregation of adsorbed species occurs. However, in view of the wealth of other complicating factors, the effort may well be beyond a point of diminishing returns unless a very simple rate equation results. While the strictly kinetic problems are largely analogous in both fields, the much greater complexity of the peripheral conditions in heterogeneous catalysis leaves less room for inclusion of finer reaction-kinetic detail.

Summary

Homogeneous catalysis can be classified into single-species and complex catalysis, although the distinction is not always clear-cut. In the former, a single molecule or ion acts as the catalyst; in the latter, the catalyst is a system of several species that interconvert into one another and differ in their catalytic properties. A further complication arises if significant fractions of the total catalyst material may be present in the form of reaction intermediates rather than free catalyst. If so, the concentration of the free catalyst is not known and may vary with conversion, and rate equations that instead contain the known, total amount of catalyst material are needed.

In single-species catalysis, the rate laws for noncatalytic reactions apply, the only difference being that the catalyst appears as both a reactant and a product. In catalysis by highly concentrated acids, anomalies may appear: Protonated species other than H_3O^+ (or protonated solvent in non-aqueous media) may arise and act as additional catalysts; this can be accounted for with the Hammett acidity function. Also, the rates of reactions such as hydration and hydrolysis may decrease with further increase in acid concentration because of reduced availability of free water as reactant.

In acid-base catalysis, both an acid (or base) and its conjugate base (or acid) take part in different reaction steps and are eventually restored. Such reactions are first order in acid (or base) if the link-up with that species controls the rate, or first order in H^+ (or OH^-) if a subsequent step involving the conjugate base (or acid) does so. Traditionally, the first alternative is called "general" acid or base catalysis; the second, "specific" acid or base catalysis. However, this distinction is not always applicable as there may be no clear-cut rate-controlling step, and reversibility of later steps may produce a more complex behavior.

Many reactions of organic chemistry are catalyzed by complexes of transition-metal ions, most notably those of Group VIII. Here, the catalyst is a system of different complexes that are linked by ligand-exchange and ligand-dissociation equilibria and differ in their catalytic properties, occasionally with rather counterintuitive results such as a decrease in rate with increase in pressure in a reaction in which gas is consumed.

For historical interest and to illustrate a general facet of systems with arbitrary distribution of catalyst material over free catalyst and reaction intermediates, the classical models of enzyme catalysis are briefly reviewed. They show "saturation kinetics:" An increase in reactant concentration causes a shift of catalyst material from free catalyst to an intermediate, so that the rate has an asymptotic limit that can at most be approached even at the highest reactant concentrations.

A general formula for single catalytic cycles with arbitrary number of members and arbitrary distribution of catalyst material has been derived by Christiansen. Unfortunately, the denominator of his rate equation for a cycle with k members contains k^2 additive terms. Such a profusion makes it imperative to reduce complexity. If warranted, this can be done with the concept of relative abundance of catalyst-containing species or the approximations of a rate-controlling step, quasi-equilibrium steps, or irreversible steps, or combinations of these (the Bodenstein approximation of quasi-stationary states is already implicit in Christiansen's mathematics). In some fortunate instances, the rate equation reduces to a simple power law.

If significant fractions of the catalyst material may be bound in the form of reaction intermediates, the rules for reaction orders in noncatalytic simple pathways no longer apply. However, if one of the cycle members—the free catalyst or an intermediate—is a *macs* (most abundant catalyst-containing species, containing practically all of the catalyst material), the rules for noncatalytic pathways can be adapted: The rate equation and reaction orders for the cycle are the same as for an imaginary equivalent linear pathway that starts and ends with *macs*. A cycle member that contains only an insignificant fraction of catalyst material is a *lacs* (low-abundance catalyst containing species), and the denominator terms it contributes can be dropped.

In practice, catalytic networks often involve more than a single cycle. In a very common type of network, a linear pathway is attached to the cycle. This is so, for instance, in reactions with ligand-deficient catalysts. Here, the coordinatively saturated "catalyst" has no activity but, rather, must first lose one of its ligands to provide access for a reactant. Other examples of cycles with attached pathways include systems with inhibition, activation, decay, and poisoning. Also, the network may consist of two or more cycles with a common member or pathway. This situation is typical for reactions yielding different isomeric products. The Christiansen formula is extended to cover such cases.

In some reactions, the rate increases rather than decreases as conversion progresses. This is loosely called autocatalysis, although no genuine catalysis may be involved. The acceleration may stem from promotion by a product or major early intermediate, or from consumption of a reactant that functions as inhibitor. In product-promoted reactions, the kinetics order with respect to a product (or early intermediate) is positive. This causes the rate to increase to a maximum and then to decline as the effect of consumption of the reactant or reactants begins to overcompensate that of promotion by the product. In reactant-inhibited reactions, the order with respect to a reactant is negative. The rate may increase until the respective reactant is used up and, in some cases, may theoretically approach infinity. The latter behavior is, of course, physically impossible, and another mechanism or event necessarily takes over.

A number of the concepts used have parallels in heterogeneous catalysis. For context, the analogies are briefly reviewed.

Examples include acetal hydrolysis, base-catalyzed aldol condensation, olefin hydroformylation catalyzed by phosphine-substituted cobalt hydrocarbonyls, phosphate transfer in biological systems, enzymatic transamination, adiponitrile synthesis via hydrocyanation, olefin hydrogenation with Wilkinson's catalyst, and osmium tetroxide-catalyzed asymmetric dihydroxylation of olefins.

References

General references

G1.　J. P. Collman, L. S. Hegedus, J. R. Norton, and R. G. Finke, *Principles and applications of organotransition metal chemistry*, University Science Books, Mill Valley, 2nd ed., 1987, ISBN 0935702512.

G2.　K. A. Connors, *Chemical kinetics; the study of reaction rates in solution*, VCH Publishers, New York, 1990, ISBN 3527218223.

G3.　J. Falbe, *Carbon monoxide in organic synthesis*, Springer, Berlin, 1970.

G4.　A. A. Frost and R. G. Pearson, *Kinetics and mechanism: a study of homogeneous chemical reactions*, Wiley, New York, 2nd. ed., 1961.

G5.　B. C. Gates, *Catalytic chemistry*, Wiley, New York, 1992, ISBN 0471517615.

G6.　L. P. Hammett, *Physical organic chemistry: reaction rates, equilibria, and mechanisms*, McGraw-Hill, New York, 2nd ed., 1970.

G7.　C. K. Ingold, *Structure and mechanism in organic chemistry: reaction rates, equilibria, and mechanisms*, Cornell University Press, Ithaca, 2nd ed., 1969, ISBN 0801404991.

G8.　O. Levenspiel, *The chemical reactor omnibook*, O.S.U. Book Stores, Corvallis, 1989, ISBN 0882461648.

G9.　J. W. Moore and R. G. Pearson, *Kinetics and mechanism: a study of homogeneous chemical reactions*, Wiley, New York, 3rd ed., 1981, ISBN 0471035580.

G10.　G. W. Parshall and J. D. Ittel, *Homogeneous catalysis: the application and chemistry of catalysis by soluble transition metal complexes*, Wiley, New York, 1992, ISBN 0471538299.

G11.　I. H. Segel, *Enzyme kinetics: behavior and analysis of rapid-equilibrium and steady-state enzyme systems*, Wiley, New York, 1975, ISBN 0471774251.

G12.　J. I. Steinfeld, J. S. Francisco, and W. L. Hase, *Chemical kinetics and dynamics*, Prentice-Hall, Englewood Cliffs, 2nd ed., 1999, ISBN 0137371233.

Specific references

1.　L. Wilhelmy, *Ann. Phys. Chem. (Poggendorf)*, **81** (1850) 413 and 499.

2.　Moore and Pearson (ref. G9), pp.338-340.

3.　Gates (ref. G5), Section 2.2.7.

4.　F. A. Long and M. A. Paul, *Chem. Rev.*, **57** (1957) 935.

5.　A. M. Wenthe and E. H. Cordes, *J. Am. Chem. Soc.*, **87** (1965) 3173.

6.　J. T. Edward and S. C. R. Meacock, *J. Chem. Soc.*, **1957**, 2000.

7.　L. P. Hammett and A. J. Deyrup, *J. Am. Chem. Soc.*, **54** (1932) 2721.

8.　G. A. Olah, *Friedel-Crafts chemistry*, Wiley, New York, 1973, ISBN 0471653152, p. 368.

9.   K. B. Wiberg, *Physical organic chemistry*, Wiley, New York, 1964, Section 3-8.

10.   Hammett (ref. G6), p. 267.

11.   Connors (ref. G2), Chapter 7.

12.   R. P. Bell, *J. Chem. Soc.*, **1937**, 1637.

13.   Ingold (ref. G7), pp. 999-1002.

14.   Frost and Pearson (ref. G4), Section 12.D.

15.   F. G. Helfferich and P. E. Savage, *Reaction kinetics for the practical engineer*, Course #195, AIChE Educational Services, New York, 7th ed., 1999, Section 7.4.

16.   J. L. Van Winkle, personal communication, 1967.

17.   L. H. Slaugh and R. D. Mullineaux, *J. Organomet. Chem.*, **13** (1968) 469.

18.   Falbe (ref. G3), Section I.2.

19.   E. Billig and D. R. Bryant, *Oxo reaction*, in *Kirk-Othmer, Encyclopedia of chemical technology*, 4th ed., J. I. Kroschwitz and M. Howe-Grant, eds., Wiley, New York, Vol. 17, ISBN 047152686X, 1996, p. 902.

20.   H. J. Baumgartner, personal communication, 1967.

21.   W. Rupilius, J. J. McCoy, and M. Orchin, *I&EC Prod. Res. Dev.*, **10** (1971) 142.

22.   Collman *et al.* (ref. G1), p. 91.

23.   P. Szabó, L. Fekete, G. Bor, Z. Nagy-Magos, and L. Markó, *J. Organomet. Chem.*, **12** (1968) 245.

24.   W. W. Spooncer, A. C. Jones, and L. H. Slaugh, *J. Organomet. Chen.*, **18** (1969) 327.

25.   C. A. McAuliffe and W. Levason, *phosphine, arsine, and stibine complexes of transition metals*, Vol. 1 of *Studies in inorganic chemistry*, Elsevier, Amsterdam, 1979, ISBN 0444417494, Sections III.8.1 and V.6.1.

26.   R. V. Kastrup, J. S. Merola, and A. A. Oswald, *Adv. Chem. Ser.*, **196** (1981) 43.

27.   L. Michaelis and M. L. Menten, *Biochem. Z.*, **49** (1913) 333.

28.   Segel (ref. G11), Chapter 2.

29.   V. Henri, *Lois générales de l'action des diastases*, Hermann, Paris, 1903.

30.   G. E. Briggs and J. B. S. Haldane, *Biochem. J.*, **19** (1925) 338.

31.   H. Lineweaver and D. Burk, *J. Am. Chem. Soc.*, **56** (1934) 658.

32.   C. S. Hanes, *Biochem. J.*, **26** (1932) 1406.

33.   G. S. Eadie, *J. Biol. Chem.*, **146** (1942) 85.

34.   B. H. J. Hofstee, *Nature*, **184** (1959) 1296.

35.   G. G. Hammes and D. Kochavi, *J. Am. Chem. Soc.*, **84** (1962) 2064, 2073, and 2076.

36.   J. A. Christiansen, *Z. physik. Chem.*, Bodenstein-Festband (1931) 69.

37.   J. A. Christiansen, *Z. physik. Chem.*, **B 28** (1935) 303.

38.   J. A. Christiansen, *Adv. Catal.*, **5** (1953) 311.

39.   E. L.. King and C. Altman, *J. Phys. Chem.*, **60** (1956) 1375.

40.   Segel (ref. G11), Section 9.I.

41. P. L. Brezonik, *Chemical kinetics and process dynamics in aquatic systems*, CRC Press, Boca Raton, 1994, ISBN 0873714318, Section 6.2.2.

42. M. Boudart, *AIChE J.*, **18** (1972) 465.

43. J. D. Druliner, *Organometallics*, **3** (1984) 205.

44. C. A.. Tolman, R. J. McKinney, W. C. Seidel, J. L. Druliner, and W. R.. Stevens, *Adv. Catal.*, **33** (1985) 1.

45. Parshall and Ittel (ref. G10), Section 3.6.

46. C. A. Tolman and J. P. Jesson, *Science*, **181** (1973) 501.

47. Helfferich and Savage (ref. 15), Section 6.8.

48. F. H. Jardine, J. A. Osborn, G. Wilkinson, and J. F. Young, *Chem. & Ind.*, **1965**, 560.

49. J. A. Osborn, F. H. Jardine, J. F. Young, and G. Wilkinson, *J. Chem.. Soc.*, **A 1966**, 1711.

50. C. O'Connor and G. Wilkinson, *Tetrahedron Lett.*, **1969**(18), 1375.

51. Collman *et al.* (ref. G1), pp. 530-535.

52. Gates (ref. G5), Section 2.4.2.

53. J. Halpern and C. S. Wong, *J. Chem. Soc., Chem. Commun.*, **1973**, 629.

54. J. Halpern, T. Okamoto, and A. Zakhariev, *J. Mol. Cat.*, **2** (1976) 65.

55. J. Halpern, *Trans. Am. Cristallogr. Assoc.*, **14** (1978) 59.

56. C. A. Tolman, P. Z. Meakin, D. L. Lindner, and J. P. Jesson, *J. Am. Chem. Soc.*, **96** (1974) 2762.

57. C. Daniel, N. Koga, J. Han, X. Y. Fu, and K. Morukuma, *J. Am. Chem. Soc.*, **110** (1988) 3773.

58. Collman *et al.* (ref. G1), p. 534.

59. N. A. Muhammadi and G. L. Rempel, *Macromolecules*, **20** (1987) 2362; *J. Mol. Catal.*, **50** (1989) 259.

60. e.g., see Gates (ref. G5), p. 82.

61. Segel (ref. G11), Chapters 3 and 4.

62. Segel (ref. G11), Chapter 5.

63. J. S. M. Wai, I. Markó, J. S. Svendsen, M. G. Finn, E. N. Jacobsen, and K. B. Sharpless, *J. Am. Chem. Soc.*, **111** (1989) 1123.

64. T. Göbel and K. B. Sharpless, *Angew Chem. Internat. Ed. English*, **32** (1993) 1329.

65. C. G. Hill, Jr., *An introduction to chemical engineering kinetics & reactor design*, Wiley, New York, 1977, ISBN 0471396095, Problem 8.26.

66. Moore and Pearson (ref. G9), p. 26.

67. Steinfeld *et al.* (ref. G12), Section 5.3.

68. G. G. Froment and K. B. Bischoff, *Chemical reactor analysis and design,* Wiley, New York, 2nd. ed., 1990, ISBN 0471510440, Example 1.3.1.

69. L. D. Schmidt, *The engineering of chemical reactions*, Oxford University Press, 1998, ISBN 0195105885, pp. 112-114.

70.   R. W. Missen, C. A. Mims, and B. A. Saville, *Introduction to chemical reaction engineering and kinetics*, Wiley, New York, 1999, ISBN 0471163392, pp. 189-191.

71.   S. C. Datta, J. N. E. Day, and C. K. Ingold, *J. Chem. Soc.*, **1939**, 838.

72.   J. N. E. Day and C. K. Ingold, *Trans. Faraday Soc.*, **37** (1941) 686.

73.   Frost and Pearson (ref. G4), pp. 316-320.

74.   M. Frenklach and D. Clary, *I&EC Fundamentals*, **22** (1983) 433.

Chapter 9

Chain Reactions

Chain reactions rely on the existence of highly reactive intermediates, so-called chain carriers, that take part in a repeating sequence of steps in which they are consumed and produced anew. A typical sequence of this kind, with two chain carriers X and Y converting reactants A and B to products P and Q, is

$$
\begin{aligned}
X + A &\longrightarrow P + Y \\
Y + B &\longrightarrow Q + X
\end{aligned}
\tag{9.1}
$$

Chain carrier X, consumed in the first step, re-appears as a product in the second. Likewise, Y, consumed in the second step, re-appears as a product in the first. If you will (and belong to those who still learned FORTRAN in school), a chain reaction is a "reaction with a DO loop."

Like a DO loop in FORTRAN, the chain must be started and eventually be terminated. It is started by an additional reaction that serves as source of chain carriers. This is called *initiation*. It is terminated by a reaction that, once in a while, consumes chain carriers without generating others. This is called *termination*. The steps of the DO loop are called *propagation*. Typically, chain carriers are species (atoms or free radicals) with unpaired electrons. They resemble catalysts in that they arise again after having been consumed, but differ in having extremely short life spans that prevent their isolation. Moreover, the initiation and termination steps and their kinetic implications set chain reactions apart from catalysis as well as from any other kinds of chemical reactions.

9.1. General Properties

Once a chain has been started, the propagation cycle typically runs through many repetitions before being terminated. This makes chain reactions very sensitive: A minute amount of an initiator that generates chain carriers can set off a massive and fast reaction.

Chain reactions are also quite sensitive to any substances that act as free-radical traps. The presence of a very small amount of such a substance can effectively cause a chain reaction to "flame out."

Another typical feature occasionally encountered in chain reactions is a time delay upon initiation: When the initiator is added to the reactants, the reaction may start only after an *induction period* rather than immediately. This happens, for example, if a trace contaminant acting as inhibitor is present: The contaminant catches and consumes the chain carriers produced by the initiator until it has been destroyed in the process.

Perhaps the most striking feature of chain reactions is that some of them can result in detonation. If the propagation mechanism includes a step or steps that produce more chain carriers than they consume, the reaction is self-accelerating. This is called *chain branching*. The result may be an event much more violent than a thermal explosion, in which self-acceleration stems from temperature increase owing to the inability of heat transfer to keep up with the heat production of the reaction. The detonation of a nuclear bomb can be viewed as a chain reaction with neutrons as carriers and with chain branching.

As highly reactive species with very short life spans, the chain carriers remain at trace level (except in a detonation). The Bodenstein approximation of quasi-stationary behavior thus is applicable to them. In fact, it was first introduced by Bodenstein and his school in their study and mathematical analysis of chain reactions [1]. Its validity for chain carriers will be taken for granted throughout this chapter.

9.2. Initiation

Some chain reactions start spontaneously, without prodding by an added initiator because a reactant does that job by itself. The hydrogen-bromide reaction, in which thermal dissociation of Br_2 initiates the reaction at sufficiently high temperatures, is an example (see Example 9.1 in next section). In other chain reactions, an initiator need not be added because it forms spontaneously from some intermediate or product. Such a reaction may be quite unintended. For example, upon prolonged standing in contact with air, hydrocarbons accumulate peroxides. The decomposition of the latter into free radicals can be set off by trace contaminants (e.g., transition-metal ions, see below) or even the friction of a ground-glass stopper being turned as a flask is opened or closed. This can trigger an autocatalytic reaction that leads to an explosion.

Most chain reactions have to be jump-started by an intentionally added initiator. For liquid-phase reactions, the most commonly used initiators are benzoyl peroxide and 2,2'-azo-*bis*-isobutyronitrile (AIBN), unstable compounds that readily decompose into free radicals:

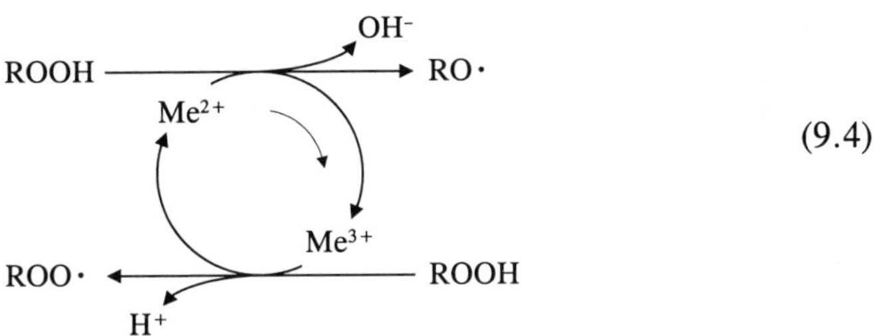

(dots indicate unpaired electrons). The initiator molecule produces two free radicals, $R\cdot$, that can act as chain carriers (and possibly an additional inert species such as N_2):

$$\text{in} \longrightarrow 2\,R\cdot + \ldots \tag{9.2}$$

(in = initiator). The rate of chain-carrier production accordingly is

$$r_{\text{init}} = 2k_{\text{init}} C_{\text{in}} \tag{9.3}$$

At low temperature, decomposition of a hydroperoxide initiator may require promotion. Transition-metal ions such as Fe^{2+}, Co^{2+}, Ag^+, etc., that can easily switch valence states can serve this purpose. A likely mechanism of such promotion by an ion Me^{2+} is the Haber-Weiss redox cycle [2]:

$$\tag{9.4}$$

Some chain reactions can be initiated photochemically [3]. In fact, most of the early work on kinetics of chain reactions was done with photochemical initiation. For example, in the hydrogen-bromide reaction (see next section), initiation $Br_2 \longrightarrow 2\,Br$ can be achieved with ultraviolet light. Such initiation allows the reaction to be conducted at a lower temperature at which thermal initiation is ineffective. This may be an advantage in an industrial process, and also offers opportunities for elucidation of reaction mechanisms.

Lastly, some gas-phase chain reactions, among them the hydrogen-oxygen reaction, can be triggered by an electric discharge.

9.3. Reactions with two chain carriers: the hydrogen-bromide reaction

The simplest chain reactions are those with two alternating chain carriers as in the sequence 9.1, one or both generated by initiation, and with termination by binary *coupling* (also called *recombination*) of chain carriers. The propagation steps may or may not be reversible. In general terms the network can be written:

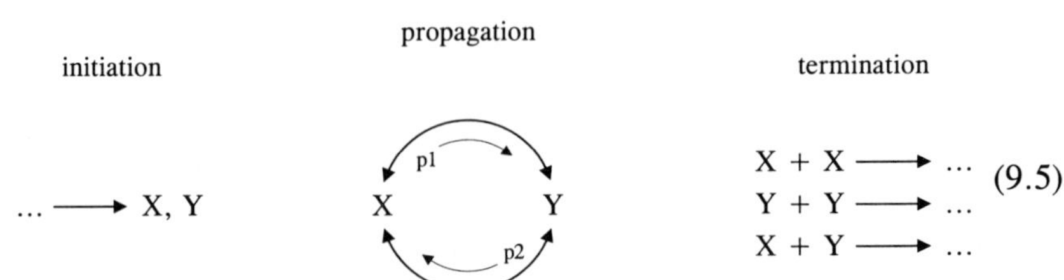

$$(9.5)$$

(reactants and products other than chain carrier not shown, and to be accounted for in the respective λ coefficients; see Section 6.2).

In such a network, the rates of the two propagation steps are

$$r_{p1} = \lambda_{p1} C_X - \lambda_{-p1} C_Y$$
$$r_{p2} = \lambda_{p2} C_Y - \lambda_{-p2} C_X \tag{9.6}$$

(indices p1 and p2 are used to avoid ambiguity; minus signs in indices indicate reverse steps). The Bodenstein approximation for X gives

$$r_X = (r_X)_{\text{init}} + C_Y(\lambda_{-p1} + \lambda_{p2}) - C_X(\lambda_{p1} + \lambda_{-p2}) + (r_X)_{\text{trm}} \cong 0 \tag{9.7}$$

Here, $(r_X)_{\text{init}}$ and $(r_X)_{\text{trm}}$ are the rates of production and elimination of X by initiation and termination, respectively (the termination rate is negative).

Characteristic of most chain reactions is that the propagation cycle is repeated many times between initiation and termination. If so, the initiation and termination terms (first and last in eqn 9.7) are small compared with the propagation terms. The so-called *long-chain approximation* [4] disregards them.   Equation 9.7 can then be used to express the concentration of one chain carrier as a function of that of the other:

$$C_Y \cong C_X \left[\frac{\lambda_{p1} + \lambda_{-p2}}{\lambda_{-p1} + \lambda_{p2}} \right] \tag{9.8}$$

With this substitution, eqns 9.6 yield

$$r_{p1} \cong r_{p2} \cong C_X \left[\frac{\lambda_{p1}\lambda_{p2} - \lambda_{-p1}\lambda_{-p2}}{\lambda_{-p1} + \lambda_{p2}} \right] \tag{9.9}$$

Equation 9.9 involves the still unknown chain-carrier concentration C_X. This concentration, and with it the final form of the rate equation, depends on the type of termination. In principle, there are three potential mechanisms of termination by coupling: $X + X \longrightarrow ...$, $Y + Y \longrightarrow ...$, and $X + Y \longrightarrow ...$. In reactions of two free radicals with one another to yield an inert product, no bonds need to be broken as the odd electrons link up, and the great majority of collisions are successful (possible exceptions will be discussed later). The activation energies then are almost nil and the rate coefficients are very similar. The relative rates of the three possible terminations, proportional to C_X^2, C_Y^2, and $C_X C_Y$, thus depend very strongly on the relative concentrations of the chain carriers: If X outnumbers Y by, say, a factor 100, then the rate of X + X termination is roughly 100 times as high as that of X + Y, and $(100)^2 = 10,000$ times as high as that of Y + Y.

> Termination by coupling tends to be dominated
> by the most plentiful chain carrier.

If the concentrations of X and Y are comparable, all three mechanisms contribute. In general, the overall termination rate $r_{trm} \equiv (r_X)_{trm} + (r_Y)_{trm}$ (consumption of all chain carriers) thus is seen to be:

	termination	termination rate
if $C_X \gg C_Y$	X + X	$-2k_{cX}C_X^2$
if $C_X \ll C_Y$	Y + Y	$-2k_{cY}C_Y^2$
if $C_X \approx C_Y$	X + X, X + Y, Y + Y	$-2(k_{cX}C_X^2 + k_{cY}C_Y^2 + k_{cXY}C_XC_Y)$

$$(9.10)$$

where k_{cX}, k_{cY}, and k_{cXY} are the respective rate coefficients (c stands for coupling; the rates are negative because they refer to consumption of chain carriers, and the factor 2 reflects the fact that two carriers are consumed.)

The listings 9.10 presume that the reactions of chain carriers with one another proceed unhindered to stable products rather than being blocked or producing substances that in short order decompose again into chain carriers. This is not always the case. Larger radicals may react more sluggishly. Others, among them benzyl, are stabilized to some extent by resonance. Also, for example, two oxy radicals, RO·, react with one another to form a labile peroxide, ROOR, that soon reverts to the two radicals. In cases like the latter:

> If the most plentiful radical cannot terminate the
> chain by reaction with itself, it will do so by reaction
> with whichever radical is the next most plentiful.

In reactions with only two radicals, X and Y:

$$\text{termination:}\quad X + Y \qquad\qquad \text{rate:}\quad -2k_{cXY}\,C_X C_Y \qquad (9.11)$$

This is the only situation in which "mixed" termination (i.e., coupling of two different chain carriers) can dominate termination.

Equation 9.9, derived with the long-chain approximation, shows the rates of the two propagation steps under quasi-stationary conditions to be equal: Each of the two steps consumes as many chain carriers as the other produces. Since the total number of chain carriers is not altered by the propagation steps, the overall rates of chain carrier production and consumption (of X plus Y) must be equal in magnitude:

$$r_{init} + r_{trm} \;\cong\; 0 \qquad (9.12)$$

where $r_{init} \equiv (r_X)_{init} + (r_Y)_{init}$ is the overall initiation rate (production of all chain carriers). The equality applies to the total chain-carrier population, but by no means necessarily to each chain carrier separately. For instance, in principle, initiation could produce X while termination eliminates Y (granted long chains, the initiation and termination terms in eqn 9.7 are both small compared with the two propagation terms, but are not necessarily equal to one another).

Equation 9.12 allows C_X to be calculated from the initiation rate given by eqn 9.3, the appropriate termination rate given by eqn 9.10 or 9.11, and elimination of C_Y with eqn 9.8 if necessary:

termination   concentration of chain carrier X

$$X + X \qquad\qquad C_X \cong (k_{init}/k_{cX})^{1/2}\,C_{in}^{1/2}$$

$$Y + Y \qquad\qquad C_X \cong \left[\frac{k_{init}}{k_{cY}}\right]^{1/2}\left[\frac{\lambda_{-p1} + \lambda_{p2}}{\lambda_{p1} + \lambda_{-p2}}\right] C_{in}^{1/2}$$

$$X + Y \qquad\qquad C_X \cong \left[\frac{k_{init}}{k_{cXY}}\right]^{1/2}\left[\frac{\lambda_{-p1} + \lambda_{p2}}{\lambda_{p1} + \lambda_{-p2}}\right]^{1/2} C_{in}^{1/2} \qquad (9.13)$$

$$\text{all three}\qquad C_X \cong \left[\frac{k_{init}\,C_{in}}{k_{cX} + k_{cY}\left[\dfrac{\lambda_{p1} + \lambda_{-p2}}{\lambda_{-p1} + \lambda_{p2}}\right]^{2} + k_{cXY}\left[\dfrac{\lambda_{p1} + \lambda_{-p2}}{\lambda_{-p1} + \lambda_{p2}}\right]}\right]^{1/2}$$

The rates $r_p = r_{p1} = r_{p2}$ of the propagation steps can now be calculated from eqn 9.9 with the appropriate substitution for C_X from eqns 9.13:

termination

X + X

$$r_p \cong \left[\frac{k_{init}}{k_{cX}}\right]^{1/2} \left[\frac{\lambda_{p1}\lambda_{p2} - \lambda_{-p1}\lambda_{-p2}}{\lambda_{-p1} + \lambda_{p2}}\right] C_{in}^{1/2} \tag{9.14}$$

Y + Y

$$r_p \cong \left[\frac{k_{init}}{k_{cY}}\right]^{1/2} \left[\frac{\lambda_{p1}\lambda_{p2} - \lambda_{-p1}\lambda_{-p2}}{\lambda_{p1} + \lambda_{-p2}}\right] C_{in}^{1/2} \tag{9.15}$$

X + Y

$$r_p \cong \left[\frac{k_{init}}{k_{cXY}}\right]^{1/2} \left[\frac{\lambda_{p1}\lambda_{p2} - \lambda_{-p1}\lambda_{-p2}}{\left[(\lambda_{-p1} + \lambda_{p2})(\lambda_{p1} + \lambda_{-p2})\right]^{1/2}}\right] C_{in}^{1/2} \tag{9.16}$$

all three

$$r_p \cong \frac{k_{init}^{1/2}(\lambda_{p1}\lambda_{p2} - \lambda_{-p1}\lambda_{-p2}) C_{in}^{1/2}}{\left(k_{cX}(\lambda_{-p1} + \lambda_{p2})^2 + k_{cY}(\lambda_{p1} + \lambda_{-p2})^2 + k_{cXY}(\lambda_{-p1} + \lambda_{p2})(\lambda_{p1} + \lambda_{-p2})\right)^{1/2}} \tag{9.17}$$

Equation 9.17, accounting for all three termination mechanisms, is the most general rate equation and contains the other three as special, but much more common cases.

If both propagation steps are irreversible, eqns 9.14 to 9.17 reduce to:

termination

X + X

$$r_p \cong (k_{init}/k_{cX})^{1/2} \lambda_{p1} C_{in}^{1/2} \tag{9.18}$$

Y + Y

$$r_p \cong (k_{init}/k_{cY})^{1/2} \lambda_{p2} C_{in}^{1/2} \tag{9.19}$$

X + Y

$$r_p \cong (k_{init}/k_{cXY})^{1/2} (\lambda_{p1}\lambda_{p2})^{1/2} C_{in}^{1/2} \tag{9.20}$$

all three

$$r_p \cong \frac{k_{init}^{1/2} \lambda_{p1} \lambda_{p2} C_{in}^{1/2}}{\left(k_{cX}\lambda_{p2}^2 + k_{cY}\lambda_{p1}^2 + k_{cXY}\lambda_{p1}\lambda_{p2}\right)^{1/2}} \tag{9.21}$$

(X is the chain carrier that acts as reactant in the first propagation step.) For reactions wity irreversible propagation steps:

- *the rate is controlled by the propagation step that consumes the terminating chain carrier* (first propagation step if termination is X + X; second, if it is Y + Y).

The rates of product formation (and reactant consumption) are seen to be of order one half in the initiator; or, if the reaction is initiated by a reactant converted in the propagation cycle, the rate equation involves exponents of one half or integer multiples of one half. For an example, see the hydrogen-bromide reaction below. This is one of the exceptions to the rule that reasonably simple mechanisms do not yield rate equations with fractional exponents. [The other exceptions are reactions with fast pre-dissociation (see Section 5.6) and of heterogeneous catalysis with a reactant that dissociates upon adsorption.]

Example 9.1. The hydrogen-bromide reaction. The elucidation of the mechanism of the gas-phase reaction

$$H_2 + Br_2 \longrightarrow 2\,HBr \tag{9.22}$$

by Bodenstein [3,5], Christiansen [6], Herzfeld [7], and Polányi [8] in 1907-1920 has been one of the most remarkable achievements of classical chemical kinetics. It led to the discovery of chain mechanisms and the introduction of the approximation of quasi-stationary states of trace intermediates by Bodenstein [1].

In the formalism used here, the reaction is

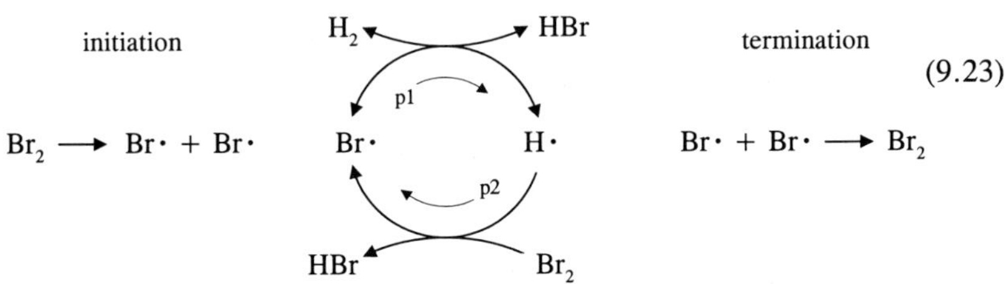

Note that the step $Br\cdot + H_2 \longrightarrow HBr + H\cdot$ is reversible, but the step $H\cdot + Br_2 \longrightarrow HBr + Br\cdot$ is not (see next section).

Since termination consumes the chain carrier that acts as reactant in the first propagation step, eqn 9.14 applies. The rates of initiation and HBr production are

$$r_{\mathrm{init}} = 2k_{\mathrm{init}}p_{Br_2}, \qquad r_{\mathrm{HBr}} = r_{p1} + r_{p2} = 2r_p$$

(because this is a gas-phase reaction, partial pressures are used in lieu of concentrations). With these rates and the substitutions

$$X = Br\cdot, \quad Y = H\cdot, \quad \lambda_{p1} = k_{p1}p_{H_2}, \quad \lambda_{p2} = k_{p2}p_{Br_2}, \quad \lambda_{-p1} = k_{-p1}p_{HBr}, \quad \lambda_{-p2} = 0$$

equation 9.14 gives

$$r_{\mathrm{HBr}} = \frac{2\left((k_{\mathrm{init}}/k_{cBr})p_{Br_2}\right)^{1/2}k_{p1}k_{p2}p_{H_2}p_{Br_2}}{k_{-p1}p_{HBr} + k_{p2}p_{Br_2}} = \frac{k_a p_{H_2}p_{Br_2}^{1/2}}{1 + k_b p_{HBr}/p_{Br_2}} \tag{9.24}$$

where

$$k_a \equiv 2K_d^{1/2}k_{p1}, \qquad k_b \equiv k_{-p1}/k_{p2}$$

Here, $K_d = k_{\mathrm{init}}/k_{cBr}$ is the dissociation constant of Br_2. A minor complication in the initiation and termination steps will be dealt with in the next section.

9.4. Identification of relevant steps

Chain carriers, be they atoms or free radicals, react with almost any molecules they encounter. As a result, their presence can give rise to a bewildering multitude of reaction steps. Usually, most or all of them occur to some extent, yet the overall behavior of the reaction is dominated by only a few. A central problem in the analysis and interpretation of chain reactions, elucidation of their mechanisms, and establishment of their proper rate equations therefore is to identify which of all the potential steps are relevant. Of course, this problem also exists to some degree in other reactions, but it is aggravated by the high reactivity of the chain carriers. Fortunately, at least as long as the reacting species are small, that same high reactivity also opens the door to an approach not generally applicable to other reactions. The key is the estimate of rate-coefficient ratios and relative rates, as will be discussed in this section and illustrated with an application to the hydrogen-bromide reaction.

Ratios of rate coefficients. Unlike most other kinetic quantities, rate-coefficient ratios of steps involving free radicals as reactants can be estimated on the basis of thermochemical data. The estimates are crude, no better than giving approximate orders of magnitude, but this often suffices. However, the estimates are reliable only if the molecular species involved are indeed small. The basis of such estimates is as follows:

Bimolecular collisions of chain carriers (atoms or free radicals) with others or with small molecules have a very high probability of resulting in reaction, provided the collision partners carry energy at least equaling the activation energy. This is to say that the pre-exponential factor A in the Arrhenius equation 2.3 for the rate coefficient is of the same or a similar order of magnitude for all of them. As a result, the ratio of two rate coefficients k_1 and k_2 with activation energies E_{a1} and E_{a2} is mainly given by the ratio of the exponential factors:

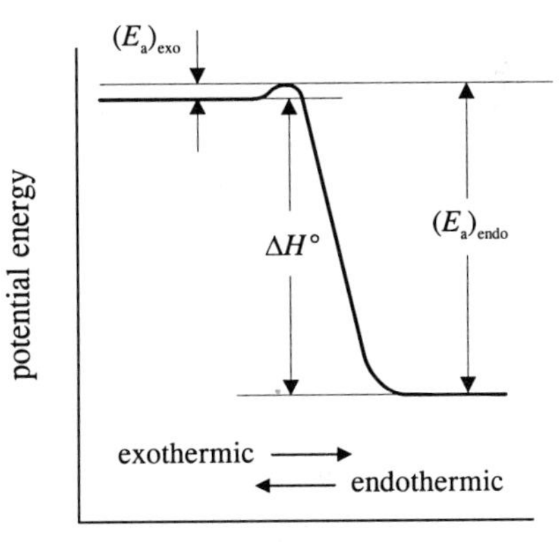

Figure 9.1. Activation energies and enthalpy of reactions of free radicals.

$$\frac{k_1}{k_2} \approx \frac{\exp(-E_{a1}/RT)}{\exp(-E_{a2}/RT)} = \exp\left[\frac{E_{a2} - E_{a1}}{RT}\right] \tag{9.25}$$

Also, the energy "hump" that such a reaction must overcome is only slightly higher than the energy level on its higher side (see Figure 9.1). As a result:

- *The activation energies of* exothermic *bimolecular reactions of free radicals are low* (usually only a few kJ mol^{-1}); experience shows them to be lowest for the most strongly exothermic reactions.

- *The activation energies of endothermic bimolecular reactions of free radicals are only marginally higher than the respective reaction enthalpies, $\Delta H°$; the difference is found smallest for the most strongly endothermic reactions.*

These regularities combined with the approximation 9.25 allow the following order-of-magnitude estimates of ratios k_1/k_2 of bimolecular rate coefficients of two reaction steps 1 and 2 with enthalpies $\Delta H_1°$ and $\Delta H_2°$ to be stated:

if $\Delta H_1° < \Delta H_2° < 0$ (both steps exothermic)	k_1/k_2 about 1 to 10 (more highly exothermic step faster)	(9.26)
if $\Delta H_1° < 0 < \Delta H_2°$ (step 1 exothermic, step 2 endothermic)	$k_1/k_2 \approx \exp(\Delta H_2°/RT)$ (exothermic step faster)	(9.27)
if $0 < \Delta H_1° < \Delta H_2°$ (both steps endothermic)	$k_1/k_2 \approx \exp[(\Delta H_2° - \Delta H_1°)/RT)]$ (less highly endothermic step faster)	(9.28)

In practice, the second and third estimates tend to be on the high side. Also, all of them become unreliable if the steps involve large species with many internal degrees of freedom, over which energy can be distributed.

The procedure underlying the approximations 9.25 to 9.28 is closely related to, but not identical with, those of Evans and Polányi [9] and Semenov [10] for reactions of homologous reactants [11,12]. The empirical Polányi equation for the activation energy is

$$E_a \approx \alpha\Delta E + C$$

where α (with $0 < \alpha < 1$) and C are constants, and ΔE is the difference in bond energies of the bonds formed and broken. ΔE is approximately equal to the standard reaction enthalpy, $-\Delta H°$, and for homologous reactants the values of α and C can be taken as very similar. With these assumptions:

$$E_{a1} - E_{a2} \approx \alpha(\Delta E_1 - \Delta E_2) \approx \alpha(\Delta H_2° - \Delta H_1°) \tag{9.29}$$

Semenov [10] has suggested $\alpha \approx 0.25$ for exothermic reactions and $\alpha \approx 0.75$ for endothermic ones. Since the pre-exponential factors should be very similar for homologous reactants, the ratio of the rate coefficients should be

$$k_1/k_2 \approx \exp\left(\alpha(\Delta H_2° - \Delta H_1°)/RT\right)$$

differing somewhat from the approximations 9.26 to 9.28. However, the procedure was not specifically designed for reactions of free radicals and is not directly applicable to comparisons of exothermic with endothermic steps as needed here.

More elaborate and more reliable procedures that can be used for estimates of rate coefficients of free-radical reactions are the bond energy-bond order method (*BEBO)* of Johnston and Parr [13] and the curve-crossing approach of Pross [14].

Because of its relative simplicity, the hydrogen-bromide reaction (Example 9.1 in the preceding section) serves well to illustrate the application of the estimates to the identification of relevant steps.

The rate equation 9.24, originally established empirically, was attributed to the mechanism 9.23. An obvious first question is, why is the reaction initiated by dissociation of Br_2 and not of H_2 or, once product has been formed, of HBr? Indeed, thermal dissociation of H_2 and HBr also contributes chain carriers, but not to an extent large enough to affect the reaction. At atmospheric pressure and, say, 500 K as a typical reaction temperature, the degrees of dissociation of gaseous Br_2, HBr, and H_2 are of the orders of 10^{-9}, 10^{-19}, and 10^{-22}, respectively. The first two of these values show that thermal dissociation of HBr as a source of bromine atoms is negligible compared that of Br_2 under any conditions of interest. This leaves the possibility that hydrogen atoms from thermal dissociation of H_2 or HBr, although in a minuscule minority, could dominate the behavior thanks to their much higher reactivity. As will be seen, however, they are vastly outnumbered by hydrogen atoms generated by the propagation cycle and so play only an insignificant role. Application of eqn 9.8 to the (quasi-stationary and long-chain) cycle yields

$$\frac{p_H}{p_{Br}} \cong \frac{k_{p1}p_{H_2} + k_{-p2}p_{HBr}}{k_{-p1}p_{HBr} + k_{p2}p_{Br_2}}$$

(irreversibility of the second step cannot yet be taken for granted). The two coefficients in the numerator are for endothermic steps; those in the denominator, for exothermic ones (see Table 9.1). Dividing numerator and denominator by either of the latter two coefficients and using the approximation 9.27 with $\Delta H°$ values from Table 9.1 one finds

Table 9.1. Approximate standard enthalpies of steps in hydrogen-bromide reaction.

steps	reaction enthalpy $\Delta H°$
$H_2 \rightarrow H\cdot + H\cdot$	$+ 436\ kJ\ mol^{-1}$
$Br_2 \rightarrow Br\cdot + Br\cdot$	$+ 190$
$HBr \rightarrow H\cdot + Br\cdot$	$+ 360$
$Br\cdot + H_2 \rightarrow HBr + H\cdot$	$+ 67$
$H\cdot + Br_2 \rightarrow HBr + Br\cdot$	$- 170$

$$\frac{p_H}{p_{Br}} \approx \frac{10^{-7}p_{H_2} + 10^{-18}p_{HBr}}{p_{HBr} + p_{Br_2}}$$

Although seen to be outnumbered more than a million-fold by bromine atoms at any compositions of interest, hydrogen atoms from propagation are still more than three orders of magnitude more plentiful than their brothers from dissociation of H_2 or HBr. Accordingly, the contribution of the latter to initiation is insignificant.

This examination also illustrates another facet of chain reactions: Granted quasi-stationary conditions and long chains, the concentrations of the chain carriers are coupled, in reactions like 9.5 through the requirement that the rates of the two propagation steps must be equal. Therefore, only one of the two chain carriers can be in dissociation equilibrium with its source, the other gets boosted by the propagation cycle to a higher than thermal concentration.

The answer to a second question, why the step $Br \cdot + H_2 \longrightarrow HBr + H \cdot$ is reversible while $H \cdot + Br_2 \longrightarrow HBr + Br \cdot$ is not, is now easy to give. Both reverse steps compete for HBr. The first of them is exothermic ($- 67$ kJ mol^{-1}) and so occurs at almost every collision; the second is strongly endothermic ($+ 170$ kJ mol^{-1}) and therefore at a great disadvantage. Estimated with the approximation 9.27, the ratio of the two rate coefficients is of the order of 10^{18} at 500 K. Thus, even though $Br \cdot$ outnumbers $H \cdot$ more than a million-fold, the rate of $H \cdot + HBr \longrightarrow H_2 + Br \cdot$ is still more than ten orders of magnitude higher than that of $Br \cdot + HBr \longrightarrow Br_2 + H \cdot$, whose contribution to HBr consumption accordingly remains insignificant.

The fact that, under typical reaction conditions, $Br \cdot$ outnumbers $H \cdot$ by six or more orders of magnitude also explains why coupling of two $Br \cdot$ alone controls termination, despite the fact that much more energy is gained by coupling of two $H \cdot$ or of $H \cdot$ and $Br \cdot$. All these couplings being highly exothermic, their enthalpies are not a relevant factor, and the much greater abundance of $Br \cdot$ alone decides the issue (see preceding section).

A high bond energy is of no great help in coupling of small radicals. For lack of other effective internal degrees of freedom, much of the released energy must be stored as bond-stretching vibration, making the molecule apt to break apart again in short order. Indeed, even $Br \cdot$ recombination occurs primarily in ternary collisions $Br \cdot + Br \cdot + M \longrightarrow Br_2 + M$, where the collision partner M is some other molecule (or the wall of the reaction vessel) that serves to absorb a substantial portion of the released energy [15]. The algebraic form of the rate equation is not affected because thermodynamic consistency requires M to participate in dissociation if it does so in recombination, and the effects cancel since the two rate coefficients appear only as the ratio k_{init}/k_{cBr}.* Evidence for the ternary mechanism has been provided by experiments with initiation by ultraviolet light at temperatures low enough for thermal dissociation to be negligible in comparison [3,17,18].

---

* It has been said that only termination, but not dissociation, involves a collision partner M and that the ratio k_{init}/k_{cBr} in the rate equation does not equal the dissociation equilibrium constant because the two coefficients are "not linked by detailed balancing" [16]. However, this argument is without merit. In the absence of H_2 (or any other species with which $Br \cdot$ can react), thermodynamic consistency and microscopic reversibility clearly require M to participate in dissociation if it does so in recombination. The addition of any species such as H_2 that takes no part in the dissociation step may cause the system to deviate from thermodynamic dissociation equilibrium, but can obviously not alter the *mechanism* of dissociation.

Activation energies and consistency checks. The discussion so far has shown how the approximations 9.25 to 9.28 can be used to identify which of all possible steps are relevant and which are not. In addition, the approximations provide a means of checking the rate coefficients for consistency.

Taking once more the hydrogen-bromide reaction as an example: The one-plus form of the rate equation 9.24 contains two phenomenological coefficients, k_a and k_b. The first is given by

$$k_a \ \equiv \ 2\,(k_{init}/k_{cBr})^{1/2}\,k_{p1}$$

and thus should have the activation energy

$$(E_a)_a \ = \ \frac{1}{2}\big((E_a)_{init} - (E_a)_{cBr}\big) + (E_a)_{p1} \tag{9.30}$$

With $(E_a)_{init}$ and $(E_a)_{p1}$ each a little more than the respective $\Delta H°$ values of $+190$ and $+67$ kJ mol^{-1} (see Table 9.1) and $(E_a)_{cBr}$ (highly exothermic) no more than a few kJ mol^{-1}, the activation energy of k_a should be about 165 to 170 kJ mol^{-1}, which is in quite satisfactory agreement with the observed value of 175 kJ mol^{-1}.

The other coefficient is given by

$$k_b \ \equiv \ k_{-p1}/k_{p2}$$

Both k_{-p1} and k_{p2} are coefficients of exothermic steps, the second (p2) being more exothermic than the first. According to the approximation 9.26, the coefficient ratio should thus be roughly in the range between 0.1 and 1 and should depend little on temperature. The observed value of k_b is 0.1 and is essentially temperature-independent, again in very satisfactory agreement with expectation.

Interestingly, because of the factor one half in the first term of eqn 9.30, the estimated overall activation energy of the reaction (ca. 170 kJ mol^{-1}) turns out to be lower than the activation energy of initiation (190 kJ mol^{-1}). Since this factor appears whenever termination is second order in chain carriers, as is true with very few exceptions, and since the activation energies of the propagation steps often are relatively low, such behavior is quite common [19].

Some rules of thumb. A few regularities that have their roots in thermodynamics are worth mentioning. They can serve as rough guidelines. Exception must be expected, especially if the propagation steps have different molecularities and if the reaction involves a large entropy change or several reactants at very different concentrations. (This excludes, for example, thermal cracking of hydrocarbons.)

The propagation cycle can be viewed as driven by the decrease in free energy that accompanies conversion of reactants to products. Barring overriding entropy effects, two conclusions can be drawn immediately:

- *At least one of the two propagation steps must be exothermic.*

- *If one propagation step is endothermic, its standard-enthalpy variation must be smaller than that of the exothermic step.*

If this were not so, equilibrium would be unfavorable and conversion would remain minimal.

If both forward steps are exothermic, both reverse steps are endothermic and so will have smaller rate coefficients. In the competition for each of the two chain carriers, the respective forward step, being exothermic, is bound to win out. Thus:

-     *If both propagation steps are highly exothermic, both are irreversible.*

If one forward step is endothermic, it competes for a chain carrier with the endothermic reverse of the other forward step, but the latter involves the larger free-energy variation and so is disadvantaged. On the other hand, the exothermic forward step competes for the other chain carrier with the exothermic reverse of the endothermic forward step; both being exothermic, they will have low activation energies and may or may not occur with comparable ease. Accordingly:

-     *If one propagation step is endothermic and the other is exothermic, the endothermic step may be reversible, the exothermic step is not.*

The likelihood of reversibility of the endothermic step is greater, the smaller the free-energy decrease accompanying the overall reaction. This is because the $-\Delta H°$ values of the two competing exothermic steps will then be more similar, and so will be their activation energies and rates.

Lastly, the termination mechanism can be conjectured on the basis of thermochemical data. The propagation step with the larger drop in free energy, or the exothermic step if one is endothermic, is apt to be "faster."* The chain carrier consumed by this step is depleted while the other accumulates. Since termination normally is controlled by coupling of the more plentiful chain carrier:

-     *The chain carrier produced by the more highly exothermic step (or by the exothermic step if the other is endothermic) controls termination.*

(However, see the preceding section for exceptions to control by the most plentiful chain carrier.)

While certainly not valid without exceptions and no substitute for a thorough understanding of the reaction at hand, these rules can serve well for preliminary orientation and as working hypotheses where their stated premises are valid.

9.5. Transmission of reactivity: indirect initiation, chain transfer

So far we have taken for granted that initiation produces a chain carrier and that termination occurs by coupling of the chain carriers. Such behavior is the norm, but there are exceptions, owing to the ability of free radicals to transmit their

---

* "Fast" as used here refers to how soon a reactant is likely to react, not to reaction rate (see Section 4.1). Granted quasi-stationary behavior and long chains, the *rates* of the two propagation steps are equal (see Section 9.3).

reactivity to other species. Specifically, initiation may be indirect in that the free radical it produces is not a chain carrier, but reacts with another molecule to form a chain carrier. More importantly, the kinetic chain may be broken by reaction of a chain carrier with another molecule, producing a radical that may or may not start a different chain. This is called *chain transfer*.

Indirect initiation. The typical step sequence of indirect initiation is

$$
\begin{array}{lll}
\text{initiation} & \text{in} \longrightarrow 2\,R\cdot \\
\text{transmission} & R\cdot + S \longrightarrow X + \ldots
\end{array}
\tag{9.31}
$$

where X is a chain carrier but $R\cdot$ is not, S may be a solvent molecule, and another molecule may or may not be produced in the second step. With $R\cdot$ at trace level, the rates of the two steps are equal, and the slow first is rate-controlling:

$$
r_{\text{init}} = 2k_{\text{init}} C_{\text{in}}
\tag{9.3}
$$

as for single-step initiation. The fact that initiation may be indirect has no effect on the rate. [If $R\cdot$ is the most plentiful radical, it will recombine in addition to transmitting; this recombination does not reduce the chain-carrier population, which then is kept in balance by coupling of $R\cdot$ with the next most plentiful chain carrier.]

Examples of indirect initiation will be encountered later in this chapter in the Rice-Herzfeld mechanisms and hydrocarbon autoxidation (see next section). Also, initiation of free-radical polymerization usually is a two-step process (see Section 10.3).

Chain transfer. While indirect initiation remains without effect on the form of the rate equation, chain transfer may profoundly affect kinetics because it may contribute an additional and possible dominant termination mechanism.

$$
\text{chain transfer} \qquad X + S \longrightarrow \ldots
\tag{9.32}
$$

Here, X is any chain carrier, and S may but need not be a solvent molecule.

The product may be, or may include, a free radical that starts a new chain. In such cases, chain transfer does not decrease the free-radical population. If the new chain is of the same kind, the reaction continues at its pace. In the rare instances in which it starts a different kind of chain, two parallel reactions, each with its own termination mechanism, must be considered.

Alternatively, chain transfer to a molecule such as carbon tetrachloride can produce radicals of low reactivity, thereby contributing to the termination of the kinetic chain. Species S then acts as a retardant. Assuming for simplicity $X+X$ coupling as the normal terminatiom, the net termination rate in such cases is

$$
r_{\text{trm}} = -(2k_{\text{cX}} C_X^2 + k_{\text{chXS}} C_X C_S)
\tag{9.33}
$$

where k_{chXS} is the rate coefficient of chain transfer from X to S.

If chain transfer is the dominant termination mechanisms, the equality of initiation and termination rates according to eqn 9.12 leads to a chain-carrier population that is proportional to the initiator concentration. For example, for chain transfer by X, the radical that is produced by initiation:

$$C_X \cong \frac{2k_{init} C_{in}}{k_{chXS} C_S} \tag{9.34}$$

With eqns 9.6, a free-radical population proportional to the initiator concentration gives a chain-reaction rate that is first order instead of half order in initiator. However, for chain transfer to outrun coupling, its rate must be high, and species S then acts essentially as an inhibitor (see also Section 9.7).

If the rates of termination by X+X coupling and chain transfer by X are comparable, one finds

$$C_X \cong \left[\left(\frac{k_{chXS} C_S}{4k_{cX}} \right)^2 + \frac{k_{init} C_{in}}{k_{cX}} \right]^{1/2} - \frac{k_{chXS} C_S}{4k_{cX}} \tag{9.35}$$

and the chain-reaction rate becomes of order between one half and one in initiator.

Equations for the chain-carrier concentrations at the various other possible combinations of initiation, coupling, and chain transfer involving the other chain carrier as well are more lengthy, but the conclusions as to kinetic behavior and reaction order with respect to the initiator are qualitatively the same.

Chain transfer is of particular interest in free-radical polymerization, where it affects not only the polymerization rate, but also the molecular weight of the product (see Section 10.3).

9.6. Reactions with more than two free radicals

So far, only a very simple type of chain reaction has been considered, that with two chain carriers generated by initiation and with a two-membered propagation cycle. The high reactivity of free radicals in general, however, can often lead to a much more complex behavior. Mechanisms with as many as 86 different elementary steps have been proposed as early as 1978 [20], and with the advent of cheap, fast, and easy-to-use computers there is now no lack of conjectured networks of even larger sizes, especially in petroleum processing and combustion engineering. The possible combinations of steps and topologies of networks stagger the imagination and make a comprehensive coverage at this place out of question. Instead, a few examples will be given and one relatively simple specific case will be discussed in detail in order to illustrate principles as well as a way of deriving rate equations.

It would be relatively easy to extend the mathematics in the preceding sections to reactions with single propagation cycles of more than two members (provided there is no chain branching). The forward and reverse reactions of the cycle could simply be expressed with clockwise and counter-clockwise Λ segment coefficients (see Section 6.3). However, there is little point in writing such equations because the high reactivity of the chain-carrying free radicals in real systems makes linear pathways of any length a rarity. In particular, many chain reactions involve hydrogen or oxygen atoms, and these react with almost any molecule they encounter. The so-called Rice-Herzfeld mechanisms of thermal degradation of organic compounds may serve as a typical example.

9.6.1. Rice-Herzfeld mechanisms: thermal cracking

Thermal cracking of organic substances is an important reaction in the petroleum industry and has been extensively studied for over seventy years. At least for simple alkanes, the decay is first order in good approximation and therefore was long believed to occur in a single, unimolecular step [21]. However, in the 1930s, Rice and coworkers [22-24] established the presence of free radicals under the conditions of the reaction by means of the Paneth mirror technique [25,26]. This observation led Rice and Herzfeld to propose a chain mechanism [22,27,28]. Extensive later studies proved the essential features of their mechanism to be correct not only for hydrocarbons, but also for many other types of organic substances.

The principal features of what has come to be called *Rice-Herzfeld mechanisms* are [21,29]:

- *initiation*: rupture of a weak bond in the reactant molecule, generating two free radicals;
- *transmission*: abstraction of a hydrogen atom from a reactant molecule by one of the free radicals, to yield a product and the first chain carrier;
- *first propagation step*: decay of this chain carrier into a product molecule and another, second chain carrier (which may or may not be $H\cdot$);
- *second propagation step*: hydrogen abstraction from another reactant molecule by the second chain carrier to yield a product molecule (H_2 if the second carrier is $H\cdot$) and another specimen of the first chain carrier;
- *termination*: coupling of the most plentiful radical.

As an example, thermal cracking of ethane will be examined here in detail.

Example 9.2. Thermal cracking of ethane. Thermal cracking of ethane at temperatures in the range of 800 to 1200 K and ambient or lower pressures yields mainly ethene and hydrogen:

$$C_2H_6 \;\longrightarrow\; C_2H_4 + H_2 \tag{9.36}$$

but small amounts of methane, butane, and propene are also formed. The reaction is generally held to be essentially first order in ethane [30]:

$$-r_{CC} \;\cong\; k_{app}\,p_{CC} \qquad\qquad (9.37)$$

(CC = ethane). Rice and Herzfeld [27] proposed the mechanism

initiation	$C_2H_6 \longrightarrow 2\,CH_3\cdot$	rate $2k_{init}p_{CC}$
transmission:	$CH_3\cdot + C_2H_6 \longrightarrow CH_4 + C_2H_5\cdot$	

propagation:

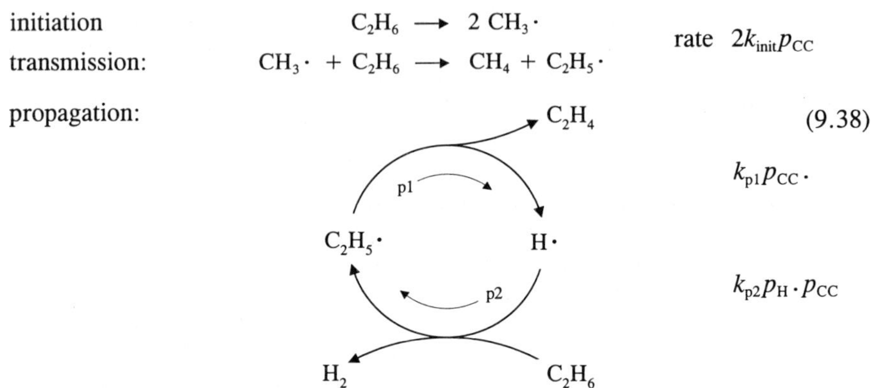

$$ (9.38) $$

$$k_{p1}p_{CC}\cdot$$

$$k_{p2}p_H\cdot p_{CC}$$

(CC· = $C_2H_5\cdot$). Initiation breaks the ethane carbon-carbon bond, which is weaker than the carbon-hydrogen bonds. The most plentiful free radical under most conditions of interest is $C_2H_5\cdot$ [30], so that coupling of two of these to butane should be the dominant termination mechanism, probably accompanied to a small extent by disproportionation to ethene and ethane [31,32]:

$$\text{termination} \qquad 2\,C_2H_5\cdot \begin{cases} \longrightarrow C_4H_{10} \\ \dashrightarrow C_2H_4 + C_2H_6 \end{cases} \qquad (9.39)$$

Propene, formed in small amounts at high degrees of conversion, is believed to arise from a side reaction $C_2H_5\cdot + C_2H_4 \longrightarrow C_3H_6 + CH_3\cdot$ in several steps with butyl and hexyl radicals as intermediates [33]. Other side reactions at high conversion are $CH_3\cdot + H_2 \longrightarrow CH_4 + H\cdot$ and $H\cdot + C_2H_4 \longrightarrow C_2H_5\cdot$ [34]. These reactions leave the number of radicals unchanged, and so have little effect on the algebraic form of the rate equation for ethane disappearance and the reaction order.

Taken at face value, the mechanism 9.38 with termination 9.39 leads to a reaction order of one half in ethane, in seeming contradiction to experimental observation. It is intriguing to trace the efforts at reconciliation.

Rice and Herzfeld [27], at a time when still little was known about free radicals and chain reactions, had tried to account for the observed first-order behavior by postulating a "mixed" termination $C_2H_5\cdot + H\cdot \longrightarrow C_2H_6$. However, since $C_2H_5\cdot$ outnumbers $H\cdot$ by several orders of magnitude under typical reaction conditions, this assumption proved untenable [30]. Thereupon Küchler and Theile [35] suggested that initiation is bimolecular in ethane; provided termination occurs without a collision partner, the overall rate then is first order in ethane. A case can be made that the butane molecule, formed by termination, has enough internal degrees of freedom to carry off the recombination energy without help by a partner even if ethane needs one for dissociation. This explanation had to be abandoned when methane formation, at least initially due exclusively to initiation, was found to be first order in ethane [33,36]. To save the day, Quinn [33] invoked Lindemann's theory of unimolecular decay [37] and applied it to the first propagation step, $C_2H_5 \cdot \longrightarrow C_2H_4 + H \cdot$. According to Lindemann, activation by binary collision must precede unimolecular decay and becomes rate-controlling at very low pressure. At start, ethane is the only available collision partner. With ethane in that role:

$$C_2H_5 \cdot \ + \ C_2H_6 \ \longrightarrow \ (C_2H_5 \cdot)_{\text{activated}} \ + \ C_2H_6$$

The overall rate of ethane consumption then is of order one-and-a-half in ethane if the rate of $C_2H_5 \cdot \longrightarrow C_2H_4 + H \cdot$ is controlled by activating collision, and of order one half if controlled by decay of the activated radical. According to Quinn, first-order behavior was observed because the reaction was studied in the "fall-off" range of pressure, that is, where rate control of $C_2H_5 \cdot$ decay shifts from one step to the other. Indeed, at very low pressures the initial rate varies with $(p_{CC}^{\circ})^{1.5}$ [31].

Quinn studied *initial* rates—i.e., in the absence of reaction products—in a limited pressure range of 60 to 230 Torr. His hypothesis can explain the dependence on initial pressure he observed, but not what is normally defined as first-order behavior, namely, a rate proportional to the reactant concentration or partial pressure *in the course of the reaction* in the presence of products formed. This is because ethene (and, for that matter, almost any other molecule with the possible exception of H_2) can also serve as activating collision partner. Indeed, addition of inerts has been found to boost the rate [35]. Since one mole of ethane produces approximately one mole of ethene, the concentration of potential collision partners is $p_{C=C} + p_{CC} \cong p_{CC}^{\circ}$ and remains essentially unchanged, so that there is no effect on the form of the rate equation and the reaction order (for simplicity, this assumes ethene to be as effective a collision partner as is ethane, and H_2 to be ineffective.) Nevertheless, textbooks to this day accept Quinn's explanation, if not Rice and Herzfeld's.

First-order behavior (as normally defined) at any pressure can be rationalized if the first propagation step is made reversible. This is not unreasonable because the step in the forward direction is strongly endothermic $(+159\ \mathrm{kJ\ mol^{-1}})$, so its reverse should make itself felt long before the reverse of the overall reaction becomes noticeable. The rate of this reverse step is proportional to a product, so that the retardation it exerts increases with progressing conversion. This translates into a higher apparent reaction order. Quantitatively, the mechanism 9.38 with termination 9.39 and reversible first step gives a rate equation of the form

$$-r_{CC} \ \cong \ \frac{k_a p_{CC}^{1/2}}{1 + k_{-p1} p_{C=C}/k_{p2} p_{CC}} \tag{9.40}$$

(for derivation, see farther below). Since both steps $-p1$ and $p2$ are exothermic and bimolecular, the ratio of their rate coefficients should not be far from unity (see Section 9.4). Setting $k_{-p1}/k_{p2} = 0.5$ one finds a behavior within 1% of first order up to over 40% conversion (see farther below). However, the initial rate now is proportional to the square root of initial pressure, at odds with Quinn's experimental results and therefore possible only at pressures above the "fall-off" range, i.e., where activating collision no longer affects the rate. Also, the acceleration by added inerts remains unexplained. Moreover, step $-p1$ being more exothermic than step p2, the ratio of their rate coefficients is expected to be larger than unity. In and below the fall-off range, if Quinn's hypothesis of an activating collision partner for $C_2H_5\cdot$ decay is accepted, a factor $(p_{CC}^\circ)^n$ with $0 < n < 1$ appears in the numerator and the second denominator term of eqn 9.40 and can produce the sought-for pressure dependence of the initial rate; overall first order in ethane now requires $k_{-p1}(p_{CC}^\circ)^n/k_{p2} \approx 0.5$, a ratio that is more believable, but confined to a pressure range around $p_{CC}^\circ \approx (k_{p2}/k_{-p1})^{1/n}$.

So far we have taken for granted that the reaction is conducted at constant volume, as in the kinetic studies by Küchler [35], Laidler [38], Quinn [39], and Lin [36]. In a plug-flow reactor as used in some other work [30], the gas expands as the mole number doubles when ethane forms ethene and hydrogen. Failure to correct for expansion would let the reaction order seem farther from first (see Section 3.3.4) and so cannot help to explain unexpected first-order behavior. Expansion keeps reducing the concentrations of the collision partners as conversion progresses. As a detailed calculation shows, this can produce an apparent reaction order close to one up to moderate conversion in and below the fall-off range, where activating collisions affect the rate. However, this effect alone cannot explain first-order behavior at constant pressure above the fall-off range, nor at constant volume at any pressure.

None of the explanations described here is entirely satisfactory, and no other simple ones come to mind. Reaction behavior appears to be more complex than the original Rice-Herzfeld network 9.38 suggests [40].

Derivation of eqn 9.40 and apparent reaction order.   Indirect initiation supplies $C_2H_5\cdot$, the chain carrier that dominates termination, so that eqn 9.14 applies. The substitutions are $\lambda_{p1} = k_{p1}$, $\lambda_{p2} = k_{p2}p_{CC}$, $\lambda_{-p1} = k_{-p1}\,p_{C=C}$, $\lambda_{-p2} = k_{-2p}p_{H_2}$, $X = C_2H_5\cdot$ $(CC\cdot)$, and in $= C_2H_6$ (CC), and give

$$-r_{CC} = \left[\frac{k_{init}\,p_{CC}}{k_{cCC\cdot}}\right]^{1/2}\frac{k_{p1}k_{p2}\,p_{CC} - k_{-p1}k_{-p2}\,p_{C=C}\,p_{H_2}}{k_{p2}\,p_{CC} + k_{-p1}\,p_{C=C}} \qquad (9.41)$$

Since step $-p1$ is strongly exothermic while step $-p2$ is endothermic (-159 vs. $+22$ kJ mol^{-1}), it can be assumed that k_{-p1} is significantly larger than k_{-p2}. The second denominator term then makes itself felt before the second numerator term does, that is, already at conversions so low that the reverse of the overall reaction is still negligible. Without the second numerator term, eqn 9.41 in one-plus form and with $k_a = (k_{init}/k_{cCC\cdot})^{1/2}k_{p1}$ gives eqn 9.40.

To establish the apparent (power-law) reaction order at low conversion, eqn 9.41 without the second numerator term must be integrated. For this purpose, p_{CC} and $p_{C=C}$ are expressed in terms of the fractional conversion of ethane; at constant volume:

$$p_{CC} = p_{CC}^o(1 - f_{CC}), \qquad p_{C=C} = p_{CC}^o f_{CC}$$

so that

$$-r_{CC} = \frac{k_a(p_{CC}^o)^{1/2}(1 - f_{CC})^{3/2}}{1 + k_b f_{CC}} \qquad \left(k_b \equiv k_{-p1}/k_{p2} - 1\right) \qquad (9.42)$$

The general relationship between reaction time t and rate $-r_A$ at constant volume is [41,42]

$$t = p_A^o \int_{f_A=0}^{f_A} -r_A^{-1}\, df_A$$

Integration with A = ethane and eqn 9.42 for the rate $-r_A$ gives

$$t = \frac{(p_{CC}^o)^{1/2}}{k_a}\int_{f_A=0}^{f_A}\frac{1 + k_b f_{CC}}{(1-f_{CC})^{3/2}}\, df_{CC} = \frac{2(p_{CC}^o)^{1/2}}{k_a}\left[\frac{(1 + 2k_b)[1 - (1 -f_{CC})^{1/2}] - k_b f_{CC}}{(1 -f_{CC})^{1/2}}\right]$$

With $k_b \approx -0.5$ (i.e., $k_{-p1}/k_{p2} \approx 0.5$) the time dependence of fractional conversion is within 1% of $t = -\ln(1 - f_{CC})$ (constant volume, first order) up to $f_{CC} = 0.42$.

Thermal cracking of ethane is an excellent example of an intricate mechanism that leads to a kinetic behavior obeying a simple, first-order rate law in good approximation over a fairly wide range of conditions. It also serves to show how easily such a deceptively simple rate law is misinterpreted. Moreover, the example illustrates an important general point:

A reverse propagation step with rate proportional to a product concentration produces an apparent overall reaction order that is higher than without the reverse step. This can happen even at conditions under which the overall reaction is irreversible.

The cause is the increasing retardation by the reverse step with progressing conversion as the product builds up [43]. This retardation can become effective even if the other propagation step and therefore the overall reaction are irreversible, or at a conversion so low that the reverse overall reaction is still insignificant.

A comment on deducing mechanistic details of Rice-Herzfeld-type reactions from apparent reaction orders is called for. Usually, a termination mechanism giving the desired result is postulated or, failing that, a collision partner in initiation, termination, or propagation steps is invoked. A formal scheme relating overall reaction orders to such mechanistic features, developed as early as 1948 by Goldfinger *et al.* [44], is quoted to this day in some textbooks. However, uncritical application can easily result in misinterpretation. The scheme implies that a collision partner in the initiation step must be another reactant molecule although most other molecules could serve just as well, it glosses over the requirements of thermodynamic consistency and that "mixed" termination is possible only under exceptional circumstances, it does not account for the effects of reverse steps, nor does it address the need for a volume correction in gas reactions at constant pressure and with change in mole number. In fact, an unusual apparent reaction order in an empirical rate equation may very well stem such ignored facets, and the mechanism may be contrary to what one is led to believe when taking the Goldfinger scheme at face value. Ethane cracking is a case in point.

Higher hydrocarbons. Thermal cracking of higher hydrocarbons is believed to occur with Rice-Herzfeld-type mechanisms [45,46]. Of course, with more carbon atoms in the molecule, more free radicals of different carbon numbers appear and produce a greater variety of products. As a still relatively simple example, the network of principal steps in cracking of *n*-butane is [47,48]:

initiation $\qquad\qquad\qquad\qquad C_4H_{10} \longrightarrow 2\, C_2H_5\cdot$

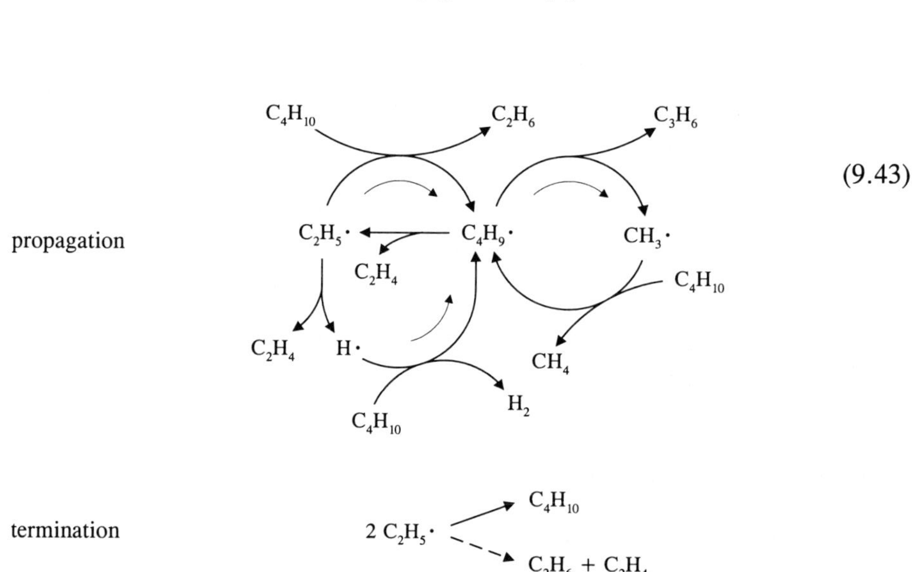

propagation

$$(9.43)$$

termination $\qquad\qquad\qquad 2\, C_2H_5\cdot \quad \begin{matrix} C_4H_{10} \\ C_2H_6 + C_2H_4 \end{matrix}$

There are three propagation cycles, all of which have the butyl radical, $C_4H_9\cdot$, in common. Even this network is grossly simplified in that it omits, among other steps, any reactions of products and hydrogen abstraction from alkanes other than butane as well as the presence of propyl radicals, which can arise from a step $C_4H_{10} \longrightarrow C_3H_7\cdot + CH_3\cdot$ and activate their own propagation cycle.

Rates of thermal cracking are first-order in good approximation for propane, butane and still higher hydrocarbons [21]. This is remarkable because chain mechanisms with initiation by break-up of a reactant normally result in reaction orders of one half or one-and-a-half, depending on which radical is consumed by termination. First-order behavior can result from "mixed" termination, which, however, can in most cases be ruled out as dominant mechanism (see Section 9.3). A more probable explanation is a combination of effects: that key hydrocarbon radicals participate in several steps of different molecularities, that some steps are reversible, and that some unimolecular ones require collision partners.

As the complexity of the reaction of even as simple a molecule as *n*-butane demonstrates, the number of steps increases steeply with carbon number. It becomes almost astronomical for complex mixtures of higher hydrocarbons as encountered in industrial petroleum processing. A more promising approach here is *discretization*, pioneered by Froment [49], a method of modeling in terms of bonds formed and broken, regardless of the exact structure of the respective molecules. The guiding idea is that the rate coefficients and activation energies are similar for like events—say, hydrogen abstraction by H· from an alkane chain, or recombination of two alkyl radicals—as long as the vicinity of the reaction site is the same, even if other parts of the molecule differ in size and structure (see also Section 11.3). For single higher hydrocarbons, conventional step-by-step modeling has become feasible thanks to the extensive data base on relevant rate coefficients and activation energies that is has been compiled over the last few years [50,51].

Other compounds. Rice-Herzfeld mechanisms appear to be the rule in thermal degradation of many other types of organic compounds, among them aldehydes [21,43,52-54] and ketones [21,55]. Many of these reactions are approximately first order. Decomposition of acetaldehyde, quite extensively studied, is of order one-and-a-half, easily explained with a Rice-Herzfeld mechanism and eqn 9.18 or 9.19 [21,56]. The reaction order is found to increase toward two at high conversion [43,56]. As seen in the example of ethane cracking, such a "creeping up" of the reaction order with progressing conversion is a typical symptom of a reverse step in the propagation cycle [43].

9.6.2. Hydrocarbon oxidation

Reactions of organic compounds, especially hydrocarbons, with oxygen in the gas or liquid phase at moderate temperatures (below 150° C), are important both as industrial processes and as natural decomposition phenomena that are to be suppressed if possible. They are chain reactions, but differ from thermal cracking in that they usually requires initiation. An initiator may have been added intentionally or be present as an impurity or early minor product, possibly a hydroperoxide that had accumulated upon prolonged standing in contact with air.

A typical mechanism of oxidation of a hydrocarbon RH [57-62] is

initiation: initiator $\longrightarrow$ 2 R'·   rate $2\,k_{init}C_{in}$

transmission: R'· + O_2 $\longrightarrow$ R'OO·

                  R'OO· + RH $\longrightarrow$ R'OOH + R·

propagation:

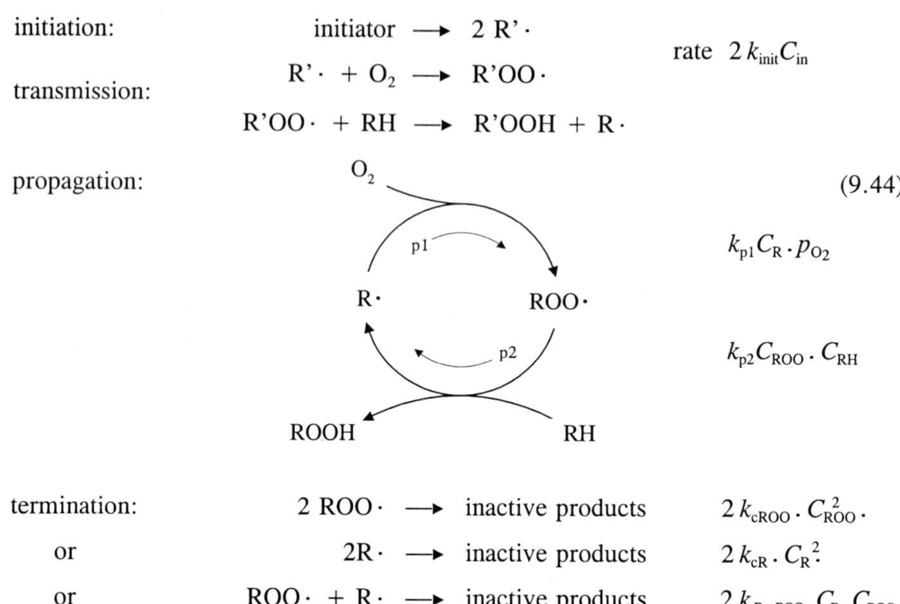

$$k_{p1}C_R \cdot p_{O_2}$$

$$k_{p2}C_{ROO} \cdot C_{RH}$$

(9.44)

termination: 2 ROO· $\longrightarrow$ inactive products $2\,k_{cROO} \cdot C_{ROO·}^2$

   or   2R· $\longrightarrow$ inactive products $2\,k_{cR} \cdot C_{R·}^2$

   or   ROO· + R· $\longrightarrow$ inactive products $2\,k_{cR·ROO} \cdot C_R \cdot C_{ROO}$

As a rule, the first propagation step is highly exothermic; the second, endothermic. If so, k_{p1} is several orders of magnitude larger than k_{p2}. As a result, the concentration of ROO· is much higher than that of R· (see eqn 9.8), even in liquid-phase reactions in which the concentration of RH is considerably higher than that of dissolved oxygen. Accordingly, coupling of ROO· is the dominant termination mechanism in most cases. At very low partial pressures of oxygen, however, ROO· is no longer as plentiful, and the other two possible termination mechanisms may also come into play. The hydroperoxide, ROOH, may decompose into free radicals that start new chains, or react further to give alcohols, ketones, and acids. Since the decomposition of the hydroperoxide gives rise to additional chain carriers, the reaction can be autocatalytic and may evolve into an explosion.

Being highly exothermic, the first propagation step in the cycle 9.44 can safely be regarded as irreversible. If the hydroperoxide were stable, the second propagation step should be reversible. However, it is unstable and likely to decompose in other ways before it has time to react with R·, which is at very low concentration. Accordingly, the assumption that both steps are irreversible is usually justified. Termination at other than very low oxygen pressures is controlled by ROO·, the free radical functioning as reactant in the second propagation step (Y in the general equations in Section 9.3). Thus, if new chains initiated by the decomposition of the hydroperoxide product can be disregarded, the rate of hydrocarbon consumption is described by eqn 9.19. At oxygen pressures so low that all three termination mechanisms contribute, the rate is given by eqn 9.21. With $r_{init} = 2k_{init}C_{in}$, $\lambda_{p1} = k_{p1}p_{O_2}$ (validity of Henry's law is assumed here), and $\lambda_{p2} = k_{p2}C_{RH}$, these equations give

$$-r_{RH} \cong (k_{init}/k_{cROO\cdot})^{1/2} k_{p2} C_{in}^{1/2} C_{RH} \tag{9.45}$$

and

$$-r_{RH} \cong \frac{k_{init}^{1/2} k_{p1} k_{p2} C_{in}^{1/2} p_{O_2} C_{RH}}{\left[k_{cR\cdot}(k_{p2}C_{RH})^2 + k_{cROO\cdot}(k_{p1}p_{O_2})^2 + k_{cR\cdot ROO\cdot} k_{p1}k_{p2}p_{O_2}C_{RH}\right]^{1/2}} \tag{9.46}$$

for normal and very low oxygen pressures, respectively.

Behavior according to eqn 9.45—orders of one half in initiator, one in the organic reactant, and zero in oxygen—is very common. If the oxygen pressure is reduced substantially, a beginning dependence on that pressure and a fall-off of the reaction order with respect to the organic reactant is observed, an indication that $R\cdot + R\cdot$ and $R\cdot + ROO\cdot$ terminations have started to contribute [58]. Termination by $R\cdot + R\cdot$ coupling alone would result in a rate that is first order in oxygen and zero order in hydrocarbon (eqn 9.18), but is unlikely even at quite low oxygen pressures. In all instances, the rate is of order one half in the initiator.

The reaction is not "clean." Hydroperoxide decomposition yields aldehyde and ketone. Moreover, at other than quite low conversion, further oxidation leads to scission of carbon-carbon bonds and formation of acids [63]. However, if a boric-acid ester or boroxine is added, secondary alcohol can be obtained in good yield (see Example 5.5 in Section 5.4).

Example 9.3. Oxidation of cyclohexane [63-65]. Air oxidation of cyclohexane to a mixture of cyclohexanol and cyclohexanone is an important step in a process for production of adipic acid and caprolactam. The reaction is carried out in the presence of a small amount of a cobalt salt (typically naphthanate or 2-ethylhexanoate) at 140 to 165° C and moderate pressure (e.g., 10 atm). The primary reaction product is cyclohexyl hydroperoxide:

$$\bigcirc + O_2 \longrightarrow \bigcirc\!\!-OOH \tag{9.47}$$

which, however, decomposes quickly. The reaction is run at low conversion to minimize degradation of the desired final products, cyclohexanol and cyclohexanone [66].

The formation of the hydroperoxide is believed to proceed with a chain mechanism much like 9.44 ($R\cdot$ being $C_6H_{11}\cdot$), except that its product acts as initiator. Cobalt functions to control the conversion of the hydroperoxide to free radicals for initiation [67], in all likelihood by a Haber-Weiss redox cycle 9.4 (see Section 9.2).

The first propagation step is highly exothermic, the second is endothermic. Cobalt also promotes the conversion of the hydroperoxide to cyclohexanol and cyclohexanone. Conditions can be adjusted so that this conversion is fairly rapid, and the reverse propagation step $C_6H_{11}\cdot + C_6H_{11}OOH \longrightarrow C_6H_{12} + C_6H_{11}OO\cdot$ then remains insignificant. A likely termination is [59-61]

$$2 \; \langle\;\rangle\text{-OO}\cdot \longrightarrow \langle\;\rangle\text{-OO-OO-}\langle\;\rangle \begin{cases} \langle\;\rangle\text{=O} \\ O_2 \\ \langle\;\rangle\text{-OH} \end{cases} \qquad (9.48)$$

The fairly harsh conditions required to break the carbon-hydrogen bond in cyclohexane cause various side reactions, and the yield to the desired end products (based on cyclohexane converted) is only about 60 to 70%, even at low conversion. A higher yield could be obtained with added borate ester or boroxine (see Example 5.5 in Section 5.5), but this would require hydrolysis of the resulting cyclohexyl ester and is not practical in a process that calls for a dry product.

Oxidation may be initiated in other ways. In *autoxidation* the hydrocarbon itself functions as initiator by reacting with oxygen to form a hydroperoxide. If so, $C_{RH}p_{O2}$ replaces C_{in} in the rate equations. The reaction orders then are between the following limits: one and a half in hydrocarbon and one half in oxygen at moderate to high oxygen pressures, and one half in hydrocarbon and one and a half in oxygen at very low oxygen pressures. Also, the reaction may be initiated photochemically, possibly in the presence of a sensitizer [68]. The rate then is proportional to the square root of the light intensity.

Because of their great importance in chemical industry, much effort has been devoted to the study of autoxidation and combustion, and a large data base of rate coefficients and activation energies of common elementary reaction steps has been compiled [51,51,60,69-71].

9.6.3. *Reactions with chain branching: the hydrogen-oxygen reaction*

As mentioned at the outset, certain chain reactions include steps in which more chain carriers are formed than consumed, and this may cause a detonation. In most cases, branching is caused by oxygen, whose atom has two unpaired electrons.

In a reaction with chain branching, there is competition for chain carriers between the branching step or steps and the termination step. As the chain-carrier population increases after initiation, production and elimination of chain carriers may or may not reach a balance: Termination may be unable to keep pace with production; the chain-carrier population then starts to grow exponentially, and a detonation ensues. The essential features of this process are best shown with a specific example, that of the hydrogen-oxygen reaction.

Example 9.4. The hydrogen-oxygen reaction [72-75]. The hydrogen-oxygen reaction

$$2\,H_2 + O_2 \longrightarrow 2\,H_2O \qquad (9.49)$$

is one of the most interesting and most thoroughly studied reactions with chain branching. Although only two elements—hydrogen and oxygen—are involved, they form a large number of molecular and free-radical species, and these can undergo many steps that are interconnected in a labyrinthine fashion. An earlier volume of this series [76] starts (under the title *Minima minimorum*) with a listing of thirty such steps.

At low pressure and moderate temperature (say, 400 to 700 K), the propagation cycles can be represented in a simplified fashion by a network of just three interlocking steps:

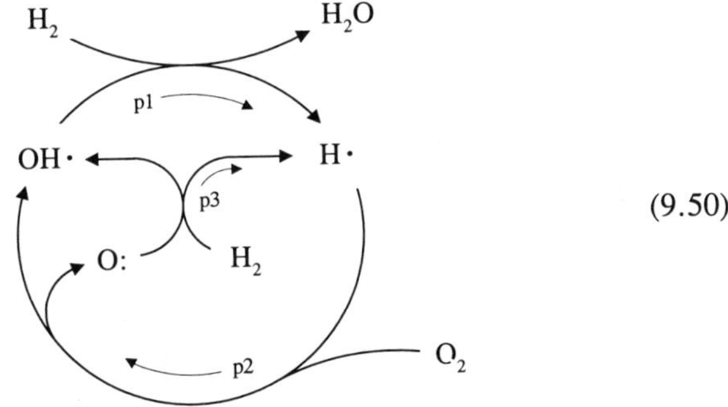

$$(9.50)$$

Two of the three steps involve chain branching: $H\cdot + O_2 \longrightarrow O: + OH\cdot$ and $O: + H_2 \longrightarrow OH\cdot + H\cdot$ Initiation can be triggered by an electric discharge or occur by dissociation

$$H_2 + M \longrightarrow H\cdot + H\cdot + M$$

or reaction

$$H_2 + O_2 + M \longrightarrow OH\cdot + OH\cdot + M$$

where M is a collision partner or the wall of the reaction vessel. Possible termination mechanisms are adsorption of $H\cdot$ at the vessel wall or reaction of $H\cdot$ with oxygen:

$$H\cdot \longrightarrow H_{adsorbed} \qquad\qquad \text{rate} \quad -k_{adsH}\,p_H \qquad (9.51)$$

$$H\cdot + O_2 + M \longrightarrow HOO\cdot + M \qquad\qquad -k_{cH\cdot O_2}\,p_H\,p_{O_2}\,p_M \qquad (9.52)$$

The hydroperoxy radical, $HOO\cdot$, is fairly unreactive. It builds up to relatively high concentrations and is then likely to react with itself according to

$$2\,HOO\cdot \longrightarrow H_2O_2 + O_2 \qquad (9.53)$$

or be deactivated at the vessel wall. Because the gas mixture lacks inert large molecules that could effectively carry off the released energy, coupling of $H\cdot$ is not competitive under usual conditions.

The question of greatest interest is whether or not a detonation will ensue once the reaction has been initiated. To answer it, an equation describing the growth rate of the chain-carrier population is sought. Because population growth is considered, the Bodenstein approximation of quasi-stationary states of chain carriers cannot be applied across the board. However, as long as the system is still remote from detonation, an approximate description can be obtained by application of that approximation to $OH\cdot$ and $O:$ only, but not to $H\cdot$. This is admissible because, in the propagation cycle, $H\cdot$ is consumed only by a highly endothermic step (ca. $+70$ kJ mol^{-1}) and thus becomes more plentiful than $OH\cdot$ and $O:$ by several orders of magnitude, with the result that on an absolute scale the growth rates of the latter two free radicals are negligible by comparison with that of $H\cdot$.

The Bodenstein approximations for $OH\cdot$ and $O:$ amount to

$$r_{O:} \;=\; k_{p2}p_H\cdot p_{O_2} - k_{p3}p_{O:}p_{H_2} \;\cong\; 0$$

$$r_{OH\cdot} \;=\; k_{p2}p_H\cdot p_{O_2} + k_{p3}p_{O:}p_{H_2} - k_{p1}p_{OH}\cdot p_{H_2} \;\cong\; 0$$

respectively, and yield

$$p_{O:} \;\cong\; k_{p2}p_H\cdot p_{O_2}/k_{p3}p_{H_2} \tag{9.54}$$

$$p_{OH\cdot} \;\cong\; (k_{p2}p_H\cdot p_{O_2} + k_{p3}p_{O:}p_{H_2})/k_{p1}p_{H_2} \;\cong\; 2k_{p2}p_H\cdot p_{O_2}/k_{p1}p_{H_2} \tag{9.55}$$

(Equation 9.54 has been used to replace $p_{O:}$ in eqn 9.55.) The growth rate of $H\cdot$ can now be calculated. Since $H\cdot$ is by far the most plentiful chain carrier, its growth rate is representative of that of the total chain-carrier population, $\Sigma\cdot$. Allowing for both termination mechanisms 9.51 and 9.52:

$$r_{\Sigma\cdot} \;\cong\; r_{H\cdot} \;\cong\; r_{init} + k_{p1}p_{OH}\cdot p_{H_2} + k_{p3}p_{O:}p_{H_2} - p_H\cdot(k_{p2}p_{O_2} + k_{adsH} + k_{cH\bullet O_2}p_{O_2}p_M)$$

and with eqns 9.54 and 9.55:

$$r_{\Sigma\cdot} \;\cong\; r_{init} + p_H\cdot(2k_{p2}p_{O_2} - k_{adsH} - k_{cH\bullet O_2}p_{O_2}p_M) \tag{9.56}$$

For a batch system at constant volume, $r_{\Sigma\cdot} = dp_{\Sigma\cdot}/dt$, and eqn 9.56 can be integrated over time:

$$p_{\Sigma}(t) \;\cong\; \int_{t=0}^{t} r_{\Sigma}\cdot dt \;\cong\; \frac{r_{init}}{B-A}\left(1 - \exp[(A-B)t]\right) \tag{9.57}$$

where

$$A \;\equiv\; 2k_{p2}p_{O_2} \qquad\qquad \text{chain–branching effect}$$

$$B \;\equiv\; k_{adsH} + k_{cH\bullet O_2}p_{O_2}p_M \qquad\qquad \text{termination effect}$$

[Even if the gas is not confined in a closed reaction vessel, the increase in mole number becomes so rapid as detonation is approached that pressure builds up, so that, for the integration of eqn 9.56 in this range, constant volume is a better approximation than constant pressure.]

Equation 9.57 shows the behavior of the gas mixture to depend critically on the relative strength of the branching and termination effects. If $B > A$, termination can keep pace with branching, and the chain-carrier population approaches a quasi-stationary level

$$ t \to \infty, \qquad p_{\Sigma} \cdot (t) \cong \frac{r_{init}}{B - A} = \frac{r_{init}}{k_{ads\,H} + k_{cH\cdot O_2} p_{O_2} p_M - 2 k_{p2} p_{O_2}} $$

The chain-branching effect, A, is seen to counteract the termination effect, B, but does not overcome it. On the other hand, if $B < A$, chain-carrier production outruns elimination by termination, and the population begins to increase exponentially as the exponential in eqn 9.57 becomes dominant. When this starts to happen, the Bodenstein approximations for $OH\cdot$ and O: and eqns 9.56 and 9.57 derived from them lose their validity. Moreover, with the accompanying rise in temperature and pressure, other steps enter the picture to produce a more complex behavior. A result is the anomalous shape of the detonation limit, shown in Figure 9.2.

Equation 9.57 provides some clues about the dependence of the detonation limit on conditions. An increase in oxygen pressure promotes branching more than termination, and so favors detonation; an addition of a heavy inert component (M) does the opposite. A large surface-to-volume ratio makes for a large value of $k_{ads\,H}$, favoring termination. The propagation step p2, in which an oxygen-oxygen bond is broken, is highly endothermic and thus has a high activation energy; accordingly, a temperature increase strongly promotes chain branching and detonation (above about 1000 K, the stoichiometric gas mixture is explosive at any pressure).

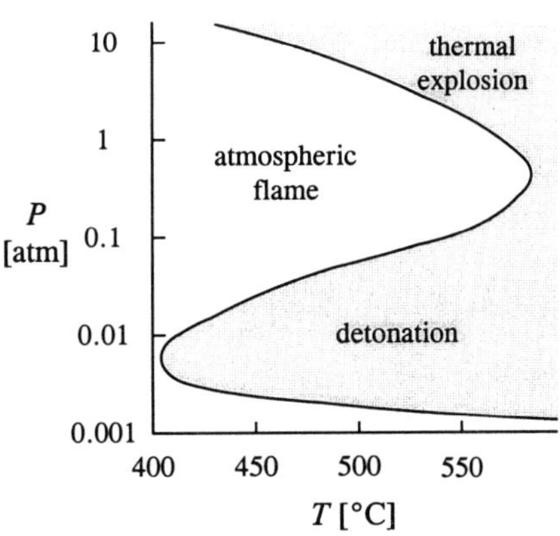

Figure 9.2. Explosion limit of 2:1 molar mixture of H_2 and O_2 as function of temperature and pressure (schematic).

The example of the hydrogen-oxygen reaction and eqn 9.57 nicely illustrate a critical facet of reactions with chain branching: the competition between the rates of excess chain-carrier production by branching and increased elimination by termination. However, reality is more complex. Systems with chain branching by their nature involve highly aggressive radicals and, therefore, a large number of possible steps of different molecularities and very different activation energies. As a result, variations in pressure and temperature, bound to occur when population growth becomes fast, are apt to produce dramatic shifts in control between various mechanisms. To some extent this is, of course, true for all types of reactions, but the potential for exponential self-acceleration caused by chain branching aggravates matters greatly.

The procedure of arriving at eqn 9.57 as an approximation for growth of the chain-carrier population is generally applicable:

(1)   use of the Bodenstein approximation for the free radicals *except* the most plentiful one in order to obtain an equation for the growth rate of the latter;

(2)   integration of this equation over time at constant volume.

As the example of the hydrogen-oxygen reaction has shown, this procedure provides clues about the sensitivity of the system and its dependence on conditions. Because of the many simplifications in its derivation, however, it can not be used to predict detonation limits.

9.7. Inhibition and induction periods

As mentioned at the outset, chain reactions, relying on free radicals as chain carriers, are sensitive toward any substances that can destroy or trap such radicals. The interference with chain propagation can assume two forms. An added substance can reduce the reaction rate to almost nil or bring it to essentially a complete and permanent stop. This is called *inhibition*. It occurs if the inhibitor catches practically all free radicals produced by the initiator. Under different conditions, an added substance or impurity can delay the start of a chain reaction for some period of time, called an *induction period*, without affecting its later course.

Inhibition [77]. An inhibitor is itself being consumed as it traps free radicals. To be effective, it must therefore be present in an excess over the initiator. In practice, this limits effective inhibition to chain reactions apt to be set off by small amounts of an initiator other than the bulk reactant. The most common application of inhibition is for protection of sensitive chemicals whose decomposition or polymerization by chain mechanisms may easily be triggered.

A typical example is the stabilization of highly reactive monomers such as styrene or methyl methacrylate by inhibitors such as hydroquinone, 4-*tert*-butyl-catechol, or TEMPO (a nitrosoxide) [78-81]. Free-radical polymerization of styrene is easily set off by impurities or the slightest amount of a free-radical producing agent and, being highly exothermic, can result in a thermal explosion. Another example is the use of antioxidants for protection of polymers against degradation by radicals produced by oxygen from air or ultraviolet radiation from sunlight [82,83].

Induction periods. An induction period typically occurs if a free-radical trap is present at a concentration much lower than that of the initiating substance (reactant or added initiator). An often cited example is the temporary inhibition of oxidation of cumene by a cresol derivative [57,84]: With twice as much inhibitor, the induction period is twice as long, and once the inhibitor is used up, the reaction is faster if cobalt is present (see Figure 9.3).

There are, however, other phenomena that can lead to a delay of a chain reaction. In autoxidation of hydrocarbons, for example,

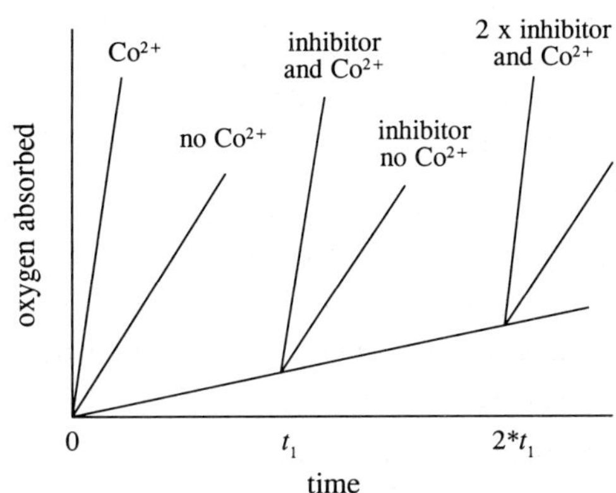

Figure 9.3. Effect of initiation (by AIBN and Co²⁺) and inhibition (by 2,6-di-*t*-butyl-*p*-cresol) on oxidation of cumene in glacial acetic acid (adapted from Moore and Pearson [77]).

the primary reaction product is a peroxide that, in turn, can act as initiator (see Example 9.3 in Section 9.6.2). Unless an initiator has intentionally been added, the reaction may start at a very low rate and then pick up speed as it produces its own initiator. Such behavior is best classified as *autocatalytic* (see Section 8.9).

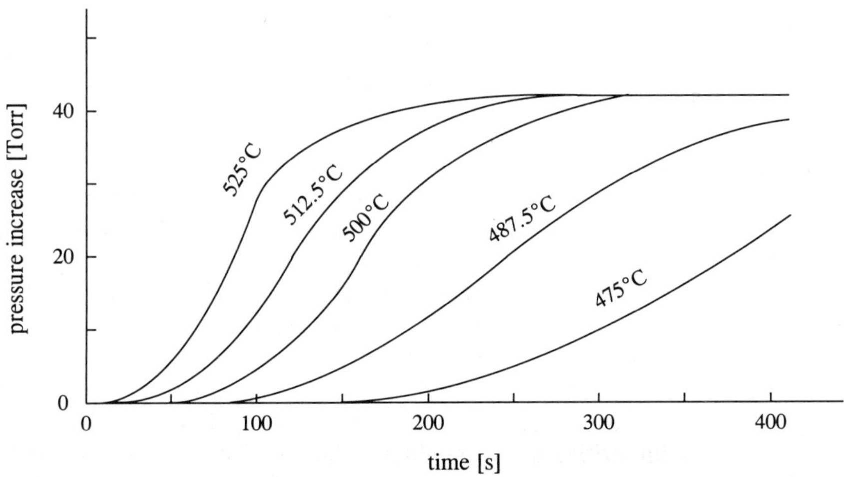

Figure 9.4. Pressure versus time in low-temperature oxidation of methane in constant volume batch (methane-to-oxygen mole ratio 2:1; adapted from Hoare [85]).

The agent responsible for autocatalytic behavior need not be the product of the reaction, it can be an intermediate. Low-temperature oxidation of methane provides an example [59,85]. The key free radical turns out to be $CH_3\cdot$, produced from methane by initiation and giving rise to other free radicals. The propagation mechanism with six interlocking steps and chain branching is such that build-up of $CH_3\cdot$ accelerates the rate (see Figure 9.4, previous page).

Yet another possible mechanism that can cause an induction period is reversible coupling (or other reaction with itself) of the most plentiful free radical to yield a product of low stability. Initially, none of that product is present. In the early stages of the reaction, its formation can be the dominant and highly effective termination, keeping the reaction at a low rate. With time, however, the metastable product builds up and approaches equilibrium with the free radicals from which it is formed. Picture the product as a reservoir into which the reaction drains free radicals until it is filled to capacity. By default, another and slower termination mechanism then takes over, and the reaction speeds up accordingly.

Summary

Chain reactions rely on highly reactive free radicals that convert reactants to products while cycling through a step sequence like a DO loop in FORTRAN, in which they are consumed and produced anew. The step sequence, called propagation, is initiated by a reaction or event that generates free radicals, and is terminated by reaction of free radicals with one another to form a stable product or products or, more rarely, their deactivation by some other mechanism. With few exceptions, termination is by reaction of the most plentiful free radical with itself. Termination usually is second-order in free radicals, and this leads to rate equations with exponents of one half or integer multiples of one half.

The steric and frequency factors of reactions of free radicals with one another or with small molecules are rather alike, and the activation energies are very low for exothermic reactions and only barely higher than the standard-enthalpy changes $\Delta H°$ for endothermic steps. In simple cases this makes it possible to use thermochemical data to identify which of the many possible steps dominate kinetics.

The free radicals generated by initiation are not necessarily the chain carriers. Rather, they may transmit their reactivity to other species, which then act as chain carriers. If so, initiation is indirect. Similarly, chain carriers may transmit their reactivity to other species, which then may or may not start new chains. This is called chain transfer. Whether initiation is direct or indirect makes little difference for kinetics. In contrast, chain transfer can have a profound effect. If a new chain of the same kind is started, the reaction continues at its pace, but chain transfer producing an unreactive radical may dominate termination, and the reaction then is first order rather than half order in initiator. If such transfer is fast, the reaction may be retarded or inhibited.

Very common in thermal degradation of organic substances, especially hydrocarbons, are so-called Rice-Herzfeld mechanisms. One of the two chain carriers abstracts a hydrogen atom from the reactant to form a product plus the other chain carrier, which then reacts to yield another product plus the first chain carrier.

Other typical chain reactions include those of hydrogen with halogen in the gas phase and oxidation of organic substances at moderate temperatures (autoxidation). A special facet of the latter reactions is that the product or an intermediate can act as initiator, and the reaction then is autocatalytic.

Certain chain reactions involve steps which generate more free radicals than they consume. This is called chain branching. The inherent self-acceleration may or may not outrun termination. If it does, a detonation results. A classical example is the reaction of hydrogen with oxygen.

Chain reaction may be inhibited by agents acting as free-radical traps. They may also exhibit induction periods, that is, delayed onset of the reaction. An induction period may be caused by a trace inhibitor that must first be destroyed by the free radicals, or actually reflect autocatalytic behavior. Yet another possible mechanism is the reversible formation of a metastable compound from free radicals, acting as a free-radical sink until filled.

Examples include the hydrogen-bromide reaction, thermal cracking of ethane and *n*-butane, oxidation of cyclohexane, and the hydrogen-oxygen reaction.

References

General references

G1. S. W. Benson, *The foundations of chemical kinetics*, McGraw-Hill, New York, 1960 (updated and corrected reprint, Krieger, Melbourne, 1982, ISBN 0898741947).

G2. M. Boudart, *Kinetics of chemical processes*, Prentice-Hall, Englewood Cliffs, 1968.

G3. R. J. Farrauto and C. H. Bartholomew, *Fundamentals of industrial catalytic processes*, Chapman & Hall, London, 1997, ISBN 0751404063, Chapter 12.

G4. G. F. Froment and K. B. Bischoff, *Chemical reactor analysis and design*, Wiley, New York, 2nd ed., 1990, ISBN 0471510440.

G5. A. A. Frost and R. G. Pearson, *Kinetics and mechanism: a study of homogeneous chemical reactions*, Wiley, New York, 2nd. ed., 1961.

G6. B. C. Gates, *Catalytic chemistry*, Wiley, New York, 1992, ISBN 0471517615.

G7. V. N. Kondratiev, in *The theory of kinetics*, Vol. 2 of *Comprehensive chemical kinetics*, C. H. Bamford and C. F. H. Tipper, eds., Elsevier, Amsterdam, 1969, ISBN 0444406743, Chapter 2.

G8. K. J. Laidler, *Chemical kinetics*, McGraw-Hill, New York, 3rd ed., 1987, ISBN 0060438622.

G9. J. W. Moore and R. G. Pearson, *Kinetics and mechanism: a study of homogeneous chemical reactions*, Wiley, New York, 3rd ed., 1981, ISBN 0471035580.

G10. G. W. Parshall and S. D. Ittel, *Homogeneous catalysis: the application and chemistry of catalysis by soluble transition metal complexes*, Wiley, New York, 1992, ISBN 0471538299.

G11.   A. Pross, *Theoretical and physical principles of organic reactivity*, Wiley, New York, 1995, ISBN 0471555991.

G12.   W. A. Pryor, *Free radicals*, McGraw-Hill, New York, 1966 (reprinted by Books on Demand, Ann Arbor, publication projected, ISBN 0608099430).

G13.   F. O. Rice and K. K. Rice, *The aliphatic free radicals*, Johns Hopkins Press, Baltimore. 1935.

G14.   N. N. Semenov, *Some problems of chemical kinetics and reactivity*, Pergamon, New York, Vol. 1, 1958.

G15.   E. W. R. Steacie, *Atomic and free radical reactions*, Reinhold, New York, 2nd ed., 1954 (2 volumes).

G16.   J. I. Steinfeld, J. S. Francisco, and W. L. Hase, *Chemical kinetics and dynamics*, Prentice-Hall, Englewood Cliffs, 2nd ed., 1999, ISBN 0137371233.

G17.   G. S. Yablonskii, V. I. Bykov, A. N. Gorban', and V. I. Elokhin, *Kinetic models of catalytic reactions*, in *Comprehensive chemical kinetics*, R. E. Compton, ed., Elsevier, Amsterdam, Vol.32, 1991, ISBN 0444888020.

Specific references

1.   M. Bodenstein, *Z. physik. Chem.*, **85** (1913) 329.

2.   F. Haber and J. Weiss, *Naturwissenschaften*, **20** (1932) 948; *Proc. Roy. Soc.*, **A 147** (1934) 332.

3.   M. Bodenstein and H. Lütkemeyer, *Z. physik. Chem.*, **114** (1925) 208.

4.   G. R. Gavalas, *Chem. Eng. Sci.*, **21** (1966) 133.

5.   M. Bodenstein and S. C. Lind, *Z. physik. Chem.*, **57** (1907) 168.

6.   J. A. Christiansen, *Kgl. Dansk Videnskab Selskab, Mat-fys*, **1** (1919) 14.

7.   K. F. Herzfeld, *Ann. Physik*, **59** (1919) 14.

8.   M. Polányi, *Z. Elektrochem.*, **26** (1920) 49.

9.   M. G. Evans and M. Polányi, *Trans. Faraday. Soc.*, **34** (1938) 3 and 11.

10.   Semenov (ref. G14), pp. 27-28.

11.   Boudart (ref. G2), Section 8.1.

12.   Moore and Pearson (ref. G9), pp. 199-201.

13.   H. S. Johnston and C. Parr, *J. Am. Chem. Soc.*, **85** (1963) 2544.

14.   Pross (ref. G11), Chapter 10.

15.   E. Rabinowitch and H. L. Lehmann, *Trans. Faraday Soc.*, **31** (1935) 689.

16.   Steinfeld *et al.* (ref. G16), Section 2.3.1.

17.   W. Jost and G. Jung, *Z. physik. Chem.*, **B 3** (1929) 83.

18.   M. Ritchie, *Proc. Roy. Soc.*, **A 146** (1934) 828.

19.   e.g., see Froment and Bischoff (ref. G4), pp. 36-37.

20.   K. M. Sundaram and G. F. Froment, *I&EC Fundam.*, **17** (1978) 174.

21.   Steacie (ref. G15), Chapter IV (includes copious references to earlier work).

22. F. O. Rice, *J. Am. Chem. Soc.*, **53** (1931) 1959; **55** (1933) 3035.

23. F. O. Rice, W. R. Johnston, and B. L. Evering, *J. Am. Chem. Soc.*, **54** (1932) 3529.

24. Rice and Rice (ref. G13), Chapter III.

25. F. A. Paneth and W. Hofeditz, *Chem. Ber.*, **62 B** (1929) 1335.

26. Rice and Rice (ref. G13), Chapter IV.

27. F. O. Rice and K. F. Herzfeld, *J. Am. Chem. Soc.*, **56** (1934) 284.

28. Rice and Rice (ref. G13), Chapter VII.

29. Froment and Bischoff (ref. G4), p.27.

30. Steacie (ref. G15), Section IV-8 (includes copious references to earlier work).

31. K. J. Laidler, and B. W. Wojciechowski, *Proc. Roy. Soc.*, **A 260** (1961) 91.

32. Laidler (ref. G8), Section 8.5.3.

33. C. P. Quinn, *Proc. Roy. Soc.*, **A 275** (1963) 190.

34. M. J. Pilling, in *Modern gas kinetics: theory, experiment and application*, M. J. Pilling and I. W. M. Smith, eds., Blackwell, Oxford, 1987, ISBN 0632016159, Chapter C5.

35. L. Küchler and H. Theile, *Z. physik. Chem.*, **B 42** (1939) 359.

36. M. C. Lin and M. H. Back, *Can. J. Chem.*, **44** (1966) 505; **45** (1967) 3165.

37. F. A. Lindemann, *Trans. Faraday Soc.*, **17** (1922) 598.

38. K. J. Laidler and B. W. Wojciechowski, *Proc. Roy. Soc.*, **A 259** (1960-61) 257.

39. J. H. Purnell and C. P. Quinn, *J. Chem. Soc.*, **1961**, 4128.

40. M. J. Pilling (ref. 35), pp. 234-235.

41. C. G. Hill, Jr., *An introduction to chemical engineering kinetics and reactor design*, Wiley, New York, 1977, ISBN 0471396095, Section 8.5.1.

42. O. Levenspiel, *Chemical reaction engineering*, Wiley, New York, 3nd ed. 1972, ISBN 047125424X, Chapter 5.

43. M. Letort, *J.Chim.Phys.*, **34** (1937) 206 and 256.

44. P. Goldfinger, M. Letort, and M. Niclause, *Contribution à l'étude de la structure moléculaire*, Victor Henri Commemorative Volume, Desoer, Liège, 1947-48, p. 283; see Froment and Bischoff (ref. G3), Table 1.4.1, p. 29.

45. D. A. Leathard and J. H. Purnell, *Ann. Rev. Phys. Chem.*, **21** (1970) 197.

46. Benson (ref. G1), p. Section XIII.9.

47. J. H. Purnell and C. P. Quinn,*Proc. Roy. Soc.*, **A 270** (1962), 267.

48. K. J. Laidler, *Chemical kinetics*, McGraw-Hill, New York, 2nd ed., 1965,, pp. 406-407.

49. e.g., see M. A. Baltanas, K. K. Van Raemsdonck, G. F. Froment, and S. R. Mohedas, *I&EC Research*, **28** (1989) 899.

50. *Landolt-Börnstein, New Series, Radical reaction rates in liquids*, H. Fischer, ed., Springer, Berlin, Part II Vol. 13 (5 subvolumes), 1983-1985, ISBN 0387126074, 0387132414, 0387117253, 0387121978, 0387136762; Vol. 18 (5 subvolumes), 1994-1997, 3540560548, 3540560556, 3540560564, 3540603573, 3540560572.

51.   A. B. Ross, W. G. Mallard, W. P. Helman, G. V. Buxton, R. E. Huie, and P. Neta, *NDRL-NIST Solution kinetics database*, Ver. 3.0, Notre Dame Radition Laboratory, Notre Dame, and National Institute of Standards, Gaithersburg, 1998.

52.   W. D. Walters, in A. L. Fries and A. Weisberger, eds., *Techniques of organic chemistry*, Interscience, NY, vol. 8, 1953, pp. 283-296.

53.   M. Letort, *Chim. et Ind.*, **76** (1956) 430.

54.   J. G. Calvert and J. T. Gruver, *J. Am. Chem. Soc.*, **80** (1958) 1313.

55.   Rice and Rice (ref. G13), Chapter VIII.

56.   Frost and Pearson (ref. G13), pp. 256-258 (includes references to earlier work).

57.   G. A. Russell, *J. Chem. Educ.*, **36** (1959) 111.

58.   Semenov (ref. G14), p. 399.

59.   Kondratiev (ref. G7), Section 3.3.2.

60.   E. T. Denisov and T. Denisova, *Handbook of antioxidants: bond dissociation energies, rate constants, activation energies, and enthalpies of reactions*, CRC Press, Boca Raton, 2nd ed., 2000, ISBN 0849390044, Chapter 2.

61.   V. A. Kritsman, G. E. Zaikov, and N. M. Emanuel', *Chemical kinetics and chain reactions: historical aspects*, English translation Nova Science, Commack, 1995, ISBN 1560721669, Section VII-2 (many references to Russian work).

62.   L. Reich and S. S. Stivala, *Autoxidation of hydrocarbons and polyolefins: kinetics and mechanisms*, Dekker, New York, 1969; reprint Books on Demand, Ann Arbor (projected), ISBN 0783707754.

63.   Gates (ref. G6), p. 67.

64.   I. V. Berezin, E. T. Denisov, and N. M. Emanuel', *The oxidation of cyclohexane*, Pergamon, Oxford, 1966.

65.   Parshall and Ittel (ref. G10), Section 10.3.

66.   J. W. Parshall, *J. Mol. Catal.*, **4** (1978) 243.

67.   C. A. Tolman, J. D. Druliner, F. J. Krusic, M. J. Nappa, W. C. Seidel, I. D. Williams, and S. D. Ittel, *J. Mol. Catal.*, **48** (1988) 129.

68.   P. D. Bartlett, ACS Symp. Ser., **69** (1978) 15.

69.   D. L.Baulch, C. J. Cobos, R. A. Cox, C. Esser, P. Frank, T. Just, J. A. Kerr, M. J. Pilling, J. Troe, R. W. Walker, and J. Warnatz, *J. Phys. Chem. Ref. Data*, **21** (1992) 411.

70.   D. L. Baulch, C. J. Cobos, R. A. Cox, P. Frank, G. Hayman, T. Just, J. A. Kerr, T. Murrells, M. J. Pilling, J. Troe, R. W. Walker, and J. Warnatz, *J. Phys. Chem. Ref. Data*, **23** (1994) 847.

71.   *Handbook of chemistry and physics*, D. R. Lide, ed.-in-chief, CRC Press, Boca Raton, 80th ed., 1999-2000, ISBN 0849304806, Section 5, pp. 505-519.

72.   Kondratiev (ref. G7), Section 3.2.

73.   R. R. Baldwin and R. W. Walker, *Essays Chem.*, **3** (1972) 1.

74.   Moore and Pearson (ref. G9), pp. 408-411.

75.   Steinfeld *et al.* (ref. G16), Section 14.2

76.   Yablonskii *et al.* (ref. G17), Section 1.1.

77. Moore and Pearson (ref. G9), pp. 398-401.

78. V. W. Bowry and K. U. Ingold, *J. Am. Chem. Soc.*, **114** (1992) 4992.

79. G. Odian, *Principles of polymerization*, Wiley, 3rd ed., 1991, ISBN 0471610208, Section 3.7b.

80. G. Moad and D. H. Solomon, *The chemistry of free radical polymerization*, Pergamon, 1995, ISBN 0080420788, Section 5.4.

81. S. S. Chen, *Styrene*, in *Kirk-Othmer, encyclopedia of chemical technology*, J. I. Kroschwith and M. Howe-Grant, eds., 7th ed. Wiley, New York, Vol. 22, 1997, ISBN 0471526916, p.956 (see p. 973).

82. Yu. A. Shliapnikov, *Antioxidative stabilization of polymers*, Taylor & Francis, London, 1996, ISBN 0748405771.

83. Denisov and Denisova (ref. 56), Chapter 9.

84. H. S. Blanchard, *J. Am. Chem. Soc.*, **82** (1960) 2014.

85. D. E. Hoare, *Combustion of methane*, in *Low temperature oxidation*, W. Jost, ed., Gordon & Breach, New York, 1968, p. 125.

Chapter 10

Polymerization

In principle, polymerization is a chemical reaction like any other.* However, in most cases it occurs under circumstances that differ from those of ordinary homogeneous reactions, with consequences for kinetic behavior. For example, although the reaction mixture may be homogeneous at start, the polymer may precipitate, and monomer diffusion into the newly formed particles may then become an important factor. Even if the reaction mixture does remain homogeneous, entanglement of polymer chains grown to great length may impede the mobility of the reactive groups at their ends. Depending on conditions, this may slow the reaction down, may accelerate it into a runaway, or may bring it to a standstill short of complete conversion. Such effects are matters of polymer reaction engineering and can be dealt with here only in outline. The present chapter is intended mainly as a guide to show how the principles, methods, and mathematics developed in earlier parts of this book can be used where they are applicable, namely, in homogeneous polymerization at no more than moderate conversion or in dilute solution (several phases may be present, provided the reaction is confined to a homogeneous liquid). Even in such cases, an accuracy and reliability as high as for ordinary homogeneous reactions cannot be expected.

As in previous chapters, the presentation remains restricted to kinetic principles and their applications in modeling. No attempt is made to review the intricacies of reactivities of monomers and their molecular causes or their effect on polymer structure and properties. For details of these, the reader is referred to excellent texts on the subject [G1-G12].

In polymerization kinetics, not only rates are of interest, but also molecular weights and their distributions. To the extent that this can be done without getting embroiled in major complications, such aspects are included.

10.1. Types of polymerization reactions

From a point of view of chemistry, a distinction can be made between *condensation* and *addition* as mechanisms of polymerization, depending on whether or not a small molecule such as H_2O or HBr is cast off when monomers link up [1]. Examples are

---

* The term polymerization is meant to include *oligomerization* (formation of low polymer).

condensation polymerization 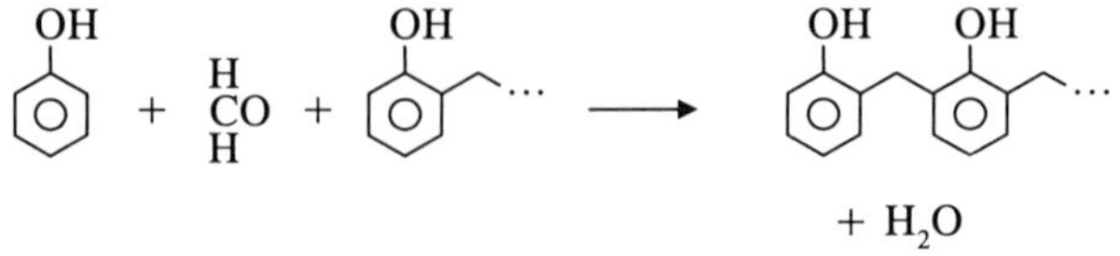 (10.1)

addition polymerization 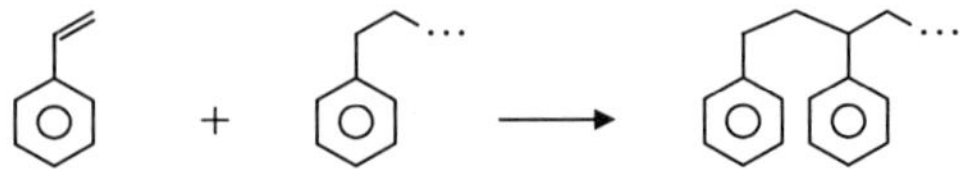 (10.2)

However, a classification into condensation and addition polymers as originally envisaged by Carothers [2] is no longer appropriate because some polymers can be synthesized by either method. For example, *Nylon*-6, a polymer with repeating units $-NH(CH_2)_5CO-$, can be made either from 6-aminocaproic acid by condensation polymerization or from caprolactam by addition polymerization.

As far as mathematics of reaction kinetics is concerned, there is no fundamental difference between condensation and addition polymerization: Whether or not a small molecule is cast off when two monomer or polymer molecules link up has little impact in this respect. On the other hand, a distinction relevant to kinetics is that between *step growth* and *chain growth* as reaction mechanisms. In step-growth polymerization, monomer or already formed polymer is added step by step to the growing polymer at its functional groups in an ordinary chemical reaction, and such link-up is the only reaction that occurs. Chain-growth polymerization differs in that a "reactive center" must first be created at a molecule, and new monomer molecules are then added successively until some other reaction or event brings the process to a halt and produces "dead" polymer that reacts no further: Chain growth must be initiated and usually also involves a termination reaction. It is sustained by a small minority of highly reactive chain carriers that act as an assembly line for churning out dead polymer molecules.

The most common type of chain-growth polymerization is *free-radical polymerization*. An initiator or a photochemical reaction produces a free radical that attaches itself to a monomer molecule, creating a group with odd-electron configuration (reactive center) at which monomer molecules are added until two such centers react with one another or, more rarely, a center is deactivated by some other process. This is a mechanism much like that of ordinary chain reactions (see Chapter 9; the term "chain" in chain growth refers to that kind of mechanisms, not to the growing molecular chain of repeating units in the polymer.)

Unlike ordinary chain reactions, chain-growth polymerization need not involve free radicals. The reactive center may instead be a carbanion or carbocation generated by intermolecular transfer of a proton or electron. Depending on the sign of the ionic charge on the chain carriers, the overall reaction is called *anionic* or *cationic polymerization*. As in free-radical polymerization, initiation is required.

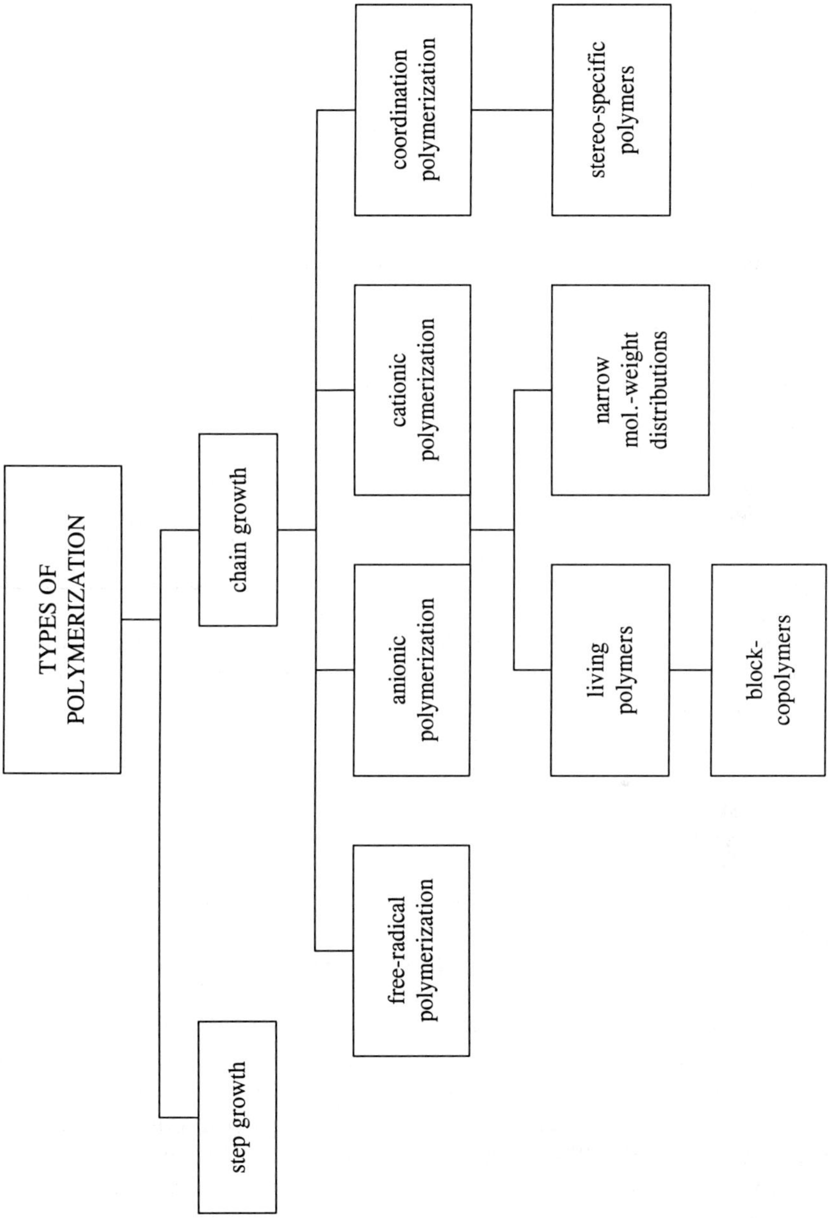

Figure 10.1. Types of polymerization (adapted from Helfferich and Savage [3]).

However, termination by reaction of two chain carriers with one another cannot occur because charges of the same sign repel one another. As a result, reactive centers may be left over when all monomer is used up (*living polymers*). This can be utilized for production of specialty polymers.

Coordination polymerization is yet another variation on the same theme. Here, polymerization is initiated by attachment of a monomer molecule to a metal complex. The polymer grows by successive insertion of monomer molecules at the metal. Growth stops when the metal complex detaches itself or the reactive center becomes deactivated by some intended or inadvertent event. Stereo-specific polymers can be produced.

10.2. Step-growth polymerization

The earliest polymers of practical use were prepared by step-growth reactions, most notable among them *Bakelite*, a phenol-formaldehyde copolymer first marketed in 1910 [4]. Its name was long almost synonymous with synthetic plastics and resins, has become generic, and is no longer restricted to phenol-formaldehyde copolymers. Most but not all step-growth polymerizations are condensations.

10.2.1. Functionality

Step growth involves reactions of functional groups with one another. For example, the functional groups in polymerization of 6-aminocaproic acid to *Nylon*-6 [5,6]

$$n \ \ H_2N \diagdown\diagup\diagdown\diagup \ COOH \ \longrightarrow \ H-\left[\begin{matrix} N \\ H \end{matrix}\diagdown\diagup\diagdown\diagup\begin{matrix} C \\ O \end{matrix}\right]_n-OH \ + \ n\text{-}1 \ \ H_2O \qquad (10.3)$$

are $-NH_2$ and $-COOH$. The monomer, carrying two groups per molecule, is said to be *bifunctional*. The functionality may be higher. For example, glycerol with its three hydroxyl groups is *trifunctional* in condensation polymerization with, say, a dicarboxylic acid or organic dihalide.

The functionality may vary with reaction conditions. For example, in base-catalyzed copolymerization of phenol and formaldehyde, both monomers are bifunctional at ambient temperature, but phenol becomes trifunctional if the temperature is raised sufficiently. Copolymerization at ambient temperature can produce essentially linear, liquid, resole-type "prepolymers" of low molecular weight. Upon acidification and heat-curing, methylene and ether crosslinks formed by the now trifunctional phenol units transform the polymer into an insoluble resin [7] (see next page). The original *Bakelite* was such a "thermosetting" product.

An additional functionality that comes into play only when the reaction conditions are changed is called *latent functionality*.

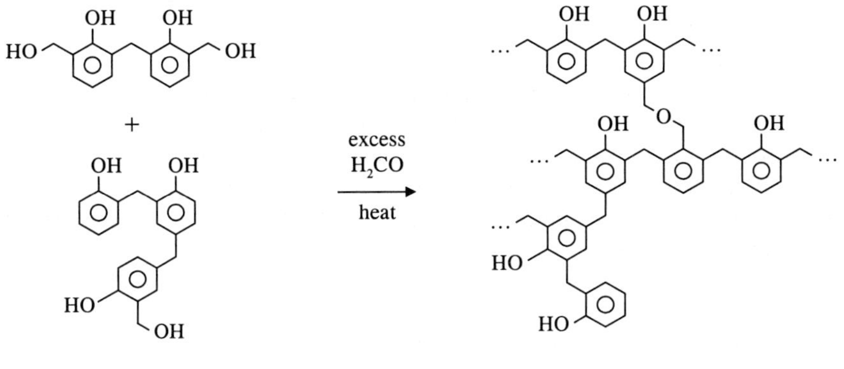

resole-type oligomers

heat-cured resin

10.2.2. Mechanism and rate

Homopolymerization. In the simplest type of step growth, a bifunctional monomer reacts successively with itself, eventually forming a polymer with a large number of repeating units. The reaction may be an addition, but more commonly is a condensation. Although condensation usually is reversible, its equilibrium is driven toward complete conversion by removal of the small and volatile cast-off molecule:

$$M \longrightarrow P_2 \longrightarrow P_3 \longrightarrow \ldots \tag{10.4}$$

where M is monomer, Q is a cast-off small molecule, and P_i is a polymer molecule with i repeating units. A reaction involving only one kind of monomer, as in 10.4, is called *homopolymerization*. Polymerization of 6-aminocaproic acid to *Nylon*-6 (reaction 10.3 farther above) is an example from industrial practice.

Since the functional end groups of the polymer molecules P_i formed are the same as those of the monomer M from which they were formed, one must expect monomer link-up per reactions 10.4 to be accompanied by link-up of polymer molecules with one another. Also, if the cast-off small molecule is not effectively removed, polymer molecules may split up again:

link-up: $\qquad P_i + P_j \longrightarrow P_k + Q \qquad (k = i + j) \qquad (10.5)$

split-up: $\qquad P_k + Q \longrightarrow P_\ell + P_m \qquad (k = \ell + m) \qquad (10.6)$

Moreover, a so-called *interchange reaction*

interchange: $\qquad P_i + P_j \longleftrightarrow P_\ell + P_m \qquad (i+j = \ell+m) \qquad (10.7)$

may scramble polymer fragments of different length. Even if normal condensation is dominant and irreversible, interchange may occur if the temperature is raised (e.g., upon further processing).

All steps in reactions 10.4 and 10.5 involve functional groups of the same kinds. One may therefore assume that all of them, except possibly the first linking of two monomers, have approximately the same rate coefficients [8] (see "short-sightedness" of reaction steps, Section 11.3). Moreover, if the reaction is run in dilute solution or stopped at reasonably low conversion, the coefficients may be assumed to remain unchanged as conversion progresses. In practice, these assumptions are usually valid, but not without exceptions [9-11], and must therefore be verified. However, even granted their validity, the wealth of simultaneous rate equations for all the participants makes it quite cumbersome to obtain concentration histories for the monomer and the individual polymers. [Analytical solutions can be obtained if condensation is irreversible and polymer-polymer link-up and inter-change per reactions 10.5 and 10.7, respectively, can be disregarded [12], but are of little more than academic interest.]

Easier to come by and just as useful is information about conversion of the functional groups. Each link-up in reactions 10.4 and 10.5 eliminates the two functional end groups that react with one another, and any interchange by reaction 10.7 leaves their number unchanged. Accordingly, the disappearance of functional groups is a bimolecular reaction and so essentially follows second-order kinetics. If the reverse reaction and polymer split-up are insignificant or suppressed, e.g., by removal or elimination of the cast-off small molecule Q as it is formed, the rate is

$$-r_\mathrm{F} \;=\; 2kC_\mathrm{F}^2 \tag{10.8}$$

where F stands for the functional groups of both kinds. Of particular interest is the fractional conversion of the functional groups, $f_\mathrm{F} \equiv 1 - C_\mathrm{F}/C_\mathrm{F}^\circ$, as a function of reaction time t or reactor space time τ:

batch:
$$f_\mathrm{F}(t) = 1 - \frac{1}{1 + 2ktC_\mathrm{F}^\circ} \tag{10.9}$$

continuous stirred tank:
$$f_\mathrm{F}(\tau) = 1 - \frac{(1 + 8k\tau C_\mathrm{F}^\circ)^{1/2} - 1}{4k\tau C_\mathrm{F}^\circ} \tag{10.10}$$

Derivation. For liquid-phase batch, where $-r_\mathrm{F} = -dC_\mathrm{F}/dt$, eqn 10.9 is obtained by integration of eqn 10.8 over time; for a continuous stirred tank, eqn 10.10 is obtained from eqn 10.8 and the material balance for the functional groups, $-r_\mathrm{F} = (C_\mathrm{F}^\circ - C_\mathrm{F})/\tau$. Equations 10.9 and 10.10 assume the reverse reaction to be negligible or suppressed, the rate coefficient to be independent of conversion, and no significant fluid-density variation to occur upon reaction.

Copolymerization. Up to this point we have considered polymerization of a single monomer that carries two different functional end groups, those of one kind reacting with those of the other. Many commercial polymers, however, are produced by condensation of two monomers, each of which carries functional groups of the same kind at both ends. An example is *Nylon*-6,6, a polymer with alternating diamine and dicarbonyl units, made from 1,6-diaminohexane and adipic acid [5]:

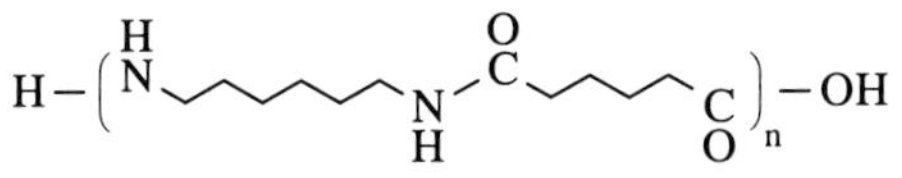

In such cases also, all linking steps are reactions of the same groups, so that the rate coefficient can once again be assumed to have approximately the same value for all. However, the two types of functional groups now are on different monomers and therefore are not necessarily present in stoichiometric amounts. For stoichiometric mixtures of monomers with different functionalities, the rate equation 10.8 and eqns 10.9 and 10.10 for fractional conversion remain valid. For nonstoichiometric mixtures, eqn 10.8 must be replaced by

$$-r_{F_A} = k C_{F_A} C_{F_B} \tag{10.11}$$

where F_A and F_B are the two different functional groups. The fractional conversion of F_A, the group that is in the minority, is now given by

batch:
$$f_{F_A} = 1 - \frac{\Delta}{C_{F_B}^o \exp(\Delta kt) - C_{F_A}^o} \tag{10.12}$$

CSTR:
$$f_{F_A} = 1 - \frac{1 + \Delta k\tau}{2k\tau C_{F_A}^o}\left[\left(1 + \frac{4k\tau C_{F_A}^o}{(1 + \Delta k\tau)^2}\right)^{1/2} - 1\right] \tag{10.13}$$

where

$$\Delta \equiv C_{F_B}^o - C_{F_A}^o$$

Derivation. Equation 10.12 is obtained from the integrated form of eqn 10.11 found in standard texts [13,14]:

$$\ln(C_{F_B}/C_{F_A}) = \ln(C_{F_B}^o/C_{F_A}^o) + (C_{F_B}^o - C_{F_A}^o)kt$$

Equation 10.13 follows from eqn 10.11 and the material balance for F_A in the CSTR.

More than two monomers may participate in copolymerization, and functionalities may be higher. Nevertheless, eqns 10.8 to 10.10 or eqn 10.11 remain applicable as long as the reverse reaction is negligible or suppressed and the rate coefficient is independent of conversion. However, the latter assumption becomes questionable as monomers of higher functionalities begin to form crosslinks (see also discussion of gel point farther below).

Cyclization. The two functional end groups of a monomer or polymer molecule might react with one another to form a cyclic compound [15,16]. An example is the formation of caprolactam as a by-product in condensation polymerization of 6-aminocaproic acid to *Nylon*-6 [5]:

$$H_2N \sim\sim\sim COOH \longrightarrow \quad \begin{array}{c} C = O \\ | \\ NH \end{array} \quad + \quad H_2O \qquad (10.14)$$

Such so-called *cyclization* can occur to the almost complete exclusion of polymerization if five- or six-membered rings are formed [17]. Smaller rings are not favored because their bond angles are strained; neither are much larger ones because a functional end group on a long chain is likely to react with an end group of another molecules before it has a chance to come close to that on the other end of its own chain (see Figure 10.2). Rarely will more than one or perhaps two of the species involved undergo cyclization to any significant extent.

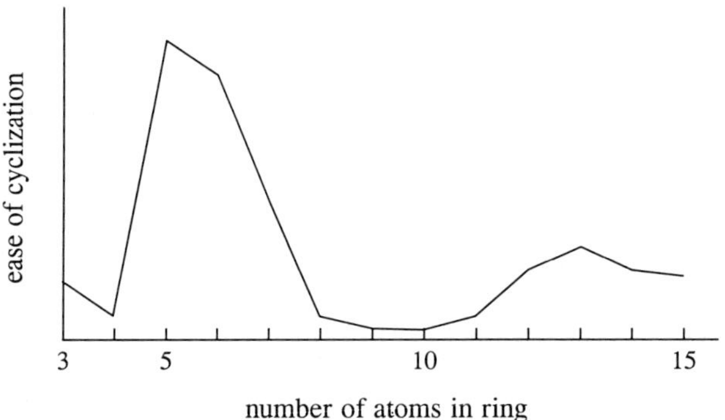

Figure 10.2. Dependence of extent of cyclization on size of ring formed (from Odian [16]).

Cyclization comes about by a reaction of the same kinds of functional groups as the link-ups of monomer or polymer molecules and, like these, consumes two functional groups. Cyclization therefore does not change the form of the rate equation 10.8 or 10.11 for consumption of functional groups. However, the degree of polymerization and the molecular weight are affected (see below).

10.2.3. Degree of polymerization and molecular weight

Number-average degree of polymerization. For step-growth polymerization, the number-average degree of polymerization, $\bar{N}$, is traditionally defined as

$$\bar{N} \equiv \frac{\text{number of monomer molecules at start}}{\text{number of molecules after polymerization}} \qquad (10.15)$$

As will be seen, it can be related in a relatively simple fashion to the fractional conversion of functional groups [18], provided cyclization is insignificant.

In step-growth polymerization of bifunctional monomers, each molecule, whether monomer or polymer, carries two functional groups: There are always half as many molecules as functional groups. Thus, when the number of unreacted groups has decreased to a fraction $1 - f_F$ of the initial, the number of molecules has decreased to that same fraction of its initial. By virtue of its definition 10.15, the number-average degree of polymerization (monomer included in averaging) is the reciprocal of this fraction:

$$\bar{N} = 1/(1 - f_F) \qquad (10.16)$$

In homopolymerization or in copolymerization of stoichiometric mixtures of bifunctional monomers, f_F can be replaced by means of eqn 10.9 or 10.10. [If monomer is excluded from the mole count, $\bar{N} = (2 - f_F)/(1 - f_F)$ (see eqn 10.85 in Section 10.5.2). At the usual high degrees of polymerization ($f_F \rightarrow 1$), the difference becomes negligible.]

> *Example 10.1. Control of molecular weight.* Assume the end use of a polymer made by step-growth homopolymerization of a bifunctional monomer requires a number-average molecular weight $\overline{MW}$ of about 10,000 and that the molecular weight of the repeating unit is 100. Accordingly, the number-average degree of polymerization should be $10,000/100 = 100$. Equation 10.16 with $\bar{N} = 100$ and solved for f_F gives a fractional conversion of functional groups $f_F = 0.990$.
>
> There is little margin for deviation because a small variation in conversion results in a large change in molecular weight: $f_F = 0.980$ would give $\bar{N} = 50$ and $\overline{MW} = 5,000$, and such a polymer may not have the required mechanical strength; $f_F = 0.995$ would give $\bar{N} = 200$ and $\overline{MW} = 20,000$, and that product may well be too stiff for processing. An easier way of controlling the molecular weight would be to add one mole percent of a monofunctional compound that reacts with and deactivates one percent of one of the functional groups, and then drive conversion of the remaining 99 percent of that group essentially to completion.

Equation 10.16 for bifunctional monomers is a special case of the more general Carothers equation [18] that is applicable to monomers with any functionalities:

$$\bar{N} = 2/(2 - \bar{n}_F f_F) \qquad (10.17)$$

where $\bar{n}_F$ is the effective average functionality. In copolymerization of mixtures with stoichiometric amounts of functional groups, it is the average functionality; in copolymerization of nonstoichiometric mixtures, groups that cannot react because they exceed the stoichiometric amount are not counted in the averaging [19].

Example 10.2. Number-average degree of polymerization in step-growth polymerization of nonstoichiometric mixture of monomers. A mixture of two moles of glycerol and five moles of phthalic acid reacts. There are 2x3 = 6 $-$OH groups from glycerol and 5x2 = 10 $-$COOH groups from phthalic acid on 2+5 = 7 molecules. Only six of the ten acid groups can react with the six $-$OH groups, the other four are not counted. The effective average functionality thus is (6+6)/7 = 1.714. According to eqn 10.17, the maximum number-average degree of polymerization that can be obtained, at complete conversion of glycerol, is 2/(2 $-$ 1.714) = 6.99.

The Carothers equation becomes invalid if cyclization occurs to a significant extent. Cyclization reduces the number of functional groups, but leaves the number of molecules unchanged. This violates the underlying premise that there are always twice as many functional groups as there are molecules.

Gel point. The Carothers equation can also be used to estimate the conversion needed to reach the so-called gel point in condensation polymerization involving monomers with functionalities higher than 2. The gel point is defined as the state of conversion at which gel formation caused by crosslinking begins to become apparent. With the assumption that this occurs when practically all molecules of the limiting monomer have reacted, the requisite fractional conversion of functional groups can be estimated with a rearranged form of the Carothers equation:

$$f_F \;=\; 2/\bar{n}_F - 2/\bar{n}_F\bar{N}$$

As the number-average degree of polymerization, $\bar{N}$, is driven as high as possible, the second term on the right-hand side becomes negligible, so that

$$(f_F)_{gel} \;=\; 2/\bar{n}_F \tag{10.18}$$

For example, in condensation of an equimolar mixture of glycerol with a trifunctional acid such as citric, $\bar{n}_F = 3$ (both monomers are trifunctional), and the gel point is reached at a fractional conversion of groups of 2/3 = 0.667.

Gelation actually begins before all molecules of the limiting monomer have reacted. The Carothers equation therefore overestimates the conversion of functional groups needed to reach the gel point. A more rigorous statistical treatment by Flory [20,21] and Stockmayer [22,23] considers the probability that a unit becomes attached to two chains [24]. This approach gives lower values for the fractional conversion at the gel point. Experimental observations suggest that the actual gel point typically falls between the estimates with Carothers' and Flory's equations [20,21,25]. A computer simulation of sol-gel distribution at high conversion has been published [26].

Molecular weight and molecular-weight distribution. The Carothers equation 10.17, where applicable, provides the number-average degree of polymerization of the reaction mixture (unreacted monomer included in the mole count). Usually, conversion of monomer is driven to a very high degree of completion and cyclization is suppressed. The number-average molecular weight of the polymer can then be obtained in good approximation from the number-average degree of polymerization simply by multiplication with the molecular weight of the structural unit (average weight if two different units alternate). However, the molecular-weight distribution is harder to come by and cannot be predicted with the same degree of accuracy. Here, the utility of mathematical theory is more in showing trends and relative magnitudes of effects than in quantitative predictions or application to design.

Since step growth is a sequential reaction, the distribution of products it yields depends on the type of reactor (see Section 5.4). Analytical solutions can be obtained only under grossly simplifying assumptions and, therefore, are of little use in practice. In principle, the simultaneous rate equations for all participants, or at least their ratios, would have to be known and solved under the respective conditions. The complications here are that polymer link-up (reaction 10.5), interchange (reaction 10.7) and, more rarely, cyclization (reactions such as 10.14) may occur. Polymer link-up shifts the distribution to higher molecular weights and broadens it, and interchange and cyclization distort the relationship between fractional conversion and molecular weight.

The most useful and most commonly employed simplified approach dates back to Flory [27,28] and is based on the premise of equal reactivity of functional groups and statistical growth. The most important application is to polymerization of bifunctional monomers and can be sketched as follows (Flory's derivation is more elaborate).

In homopolymerization or copolymerization of stoichiometric mixtures of two monomers, the probability that two functional groups have reacted to form a link is given by the fractional conversion of groups, f_F. A polymer with $j+1$ repeating units contains one more link than one with only j units. Therefore, its existence is less probable by a factor f_F than that of the latter. In view of the large number of molecules involved, the ratio of the existence probabilities is also that of the mole fractions, x_{j+1} and x_j. Accordingly:

$$x_{j+1} = f_F x_j \qquad (j \geq 1) \qquad (10.19)$$

or, for the mole fraction as an explicit function of fractional conversion:

$$x_j = f_F^{j-1} \Big/ \sum_{n=1}^{\infty} f_F^{n-1} = f_F^{j-1}(1 - f_F) \qquad (j \geq 1) \qquad (10.20)$$

[The sum converges to $1/(1 - f_F)$. Note that the mole count includes the monomer, but not the solvent, cast-off small molecules, or any inerts.]

According to this statistical approach:

> In step-growth polymerization of bifunctional monomers, the mole fractions of successive polymers (with increasing number of structural units) are in a declining geometrical progression.
>
> The factor by which the mole fractions of two successive polymers differ is given by the fractional conversion of the functional groups.

A mole-fraction distribution that is a declining geometrical progression is called a *Schulz-Flory distribution* or *most probable distribution* and is quite common [29,30]. As later examples will show, it can arise from other mechanisms as well and can therefore not be taken as evidence for step growth.

Quantitatively, the weight fraction of polymer with j structural units as a function of fractional conversion of functional groups is given by

$$ w_j \;=\; j f_F^{j-1} \Big/ \sum_{n=1}^{\infty} n\, f_F^{n-1} \;=\; j f_F^{j-1}(1 - f_F)^2 \qquad (j \geq 1) \qquad (10.21)$$

[The sum converges to $1/(1 - f_F)^2$.] This formula is for addition polymerization and requires a small correction for the weight of the cast-off small molecule in condensation polymerization.

Although the mole-fraction distributions show a monotonic decline with the number of repeating units in the polymer, the molecular-weight distributions have maxima. This is because, in the low-polymer range, the weight increase with number of repeating units overcompensates the decrease in mole fraction. With progressing conversion, the maximum shifts to higher molecular weights and flattens (see Figure 10.3).

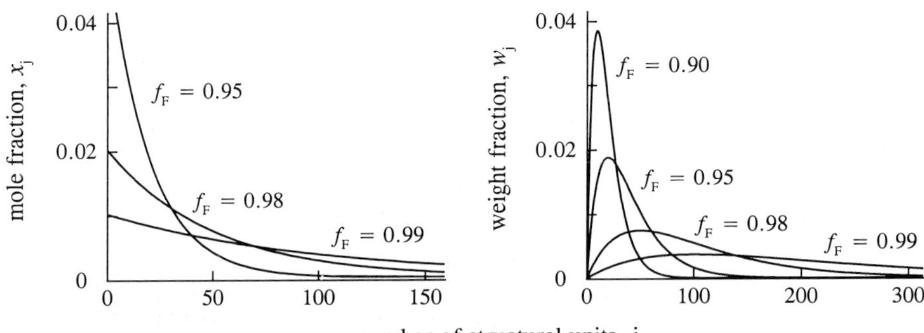

Figure 10.3. Schulz-Flory mole-fraction distribution (left) and corresponding molecular-weight distribution (right) at different degrees of fractional conversion of functional groups (adapted from Flory [27]).

Derivation of eqn 10.21. The weight of polymer with j repeating units is jN_jMW_M, where N_j is the number of moles of that polymer and MW_M is the molecular weight of the repeating unit. The total weight of the mixture (including monomer) is $N°MW_M$, where $N°$ is the number of moles of monomer at start. The weight fraction of the polymer with j repeating units is the ratio of these two weights:

$$ w_j = \frac{jN_jMW_M}{N°MW_M} = \frac{jN_j}{N°} \qquad (j \geq 1) $$

With $N_j = Nx_j$ by virtue of the definition of mole fractions (N = number of moles including monomer), $N/N° = 1-f_F$, and eqn 10.20 for x_j, this gives eqn 10.21.

10.3. Free-radical polymerization

Chain growth differs from step growth in that it involves initiation and usually also termination reactions in addition to actual growth. This makes its kinetic behavior similar to that of chain reactions (see Chapter 9). However, the chain carriers in chain-growth polymerization need not be free radicals, as they are in ordinary chain reactions. Instead, they could be anions, cations, or metal-complex adducts. While the general structure of kinetics is similar in all types of chain-growth polymerizations, the details differ depending on the nature of the chain carriers.

The most common type of chain-growth polymerization is free-radical polymerization and will be examined first.

10.3.1. Mechanism and rate *

A majority of commercial polymers are produced by free-radical polymerization. Foremost among these are polystyrene, polyethene (i.e., polyethylene), poly(vinyl chloride), poly(vinyl alcohol), poly(vinyl acetate), and poly(methyl methacrylate). In each of these, polymerization involves an olefinic double bond. However, free-radical polymerization is not restricted to such monomers.

At its simplest, the mechanism of free-radical polymerization consists of free-radical production by an initiator (*initiation*), link-up of the free radical with a monomer molecule (often considered part of initiation), addition of further monomer (*propagation*), and eventual deactivation (*termination*) of the growing polymer radicals by coupling, also called combination, that is, by link-up of two radicals with one another. This is much as in ordinary chain reactions (see Section 9.5). Free-radical polymerization of styrene may serve as an example [31,32].

---

* For an excellent coverage of chemical and structural effects and their mechanistic implications, see a recent book by Moad and Solomon [G6].

Example 10.3. Free-radical polymerization of styrene. Styrene is a highly reactive monomer. If not stabilized, it polymerizes slowly even without an initiator [33]. Commercial polystyrene is produced with peroxy or azo compounds as initiators. The mechanism of polymerization initiated by 2,2'-azo-*bis*-isobutyronitrile (AIBN) is as follows:

initiation   rate $= 2fk_{init}C_{in}$

link-up with monomer $k_{lnk}C_MC_X.$

propagation $k_pC_MC_{\Sigma P}.$

termination $-2k_{tc}C_{\Sigma P}^2.$

where in is the initiator, $X\cdot$ is the free radical produced by its decay, M is styrene monomer, and $\Sigma P\cdot$ is the total of all styrene-containing radicals (including that with only a single styrene unit). The initiation rate also involves an efficiency factor f which reflects the fact that some of the free radicals from the initiator may become deactivated by reactions with one another before they manage to initiate a kinetic chain. The factor 2 in the initiation and termination rates appears because two radicals are produced or consumed in the respective reactions. The rate equations of propagation and termination presuppose that the rate coefficients do not change with growing length of the polymer chain.

Because of their very low concentrations, quasi-stationary behavior of the free radicals can safely be assumed (Bodenstein approximation). Initiation and termination rates then are equal in absolute value:

$$2fk_{init}C_{in} = 2k_{tc}C_{\Sigma P}^2. \tag{10.22}$$

Solved for the free-radical population $C_{\Sigma P}\cdot$:

$$C_{\Sigma P}\cdot = (fk_{init}/k_{tc})^{1/2}C_{in}^{1/2} \tag{10.23}$$

The polymerization rate, which can be identified as the rate of monomer consumption, is given by

$$-r_{\mathrm{M}} \;=\; k_{\mathrm{lnk}} C_{\mathrm{M}} C_{\mathrm{X}\cdot} \;+\; k_{\mathrm{p}} C_{\mathrm{M}} C_{\Sigma P}. \tag{10.24}$$

The first term is the consumption by link-up with the free radical from the initiator, the second is the consumption by addition to the growing polymer radicals. Under practical conditions of production of high-molecular weight polymer, the first term is negligible compared with the second (long-chain approximation, see Section 9.3). If this can be assumed, elimination of $C_{\Sigma P}$. by means of eqn 10.23 gives

$$-r_{\mathrm{M}} \;=\; k_{\mathrm{p}}\,(f k_{\mathrm{init}}/k_{\mathrm{tc}})^{1/2}\, C_{\mathrm{in}}^{1/2}\, C_{\mathrm{M}} \tag{10.25}$$

making the rate first order in styrene monomer and half order in initiator. [If the first term in eqn 10.24 cannot be disregarded, the rate includes an additional term $2 f k_{\mathrm{init}}\, C_{\mathrm{in}}$, obtained with the Bodenstein approximation for the initiator radical $\mathrm{X}\cdot$, according to which the initiation and link-up rates can be taken as equal.]

Rate behavior of this kind is observed for many other olefinic monomers. As an example, Figure 10.4 shows the rate of methyl methacrylate polymerization also to be first order in monomer and about half order in initiator. However, the mechanism in Example 10.3 is by no means universal. In outline, others involve:

Termination by disproportionation.   In Example 10.3, coupling of two polymer radicals was assumed to be the only termination mechanism, as is indeed essentially true for polymerization of styrene [34]. However, various other mechanisms may contribute to termination or even dominate it. The most common of these is disproportionation, mainly observed for tertiary and other sterically hindered free radicals [35]. An example is methyl methacrylate [34] (see reaction 10.26 below). In disproportionation, two polymer radicals react with one another, transferring a

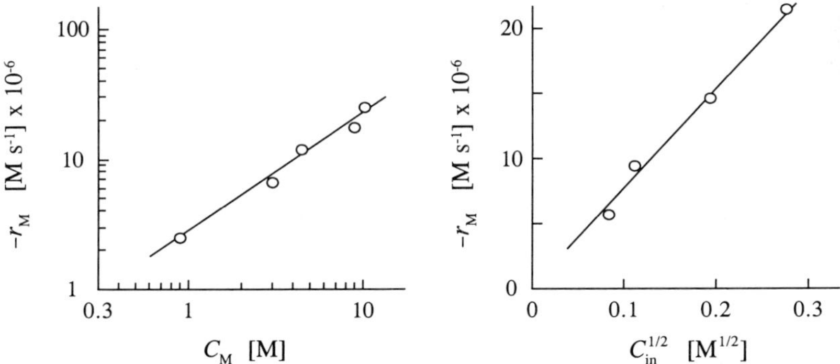

Figure 10.4. Rate of methyl methacrylate chain polymerization. Left: rate first order in monomer (redox initiator) [36]; right: rate approximately half order in initiator (benzoyl peroxide) [37].

hydrogen atom to produce a stable saturated polymer molecule and another with a carbon-carbon double bond:

$$(10.26)$$

Like coupling, disproportionation as a reaction of two polymer radicals is second order in free radicals. A contribution of disproportionation to termination thus does not alter the algebraic form of the rate equation 10.25, but the termination rate coefficient k_{tc} becomes the sum of two second-order coefficients k_{tc} and k_{td} for coupling and disproportionation, respectively. However, the degree of polymerization, the molecular weight, and the molecular-weight distribution are affected by disproportionation (see Section 10.3.4).

Chain transfer. Another mechanism of chain breaking, that is, of stopping the growth of a polymer radical, is *chain transfer* to monomer, another polymer molecule, the solvent, or some other inadvertently present or intentionally added species.*

Chain transfer to monomer in most cases predominantly yields a saturated monomer radical and a polymer molecule with double bond. Hydrogen transfer from the monomer to the polymer radical, leaving the double bond on the monomer radical, can also occur but is energetically disfavored:

$$(10.27)$$

Polymer molecules produced by the unsaturated monomer radical carry terminal vinyl groups that can react with other radicals to form radicals with reactive centers along their carbon chains. Further growth then yields branched polymer [39].

---

* In the literature there is a lack of consensus on terminology regarding "termination." We follow Kennedy and Maréchal [38]: *Termination* irretrievably ends the kinetic chain (e.g., by coupling, disproportionation, or chain transfer to produce an inactive radical); *chain breaking* ends the growth of the respective polymer radical by whatever mechanism without necessarily terminating the kinetic chain, which, upon chain transfer, may or may not continue on another molecule.

Chain transfer, whether to monomer, polymer, solvent, or an added transfer agent, breaks the kinetic chain, but does not *per se* terminate it. Unless the new radical is unreactive, chain polymerization continues, though on a different molecule. In many instances, the reactivity of the new radical is comparable to that of the old one and re-initiation of the new chain is fast. Monomer consumption then continues at its pace according to eqn 10.25, and only the degree of polymerization and the molecular weight are affected; if chain transfer is also very fast relative to propagation, only low polymer is produced (*telomerization*). On the other hand, the new radical generated by chain transfer may be unreactive. Chain transfer then decreases the rate of monomer consumption (*retardation*) or, if transfer is fast relative to propagation, polymerization may stop altogether (*inhibition*). Chain transfer producing an unreactive radical acts as another mechanism terminating the kinetic chain, in this case by a reaction that is first order rather than second order in polymer radicals (see also Section 9.5).

Deactivating chain transfer to monomer is quite common in polymerization of allyl monomers [40-42]. Allyl radicals such as that of allyl acetate are resonance-stabilized, with the result that polymerization rates and molecular weights remain low. Moreover, with chain transfer as the dominant termination mechanism, the termination rate is first order in free radicals. This lets the free-radical population become proportional to the initiator concentration and leads to a polymerization rate that is first order rather half order in initiator and zero order in monomer.

Derivation. The Bodenstein approximation of a quasi-stationary free-radical population allows the absolute values of the initiation and termination rates to be equated. With terminating chain transfer first-order in free radicals and monomer:

$$r_{init} = 2fk_{init}C_{in} = -r_{trM} = k_{trM}C_M C_{\Sigma P} \cdot$$

Solved for $C_{\Sigma P} \cdot$:

$$C_{\Sigma P} \cdot = (2fk_{init}/k_{trM})C_{in}/C_M$$

This gives a propagation rate that is first order in initiator and zero order in monomer:

$$r_p = k_p C_M C_{\Sigma P} \cdot = (2fk_{init}k_p/k_{trM})C_{in}$$

The rate of monomer consumption is the sum of the rates of propagation and link-up of the initiator radical with monomer. The latter rate equals the initiation rate (Bodenstein approximation of quasi-stationary behavior of initiator radicals). The monomer consumption rate thus becomes

$$r_{init} + r_p = 2fk_{init}(1 + k_p/k_{trM})C_{in} \tag{10.28}$$

and is also first order in initiator and zero order in monomer.

Another possible chain-breaking mechanism is *chain transfer to polymer* [43,44]. Here, a new reactive center is formed on the polymer chain of the receiving molecule, usually along its chain rather than at either end:

$$\tag{10.29}$$

New growth from such a center produces a branch.

Chain transfer may also occur to a carbon atom of the same polymer molecule five, six, or seven positions distant from the original reactive center [45]. This is called *backbiting* and is regarded as the mechanism of formation of short branches in polyethene polymerization [46,47].

Transfer agents that lead to production of unreactive radicals may be added to limit molecular weight [48-51]. Best suited are agents whose radicals are stabilized by adjacent groups or by resonance. The effectiveness of a transfer agent is characterized by its *transfer constant*, defined as the ratio of the rate coefficients of chain transfer and propagation:

$$\vartheta_{trM} \equiv k_{trM}/k_p \tag{10.30}$$

Table 10.1 lists approximate transfer constants of some common agents in polymerization of styrene, methyl methacrylate, and vinyl acetate.

Table 10.1. Approximate transfer constants of selected transfer agents in polymerization of styrene, methyl methacrylate, and vinyl acetate at 60°C (averaged and rounded values from Eastmond [52]).

transfer agent	monomer		
	styrene	methyl methacrylate	vinyl acetate
isopropylbenzene	$1*10^{-4}$	$2*10^{-4}$	0.01
isopropanol	$3*10^{-4}$	$6*10^{-5}$	0.004
chloroform	$3*10^{-4}$	$1*10^{-4}$	0.015
carbon tetrachloride	0.01	$1*10^{-4}$	40
carbon tetrabromide	50		50
n-butane thiol	25		

Dependence of rate coefficients on polymer chain length. The rate equations in Example 10.3 were derived with the assumption that the rate coefficients do not depend on the degree of polymerization of the polymer radicals and remain constant as more polymer molecules are formed. There are two major exceptions:

For most monomers, the propagation rate coefficient, k_p, is somewhat higher for the first one or two propagation steps than for later addition to longer polymer radicals [53]. This is of concern only in oligomerization, not if polymerization is carried to high molecular weight, as is the more common practice.

Potentially more troublesome is a decline in coefficient values at high conversion. Unless polymerization is carried out in dilute solution, the mixture stiffens and reactive groups have a harder time finding partners to react with. The long-chain polymer radicals become entangled with other polymer chains, and while the small molecules of monomer can still find access to radical groups on the polymer with reasonable ease, the frequency of encounters of such radical groups with one another decreases sharply [54]. Termination then is impeded, causing the reaction to accelerate (*Trommsdorff effect* or *gel effect*, see high-concentration curves in Figure 10.5) [54-58]. This calls for care in handling of large amounts of liquid monomers such as vinyl compounds, whose polymerization is strongly exothermic: An accidental initiation may result in an explosive runaway (an "unscheduled polymerization" in corporate parlance). In the absence of solvent, propagation also may come to a stand-till, so that polymerization stops short of complete conversion of monomer (see curve for 100%). Moreover, because of the arrested termination, the final polymer may still contain reactive centers.

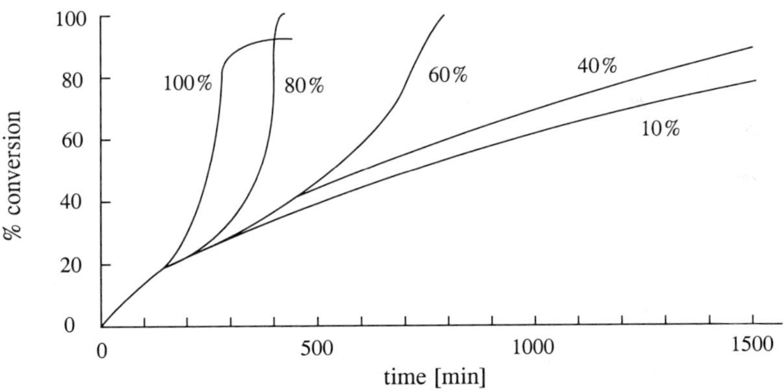

Figure 10.5. Conversion as a function of time for polymerization of methyl methacrylate at different concentrations in benzene at 50°C (adapted from Schulz and Harborth [55]).

10.3.2. Photochemical initiation.

Vinyl compounds absorb ultraviolet light in the range of 200 nm [59]. Irradiation, say, with a mercury lamp produces radicals that can initiate polymerization. The initiation rate is

$$r_{init} \; = \; \phi I_{abs} \; = \; \phi \epsilon I° C_M d \tag{10.31}$$

where I_{abs} is the intensity of absorbed light, $I°$ is the intensity of incident light, ϕ is the quantum yield (number of chains initiated per photon absorbed), ϵ is the molar absorbance, and d is the thickness of the cell (the second equality assumes the cell to be so thin that only a small fraction of the incident light is absorbed; ϕ is sometimes defined differently). With this rate of initiation and with termination by coupling, the polymerization rate becomes [60]

$$-r_M \; = \; k_p C_M \left[\frac{\phi I_{abs}}{k_{tc}} \right]^{1/2} \; = \; k_p C_M^{3/2} \left[\frac{\phi \epsilon I° d}{k_{tc}} \right]^{1/2} \tag{10.32}$$

Polymerization rate equations for other termination mechanisms are readily obtained by replacement of $2f k_{init} C_{in}$ by ϕI_{abs} or $\phi \epsilon I° C_M d$ in the respective equations for conventional initiation.

Initiation can also be achieved by irradiation of a reaction mixture to which an absorbing *sensitizer* has been added. The sensitizer then is the source of radicals and, in eqns 10.31 and 10.32, C_{sen} replaces C_M, and $C_M C_{sen}^{1/2}$ replaces $C_M^{3/2}$ (sen = sensitizer).

10.3.3. Chain length

A relatively easily determined quantity indicative of the extent of polymerization is the *kinetic chain length*, commonly defined as the average number of propagation steps occurring between initiation and termination and thus equalling the ratio of the rates of propagation and termination. However, more useful in connection with molecular weights is what we may call the *radical chain length*, v, given by the average number of propagation steps between initiation and chain breaking rather than termination, and directly related to the length of the polymer radicals:

$$v \; = \; r_p/(-r_{chbr}) \tag{10.33}$$

($-r_{chbr}$ is the total chain-breaking rate from all mechanisms.) Of course, if all chain breaking also terminates the kinetic chain, the kinetic and radical chain lengths are equal.

Regardless of the chain-breaking mechanism, the propagation rate is

$$r_p \; = \; k_p C_M C_{\Sigma P}. \tag{10.34}$$

The radical chain length depends on the mechanism of chain breaking. The most common of these is *coupling*. Here, the chain-breaking (and termination) rate is

$$-r_{chbr} = -r_{tc} = 2k_{tc}C_{\Sigma P}^2. \tag{10.35}$$

With this and eqns 10.33, 10.34, and 10.23 the radical chain length becomes

$$\nu = k_p C_M / 2k_{tc}C_{\Sigma P}. = \tfrac{1}{2}k_p(f k_{init}k_{tc})^{-1/2}C_M C_{in}^{-1/2} \tag{10.36}$$

and is seen to be inversely proportional to the free-radical population. Accordingly:

- *any attempt to boost the polymerization rate by an increase in the free-radical population—e.g., by use of more or a better initiator—comes at a sacrifice in molecular weight* [61,62].

This important conclusion becomes plausible if one considers the reaction orders of propagation and chain breaking. Propagation is first-order in free radicals whereas chain breaking is second-order (granted coupling as the chain breaking mechanism, as is most common). As a consequence, an increase in the free-radical population speeds up chain breaking more than propagation, so that growth is stopped after fewer propagation steps.

For chain breaking by *disproportionation* or any combination of coupling and disproportionation, eqns 10.35 and 10.36 also apply, but with a rate coefficient k_{td} or $t_{tc} + t_{td}$ instead of t_{tc} alone.

For chain breaking predominantly by *chain transfer to monomer*, the chain-breaking rate is

$$-r_{chbr} \cong -r_{trM} = k_{trM}C_M C_{\Sigma P}.$$

(neglecting the chain-breaking contribution from termination) and the radical chain length becomes

$$\nu \cong k_p / k_{trM} \tag{10.37}$$

Similarly, for chain-breaking predominantly by *chain transfer to solvent or an agent*, the chain-breaking rate and radical chain length are

$$-r_{chbr} \cong -r_{trS} = k_{trS}C_S C_{\Sigma P}.$$

and

$$\nu \cong (k_p / k_{trS}) C_M / C_S \tag{10.38}$$

respectively (S = recipient species). The radical chain lengths for chain transfer (eqns 10.37 and 10.38) are independent of the size of the free-radical population. Accordingly, the conclusion above does not apply. Rather, if chain transfer is the dominant chain-breaking mechanism, an increase in free radicals can accelerate polymerization without sacrifice in molecular weight. This is because both propagation and chain breaking then are first order in free radicals and so are accelerated equally by an increase in the free-radical population.

The radical chain lengths resulting from the various chain-breaking mechanisms can be summarized as follows:

chain-breaking mechanism	*radical chain length*
coupling	$\frac{1}{2}\,k_p\,(f\,k_{init}\,k_{tc})^{-1/2}\,C_M\,C_{in}^{-1/2}$
disproportionation	$\frac{1}{2}\,k_p\,(f\,k_{init}\,k_{td})^{-1/2}\,C_M\,C_{in}^{-1/2}$
chain transfer to monomer	$k_p\,/k_{trM}$
chain transfer to S (transfer agent or solvent)	$(k_p\,/k_{trS})\,C_M\,/C_S$

If several mechanisms contribute significantly to termination, their rates are additive.

10.3.4. *Degree of polymerization and molecular weight*

While the radical chain length as the ratio of the propagation and chain-breaking rates refers to conditions existing at the moment, the degree of polymerization and the molecular weight characterize a polymer that may have been produced over a span of time during which the conditions varied. Here, only some idealized situations with conditions held constant can be described in any detail.

Number-average degree of polymerization. For step-growth polymerization, The number-average degree of polymerization was defined as the ratio of the numbers of monomer molecules at start and of molecules after polymerization (eqn 10.15), and so included unreacted monomer in the molecule count. More practical for chain-growth polymerization is to exclude the unreacted monomer:

$$\bar{N}_{pol} \equiv \frac{\text{number of monomer units in total polymer}}{\text{number of polymer molecules}} \tag{10.39}$$

(subscript pol added to indicate restriction to polymer).

The number-average degree of polymerization is usually taken to be the radical chain length, or twice that length if termination is by coupling where one polymer molecule is formed from two radicals. This is permissible only for high degrees of polymerization, as becomes apparent from the fact that the minimum degree of polymerization according to the definition 10.39 is 2 (dimer being the lowest possible polymer) whereas the radical chain length as ratio of the propagation and chain-breaking rates can in principle be lower. Many complications can arise. Only two relatively simple cases will be examined here: polymerization with chain breaking by coupling and by chain transfer to an agent that produces an unreactive radical.

In *chain breaking exclusively by coupling*, the total consumption of monomer per unit time is

$$-r_{\text{M}} = r_{\text{init}} + r_{\text{p}}$$

and the number of radical chains broken per unit time is

$$-\tfrac{1}{2} r_{\text{chbr}} = -\tfrac{1}{2} r_{\text{trm}} = -\tfrac{1}{2} r_{\text{tc}} = \tfrac{1}{2} r_{\text{init}}$$

The number of polymer molecules produced per unit time is half as large because two polymer radicals are consumed when forming one polymer molecule. Accordingly, with eqn 10.33 and noting that in this case $-r_{\text{chbr}} = -r_{\text{trm}} = r_{\text{init}}$:

$$\bar{N}_{\text{pol}} = \frac{-r_{\text{M}}}{\tfrac{1}{2} r_{\text{init}}} = \frac{r_{\text{init}} + r_{\text{p}}}{\tfrac{1}{2} r_{\text{init}}} = 2 \left[1 + \frac{r_{\text{p}}}{-r_{\text{chbr}}} \right] = 2\,(1 + \nu) \qquad (10.40)$$

This formula correctly gives the limit of 2 at shortest possible chain length.

In *chain breaking exclusively by chain transfer to a deactivating transfer agent*, Tr, total monomer consumption per unit time is again the sum of the initiation and propagation rates. However, of the monomer radicals M· produced by initiation and link-up, some are lost by chain transfer to the agent; only a fraction P ends up as polymer, where P is the probability that a radical will add monomer rather than be terminated by transfer:

$$P = \frac{\text{propagation rate}}{\text{propagation rate} + \text{transfer rate}} = \frac{r_{\text{p}}}{r_{\text{p}} + (-r_{\text{trTr}})} \qquad (10.41)$$

Accordingly, the number of monomer molecules converted to polymer per unit time is

$$-(r_{\text{M}})_{pol} = P r_{\text{init}} + r_{\text{p}}$$

and the number of polymer radicals terminated per unit time, equaling the number of monomer radicals that added monomer, is

$$(-r_{\text{trTr}})_{\text{pol}} = P r_{\text{init}}$$

This gives a degree of polymerization

$$\bar{N}_{\text{pol}} = (-r_{\text{M}})_{\text{pol}} / (-r_{\text{trTr}})_{\text{pol}} = (P r_{\text{init}} + r_{\text{p}}) / P r_{\text{init}}$$

and after replacement of P with eqn 10.41 and noting that, here, $-r_{\text{chbr}} = r_{\text{init}}$:

$$\bar{N}_{\text{pol}} = 2 + r_{\text{p}} / r_{\text{init}} = 2 + \nu \qquad (10.42)$$

Here, too, the limit at shortest possible chain length is seen to be 2 as called for.

Other cases are more complex. For example, in termination by disproportionation, molecules that have undergone initiation and link-up but have been deactivated by disproportionation before managing to add further monomer must be excluded from the mole count. Also, if the concentrations of monomer and initiator decline during polymerization, as they do in batch reactors, the degree of polymerization of the eventual total product is given by the integral over time or conversion. The exception here is the case of termination by chain transfer to monomer, in which the radical chain length and degree of polymerization are independent of concentrations.

Molecular weight. The number-average molecular weight is obtained from the number-average degree of polymerization by multiplication with the molecular weight of the monomer, plus the formula weight of the end group from the initiator (a correction that is significant only at low degrees of polymerization).

Molecular-weight distribution. For low conversion or polymerization in dilute solution, approximate simple formulas can be derived. Complications arising at high conversion in bulk polymerization will be outlined later.

The probability P that a polymer radical of any length continues to grow is given by

$$P \; = \; \frac{\text{propagation rate}}{\text{propagation rate} \; + \; \text{chain--breaking rate}} \tag{10.43}$$

where chain transfers to monomer, solvent, or transfer agent are included in the chain-breaking rate. Chain transfer to polymer can be disregarded as unlikely at low conversion. With M = monomer, Tr = transfer agent, and S = solvent, and with

propagation rate $\qquad k_p C_M C_{\Sigma P\cdot}$

chain-breaking rate $\qquad \underset{\substack{\text{transfer to}\\\text{monomer}}}{k_{trM} C_M C_{\Sigma P\cdot}} \; + \; \underset{\substack{\text{transfer to}\\\text{transfer agent}}}{k_{trTr} C_{Tr} C_{\Sigma P\cdot}} \; + \; \underset{\substack{\text{transfer to}\\\text{solvent}}}{k_{trS} C_S C_{\Sigma P\cdot}} \; + \; \underset{\substack{\text{coupling and}\\\text{disproportionation}}}{2(k_{tc} + k_{td}) C_{\Sigma P\cdot}^2}$

the probability is

$$P \; = \; \frac{1}{1 + \vartheta_{trM} + \vartheta_{trTr} C_{Tr}/C_M + \vartheta_{trS} C_S/C_M + 2(\vartheta_{tc} + \vartheta_{td}) C_{\Sigma P\cdot}/C_M} \tag{10.44}$$

where the $\vartheta_i \equiv k_i/k_p$ are the respective transfer constants (see eqn 10.30).

Consider first the case of chain breaking exclusively by *disproportionation* or (terminating) *chain transfer*. Here, chain breaking converts polymer radicals into dead polymer molecules without change in the number of repeating units they contain. A polymer radical with j+1 repeating units has resulted from one more propagation step with probability P than a polymer radical with j repeating units, so the existence of the former is less probable by a factor P than that of the latter. In terms of mole fractions instead of existence probabilities:

$$(x_{j+1})_{\text{pol}} = P(x_j)_{\text{pol}} \tag{10.45}$$

where $(x_i)_{\text{pol}}$ is the mole fraction based on total polymer (monomer *not* counted) of a polymer with i repeating units. The mole fraction of the next larger polymer molecule is smaller by a factor $P < 1$. This is a Schulz-Flory distribution (declining geometrical progression; see Section 10.2.3):

- *Free-radical polymerization with chain breaking exclusively by disproportionation or chain transfer yielding unreactive radicals produces a Schulz-Flory mole-fraction distribution.*

This is as in step-growth polymerization of bifunctional monomers (eqn 10.19). The argument and result are formally the same as in that case, with the probability P taking the place of the fractional conversion f_F of functional groups. However, there is one important difference:

- In *step-growth polymerization*, the geometric factor is the fractional conversion and increases toward unity with progressing conversion. As a result, the distribution shifts toward higher molecular weights and flattens as conversion progresses. In contrast, in *free-radical polymerization*, the factor is the probability of the growth reaction and remains constant, independent of how many polymer molecules have been formed (granted the rate coefficients do not vary with monomer conversion).

This reflects a fundamental difference between step-growth and free-radical chain-growth kinetics: In step growth, conversion starts with formation of low polymer, and molecular weight increases as conversion progresses and the polymer molecules grow by link-up with monomer or other polymer molecules. In free-radical chain growth, polymer is produced by a very small family of chain carriers that quickly settle down to a quasi-stationary molecular-weight distribution and function as an assembly line that churns out dead polymer of a matching distribution. The number of polymer molecules increases with conversion, but their molecular-weight distribution remains essentially constant if monomer and initiator are replenished.

The mole and weight fractions as explicit functions of the probability of the growth reaction are

$$(x_j)_{\text{pol}} = P^{j-1}\Big/\sum_{n=1}^{\infty} P^{n-1} = P^{j-1}(1-P) \qquad (j \geq 2) \tag{10.46}$$

$$(w_j)_{\text{pol}} = jP^{j-1}\Big/\sum_{n=1}^{\infty} (n+1)P^{n-1} = \frac{jP^{j-1}(1-P)^2}{2-P} \qquad (j \geq 2) \tag{10.47}$$

[The sums converge to $1/(1-P)$ and $(2-P)/(1-P)^2$, respectively.] Unlike in step growth, the fractions are based on total polymer, to the exclusion of monomer, and are therefore written with the extra subscript pol. This difference comes about because the probability argument leading to eqn 10.43 is for molecules produced by the chain carriers and so does not apply to the monomer. It also accounts for the difference in algebraic forms between eqns 10.47 and 10.21.

For the more frequently encountered case of termination exclusively by *coupling*, eqn 10.46 indicates what fraction of polymer radicals have j repeating units, and random pairing of the members of a radical population with this distribution must then be considered. According to Flory [28,63] the result is

$$(w_j)_{pol} \;=\; \frac{j(j-1)}{2}(1-P)^3 P^{j-2} \tag{10.48}$$

This is a Poisson-type distribution, narrower than that from eqn 10.47. Any chain transfer that does *not* terminate the kinetic chain reduces the molecular weight, but does not alter the shape of the distribution, which remains given by eqn 10.48.

In practice, several or all chain-breaking mechanisms may contribute. If so, the weight fraction can be taken as the weighted average of those derived for coupling and for chain transfer plus disproportionation (eqns 10.47 and 10.48).

Molecular weights and molecular-weight distributions at high conversions are much less predictable and more dependent on the nature of the monomer. The two principal additional factors that come into play are chain transfer to polymer and the decrease of the rate coefficients for coupling and disproportionation with increasing polymer content, viscosity, and chain entanglement.

The more polymer, the greater is the chance of (non-terminating) chain transfer to polymer rather than monomer. Large polymer molecules offer larger targets for chain transfer and so are more likely recipients. This shifts growth to larger polymer and thereby broadens the molecular-weight distribution.

The slow-down of termination is most pronounced for large polymer molecules because these are more prone to chain entanglement. This also results in favored growth of large molecules and adds to the broadening of the molecular-weight distribution.

10.4. Ionic polymerization

The chain carriers in chain-growth polymerization may be anions or cations rather than free radicals. Such ionic polymerization shares many features with free-radical polymerization, but differs in one important respect: Since ions of the same charge sign repel one another, spontaneous binary termination by reaction of two chain carriers with one another cannot occur. In fact, the reaction may run out of monomer with chain carriers still intact.

There are also subtle mechanistic differences between anionic and cationic polymerization, which will therefore be examined separately.

10.4.1. Anionic polymerization

Compounds capable of forming carbanions stabilized by delocalization of the negative charge can be made to undergo anionic polymerization under appropriate conditions. Typical representatives are compounds with conjugated double bonds, such as styrene and butadiene, or with hetero-atoms that are more electronegative than carbon, among them ethene oxide and caprolactam. Historically important is *Buna S*, the first successful commercial synthetic rubber, produced by sodium-initiated copolymerization of butadiene and styrene [64]. A delicate balance of base strengths and a proper choice of solvent and reaction conditions is essential: A carbanion whose base strength is too high may deactivate itself by deprotonating a protic solvent or forming a stable complex with a polar aprotic one; a solvent with too low a dielectric constant will not sufficiently encourage ionization, while one with too high a dielectric constant will deactivate the carbanion.

Under suitable conditions, anionic polymerization is faster than free-radical polymerization and so can be conducted at lower temperatures. The main reasons are fast initiation by an ionic reaction and absence of an effective termination mechanism. However, the sensitivity to impurities is much greater and choice and control of reaction conditions are more delicate. Water, oxygen, carbon dioxide, and other substances able to react with carbanion chain carriers must be strictly excluded.

The key feature distinguishing anionic (and cationic) from free-radical polymerization is the absence of spontaneous binary termination and has already been mentioned. Unless chain transfer occurs, polymer chains keep growing until all monomer is used up. At that stage, the polymer still carries reactive centers [65] —it is said to be a "living polymer" [66,67]—, and polymerization can be started anew by addition of further monomer. Block copolymers can be synthesized from a living polymer by addition of a different monomer [68,69].

Because of the strong dependence on the nature of the solvent and initiator and the high sensitivity to impurities and reaction conditions, quantitative predictions of rates and molecular weights are difficult to make and less reliable than in free-radical polymerization.

Initiation. The three most common kinds of initiators for anionic polymerization are alkali-metal alkyls, metal amides, and elementary alkali metals.

In initiation by an *alkali-metal alkyl*, the alkyl links up with the monomer to form a carbanion, leaving the metal as a cation to compensate the negative charge. An example is the initiation of styrene polymerization by butyl lithium [70-72]:

$$\text{(10.49)}$$

This is a technique especially suited for production of living polymers.

In initiation by a *metal amide*, carried out in liquid ammonia, the amide dissociates into a metal cation and an amide anion, NH_2^-, which then adds to the monomer [73]:

$$K^+ \, NH_2^- \; + \; \overset{CH}{\underset{}{\bigcirc}} \; \longrightarrow \; \overset{H_2N}{\underset{}{HC^-}} \, K^+ \qquad (10.50)$$

Chain transfer to the ammonia solvent by deprotonation of the carbanion is common in such systems [73]:

$$\cdots HC^- \, K^+ \; + \; NH_3 \; \longrightarrow \; \cdots CH_2 \; + \; NH_2^- \, K^+ \qquad (10.51)$$

In initiation by *alkali metal*, an electron is transferred from the metal to the monomer (the metal is usually charged as a solid or colloid). Sodium-initiated butadiene polymerization provides an example [74,75]:

$$Na \; + \; H_2C \diagup \diagdown CH_2 \; \longrightarrow \; H_2\dot{C} \diagup \diagdown \ddot{C}H_2^- \, Na^+ \qquad (10.52)$$

In such systems, dimerization of the carbanion or electron transfer from another alkali metal atom can produce a di-anion that adds monomer at both ends, complicating kinetics [75].

An interesting variation is initiation by a combination of an alkali metal and an aromatic with condensed rings, e.g., naphthalene. The aromatic anion radical transfers an electron to a monomer such as styrene, which then dimerizes and grows at both ends [66,76].

Propagation. Regardless of its degree of polymerization, the carbanion and its metal counterion Me^+ can be expected to exist in a spectrum of different forms, ranging from a covalently bonded species at one extreme to separate ions at the other, with a contact ion pair and a solvent-separated ion pair as intermediates (*Winstein spectrum*) [77,78]:

$$[P \, Me] \; \longleftrightarrow \; [P^- \, Me^+] \; \longleftrightarrow \; P^- \, Me^+ \; \longleftrightarrow \; P^- + Me^+ \qquad (10.53)$$

covalently bonded   contact ion pair   solvent-separated ion pair   separate ions

(The contact and solvent-separated ion pairs are also called intimate and loose ion pairs, respectively.) Since ionic dissociation-association reactions in general are very fast, quasi-equilibrium of the species can reasonably be assumed.

Monomer is inserted at the carbon atom carrying the negative charge. The covalently bonded species tends to be unreactive, and reactivity increases sharply with progressing dissociation (left to right in 10.53). Under most conditions, the free carbanion, P^-, exists only at a much lower concentration than do the other forms: It is a *lapc* (least-abundant propagating center) constituting only a negligibly small fraction of the total of the species in 10.53 (analogous to a *lacs* in catalysis, see Section 8.5.1). However, because of its high reactivity, its contribution to the propagation rate may nevertheless be significant or even dominant. The nature of the solvent strongly affects the equilibria and thus the propagation rate and the form of the rate equation. For example, in a solvent with poor solvating power—e.g., a hydrocarbon—there is less tendency to dissociate, and free ions and a solvent-separated ion pair may not even exist.

Granted quasi-equilibrium of the various species in the reactions 10.53 at any number of repeating units in P, the relationships between the concentrations are

$$K_{01} = C_{[P^-Me^+]}/C_{[PMe]}, \qquad K_{02} = C_{P^-Me^+}/C_{[PMe]}, \qquad K_{03} = C_{P^-}C_{Me^+}/C_{[PMe]} \qquad (10.54)$$

where the K_{0i} are the respective equilibrium constants. If these do not vary as the chains grow, eqns 10.54 apply to the totals of all covalent species, contact ion pairs, etc., regardless of the lengths of their chains. The concentrations can then be expressed in terms of the total population ΣP^- of potential propagating centers:

$$C_{[PMe]} = C_{\Sigma P^-}/K, \qquad\qquad C_{[P^-Me^+]} = K_{01}C_{\Sigma P^-}/K,$$
$$C_{P^-Me^+} = K_{02}C_{\Sigma P^-}/K, \qquad\qquad C_{P^-} = K_{03}C_{\Sigma P^-}/KC_{Me^+}. \qquad (10.55)$$

where

$$\Sigma P^- \equiv [PMe] + [P^-Me^+] + P^-Me^+ + P^- \qquad (10.56)$$

$$K \equiv 1 + K_{01} + K_{02} + K_{03}/C_{Me^+} \qquad (10.57)$$

All the species in 10.53 might contribute to propagation. Leaving all possibilities open, the propagation rate can be written

$$r_P = (k_0 C_{[PMe]} + k_1 C_{[P^-Me^+]} + k_2 C_{P^-Me^+} + k_3 C_{P^-})C_M$$

$$= \left[\frac{k_0 + k_1 K_{01} + k_2 K_{02} + k_3 K_{03}/C_{Me^+}}{1 + K_{01} + K_{02} + K_{03}/C_{Me^+}}\right] C_{\Sigma P^-} C_M \qquad (10.58)$$

where the k_i are the respective rate coefficients of monomer addition.

Equation 10.58 covers all bases. Of particular interest in practice are three special and fairly common situations:

Case I:	The free anion is the *lapc* (least abundant propagating center) but, thanks to its high reactivity, provides a significant rate contribution.

If the free anion is the *lapc*, the last term in the denominator of eqn 10.58 is negligible, so that $K = 1 + K_{01} + K_{02}$. Moreover, without an added electrolyte the concentrations of P^- and Me^+ are necessarily equal. In this case, the last of eqns 10.55 gives

$$C_{P^-} = \left(K_{03}/(1 + K_{01} + K_{02})\right)^{1/2} C_{\Sigma P^-}^{1/2}$$

and the propagation rate is

$$r_p = (k_a C_{\Sigma P^-} + k_b C_{\Sigma P^-}^{1/2}) C_M \qquad (10.59)$$

where

$$k_a = \frac{k_0 + k_1 K_{01} + k_2 K_{02}}{1 + K_{01} + K_{02}}, \qquad k_b = k_3 \left[\frac{K_{03}}{1 + K_{01} + K_{02}}\right]^{1/2}$$

If only the free anion contributes to the rate, the first term in eqn 10.59 is negligible.

Case II:	As in Case I, the free anion is the *lapc* and provides a significant rate contribution. An inert strong electrolyte with same cation as that produced by the initiator has been added.

A strong electrolyte with same cation can be added to keep an overly pesky reaction in check.

Without the negligible last term in its denominator, eqn 10.58 gives

$$r_p = (k_a + k_c/C_{Me^+}) C_{\Sigma P^-} C_M \qquad \left(k_c \equiv k_3 K_{03}/(1 + K_{01} + K_{02})\right) \qquad (10.60)$$

where k_a is defined as in Case I above. A large excess of added electrolyte may suppress the rate contribution from the free anion entirely, and the second term then disappears.

Case III:	The free anion is the *lapc* and does not contribute significantly to the rate.

This may be the case in a solvent of low polarity, in which lack of dissociation keeps the free ion at too low a concentration to be effective.

If the free anion does not contribute significantly to the rate, the last term in the numerator of eqn 10.58 also becomes negligible, so that:

$$r_p = k_a C_{\Sigma P^-} C_M \qquad (10.61)$$

where k_a is as defined in Case I.

As long as the free anion is the *lapc*, the rate equations are of the algebraic forms of eqns 10.58 to 10.61, regardless of which of the other three propagating centers or combination of these contributes to the rate.

Polymerization without termination: living polymers. If care is taken, all termination reactions can be avoided. Polymerization then proceeds until all monomer is used up or the reaction is quenched by addition of a deactivating agent. Under such conditions, the rate of monomer consumption is the sum of the rates of initiation and propagation:

$$-r_M = r_{init} + r_p = k_{init} C_{in} + r_p \qquad (10.62)$$

where r_p is given by eqn 10.58 or one of its simpler special forms. The contribution from initiation is negligible except at low degrees of polymerization, and may require an effectiveness factor. Moreover, if initiation is fast relative to propagation, as is almost always true, the total population of propagating centers, ΣP^-, is equal to the amount of initiator initially added (multiplied with the effectiveness factor if called for). In the three situations discussed above:

- *Case I* (free anion is *lapc*, but contributes to the rate): The polymerization rate is first order in monomer and of order between one half and one in initially added initiator. From eqn 10.59:

$$-r_M/C_M(C_{in}^o)^{1/2} = k_b + k_a(C_{in}^o)^{1/2}$$

Accordingly, a plot of $-r_M/C_M(C_{in}^o)^{1/2}$ versus $(C_{in}^o)^{1/2}$ is linear with intercept k_b (contribution from free anion) and slope k_a (contribution from ion pair). An example is shown in Figure 10.6, left (next page).

- *Case II* (strong electrolyte with common cation has been added): The rate is first order in monomer and of order between zero and minus one in the common cation, possibly approaching zero at very large excess of the added electrolyte. From eqn 10.60:

$$-r_M/C_M C_{in}^o = k_a + k_c/C_{Me^+}$$

so that a plot of $-r_M/C_M C_{in}^o$ versus $1/C_{Me^+}$ is linear with intercept k_a (contribution from ion pairs) and slope k_c (contribution from free anion). For an example, see Figure 10.6, right.

- *Case III* (free anion is *lapc* and does not contribute significantly to the rate): The rate is first order in monomer and initially added initiator (see eqn 10.61).

If no quenching agent is added, the number-average degree of polymerization is given by the ratio of the number of monomer molecules at start and number of polymeric carbanion propagating centers, ΣP^- (also called "living ends"). Granted fast initiation, the latter equals the number of initiator molecules added at start, so that

$$(\bar{N})_{pol} \cong C_M^o/C_{in}^o \qquad (10.63)$$

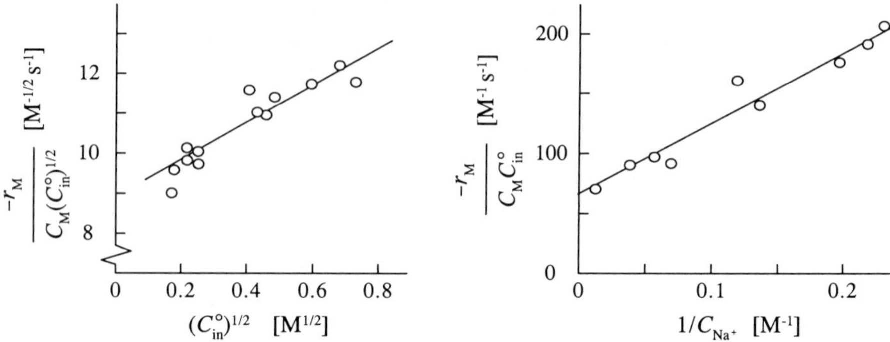

Figure 10.6. Rate of anionic polymerization of styrene initiated by sodium naphthalene in 3-methyl tetrahydrofuran at 20°C. Left: linear variation of rate with $(C^o_{in})^{1/2}$; right: inverse linear variation of rate with concentration of Na$^+$ in presence of added sodium tetraphenyl borate. (Data from Schnitt and Schulz [79].)

An effectiveness factor may have to be used if not all of the initiator produces carbanion propagating centers. If the carbanion produced by initiation dimerizes to give a di-anion, as in lithium-initiated butadiene polymerization, the degree of polymerization is twice as large as indicated by eqn 10.63.

As in other situations, the number-average molecular weight is given by multiplication of the degree of polymerization with the molecular weight of the monomer.

If initiation is very fast relative to propagation, as is usually the case, all propagating centers start growing at about the same time and grow at the same rate. If other termination reactions can be avoided, a quenching agent added at the appropriate time can stop growth when a desired molecular weight has been reached. This makes it possible to produce polymer standards of specific, highly uniform molecular weights [80,81].

- Fast initiation and suppression of termination in living polymerization can lead to a reaction behavior that is fundamentally different from that of free-radical polymerization. In the latter, the size of the chain-carrier population is dictated by a balance of initiation and termination rates. Chain carriers in quasi-stationary molecular-weight distribution convert monomer to polymer of a matching distribution. In contrast, in ionic living polymerization, the initiator is used up quickly, producing a population of propagating centers that keep adding monomer to themselves. In free-radical polymerization the number of polymer molecules increases with conversion of monomer, but the molecular-weight distribution remains essentially unchanged. In ionic living polymerization, polymer radicals keep growing to ever larger sizes as monomer is consumed, but their number does not change.

Polymerization with termination. Unless care is taken to prevent it, termination may occur by chain transfer to the solvent or impurities. Many different combinations of possibilities exist. For illustration, one example may suffice.

Example 10.4. Anionic polymerization of styrene in ammonia [73]. Polymerization of styrene in ammonia can be initiated by potassium amide and is believed to proceed as follows. Initiation consists of

$$KNH_2 \;\rightleftharpoons\; NH_2^-\;(K^+) \;\xrightarrow{\;styrene\;}\; H_2N\!-\!\overset{\cdot\cdot}{\underset{H}{C}}\!\!-\!\!\langle O \rangle$$

(see also reaction 10.50). KNH_2 dissociation is fast and at equilibrium, making the initiation rate

$$r_{init} \;=\; k_{init} C_{NH_2^-} C_M \;=\; k_{init} K_{KNH_2} C_{KNH_2} C_M / C_K. \tag{10.64}$$

where K_{KNH_2} is the dissociation constant of KNH_2.

The free carbanion remains at so low a concentration that it does not contribute significantly to the propagation rate. Accordingly, that rate is given by eqn 10.61 (Case III):

$$r_p \;=\; k_a C_{\Sigma P^-} C_M \tag{10.65}$$

where k_a is a lumped rate coefficient defined as in eqn 10.59.

There is substantial chain transfer to the ammonia solvent and some to water that is likely to be present as a trace impurity. Transfer may occur from the covalent species or either kind of ion pair or a combination of these, but the transfer rates are of the same algebraic form and so can be lumped without loss in accuracy. The rates are

$$r_{trS} \;=\; k_{trS} C_{\Sigma P^-} C_S \tag{10.66}$$

$$r_{trH_2O} \;=\; k_{trH_2O} C_{\Sigma P^-} C_{H_2O} \tag{10.67}$$

respectively. Transfer to ammonia produces a new NH_2^- anion (see reaction 10.51) which immediately initiates another chain: The kinetic chain is not terminated and the polymerization rate is not affected. Transfer to water produces an unreactive OH^- anion and so terminates the kinetic chain.

Equating the initiation rate to the rate of termination by chain transfer to water (Bodenstein approximation for the population of propagation centers) and solving for $C_{\Sigma P^-}$ one finds

$$C_{\Sigma P^-} \;=\; (k_{init} K_{KNH_2} / k_{tH_2O}) C_{KNH_2} C_M \big/ C_K . C_{H_2O} \tag{10.68}$$

With this substitution in eqn 10.65 the propagation rate becomes

$$r_p \;=\; k C_{KNH_2} C_M^2 \big/ C_K . C_{H_2O} \qquad \left(k \equiv k_b k_{init} K_{KNH_2} / k_{trH_2O} \right) \tag{10.69}$$

The rate of monomer consumption contains an additional contribution from initiation (see eqn 10.62). If the kinetic chains are long, this contribution is negligible, and the rate, then given by eqn 10.69, is first order in potassium amide, second order in styrene monomer, and of orders minus one in potassium ion and water.

The number-average degree of polymerization can be obtained from the rates of propagation (eqn 10.65) and chain breaking (sum of eqns 10.66 and 10.67) as in free-radical polymerization with termination by chain transfer to a transfer agent (see eqn 10.42):

$$(\bar{N})_{pol} = \frac{2 + r_p}{(-r_{chbr})} = 2 + \frac{k_b C_M}{k_{trS} C_S + k_{trH_2O} C_{H_2O}} \tag{10.70}$$

The molecular-weight distribution is a Schulz-Flory distribution as in eqn 10.45.

10.4.2. Cationic polymerization

With respect to kinetics and mechanism, cationic and anionic polymerization share the key features that distinguish ionic from free-radical polymerization: the absence of effective binary termination, and initiation that is fast relative to propagation. There are differences, however. Cationic polymerization is faster and may have to be conducted at temperatures below $0°$ C. Initiation may be more complex, and carbocations, while less sensitive to impurities than carbanions, are more prone to undergo a variety of chain transfer or rearrangement reactions that break the kinetic chain. As a result, high molecular weights are harder to obtain, and predictions of rates and molecular-weight distributions are less reliable than with anionic polymerization. Because of the analogies to anionic polymerization, the discussion here will be restricted to comments on differences in chemical detail.

Cationic polymerization is applied almost exclusively to monomers with olefinic double bonds. Susceptible are double bonds whose carbon atoms carry electron-donating substituents such as alkyl groups. Thus, isobutene with two methyl groups adjacent to the double bond polymerizes readily, propene with only one is sluggish, and ethene with none is inert; α-methyl styrene is more reactive than styrene; vinyl ethers are reactive, but vinyl chloride is not. The most important commercial product is *butyl rubber*, produced by copolymerization of isobutene with small amounts of isoprene, initiated by $AlCl_3$, BF_3, or $TiCl_4$ [82].

Initiation. The general mechanism of initiation of cationic polymerization is proton capture by the monomer. Brønsted acids can be used, but the choice is limited: The acid must be strong enough to be an effective proton donor, but its anion must not be so nucleophilic that it would form covalent bonds. Perchloric, sulfuric, phosphoric, and substituted sulfonic acids meet the requirements. More commonly used for production of high molecular-weight polymers are Lewis acids. Examples are BF_3, $AlCl_3$, $ZnCl_2$, $TiCl_4$, and PCl_5 among many others.

Initiation by a Lewis acid requires the presence of a proton donor (*protogen*) such as water, alcohol, or hydrogen halide. Generation of the carbocation is preceded by link-up of the initiator with the protogen, as shown here for polymerization of isoprene induced by BF_3 [83]:

$$BF_3 \; + \; H_2O \; \longleftrightarrow \; BF_3OH_2 \qquad \longrightarrow \qquad P_1^+ \; [BF_3OH]^- \qquad (10.71)$$

or, in general terms:

$$in + pr \; \longleftrightarrow \; in\cdot pr \; \overset{M}{\longrightarrow} \; P_1^+ \, A^- \qquad (10.72)$$

where in is the Lewis-acid initiator, pr is the protogen, P_1^+ is the monomeric carbocation, and A^- is the associated anion.* Usually but not necessarily, the rate-controlling step in this process is the second, and the first is at quasi-equilibrium.

Initiation can also occur spontaneously, either by so-called self-ionization [85,86] as in

$$2\,AlBr_3 \; \longleftrightarrow \; AlBr_2^+AlBr_4^- \; \overset{M}{\longrightarrow} \; AlBr_2M^+ \, AlBr_4^- \qquad (10.73)$$

or by direct reaction with monomer [87,88]:

$$TiCl_4 + M \; \longrightarrow \; TiCl_3M^+ \, Cl^- \qquad (10.74)$$

Polymerization of highly reactive monomers can also be initiated by iodine: Iodine adds to the double bond, and hydrogen iodide is then split off and acts as the protogen [89]. Other possible initiation methods include photochemical initiation [90], ionizing radiation [91], and electrolysis [92].

Termination. As in anionic polymerization, termination by coupling or disproportionation cannot occur, leaving chain transfers as the most likely mechanisms.

With regard to kinetics of chain breaking, the various chain-transfer reactions are analogous to those in free-radical polymerization (see Section 10.3.1) and need not be described again in detail. In cationic polymerization, the most common chain transfer is to monomer and leaves the polymer with a double bond while generating a monomeric chain carrier of the same structure as the original one [93].

Cationic polymerization can be effectively retarded or quenched by addition of transfer agents such as alkyl or aryl amines or phosphines that form stable quaternary cations [94]:

---

* A different notation with the protogen as "initiator" and the Lewis acid as "co-initiator" has been suggested [84], but has so far not been generally accepted.

$$\cdots\!\!\!\diagdown \quad HC^+\ A^-\ +\ NR_3\ \longrightarrow\ \cdots\!\!\!\diagdown\quad HC\!-\!\overset{+}{N}R_3\ A^- \qquad (10.75)$$

Water, alcohols (in combination with KOH), acids, and anhydrides in sufficiently high concentrations act in the same fashion [95]. It is interesting that such agents can function as the needed protogens at low concentrations, but have a retarding or quenching effect at high concentrations.

A termination unique to cationic polymerization, suggested [96] but still controversial [97], is by reaction with the counterion:

$$\cdots\!\!\!\diagdown\quad HC^+\ [BF_3OH]^-\ \longrightarrow\ \cdots\!\!\!\diagdown\quad HC\!-\!OH\ +\ BF_3 \qquad (10.76)$$

Propagation. As in anionic polymerization, a spectrum of dissociation states of the propagating centers can exist [98] (*Winstein spectrum* [77]; see carbanionic dissociation 10.53 in previous section). Accordingly, the general rate equation 10.58 and its simpler special forms 10.59 to 10.61 apply with obvious adjustments under the respective conditions. In cationic polymerization under practical conditions, dissociation may be weaker and the solvated carbocation may not exist. On the other hand, in some cases the covalently bonded species can contribute to or even dominate the propagation rate. This is referred to as *pseudo-cationic polymerization* [98-101]. It does not alter the form of the general rate equation 10.58 for propagation because the reaction orders for propagation are the same for the covalent species as for the contact and solvent-separated ion pairs. However, more complex behavior has also been observed [102,103].

Living polymers. To suppress termination in order to produce living polymers by ionic polymerization, the propagating centers must be stabilized against termination, but left reactive enough for propagation. These conflicting demands are harder to balance in cationic than in anionic polymerization, and special techniques had to be developed [104-106]. Even so, the cationic reactive ends in living polymers begin to decay slowly once all monomer is used up.

10.5. Coordination polymerization

Coordination polymerization is another variant of chain-growth polymerization. The kinetic chain starts with addition of monomer to a metal complex, propagation is by successive insertion of monomer at the metal, and termination occurs when the metal complex separates from the polymer. Inasmuch as the complex is restored upon termination, this is a catalytic process. Nevertheless, the metal complex is often called an initiator rather than a catalyst because of the analogies to other modes of chain-growth polymerization, in some processes also because the catalyst is left in the final polymer product.

By far the most important industrial coordination polymerization processes are Ziegler-Natta polymerizations of 1-olefins [107-110], most notably the production of high-density polyethene [111] and stereo-specific olefin polymers and co-polymers [108]. However, these processes employ solid catalysts, and the complex kinetics on their surfaces have no place in a book on homogeneous reactions.

The earliest Ziegler-Natta catalysts were insoluble bimetallic complexes of titanium and aluminum. Other combinations of transition and Group I-III metals have been used. Most of the current processes for production of high-density polyethene in the United States employ chromium complexes bound to silica supports. Soluble Ziegler-Natta catalysts have been prepared, but have so far not found their way into industrial processes. With respect to stereo-specificity they cannot match their solid counterparts.

Since the pioneer work of Ziegler [112] and Natta [113,114] in the 1950s, different types of soluble catalysts for coordination polymerization of monomers with olefinic double bonds have been developed. With the possible exception of some soluble Ziegler-Natta catalysts, all appear to be complex hydrides of transition metals, among these nickel, cobalt, and palladium [107]. The most important industrial processes employing homogeneous coordination catalysts are polymerizations of ethene and 1-olefins, in particular Shell's ethene oligomerization that produces straight-chain olefins in very high selectivity (see Example 10.4 farther below).

10.5.1. Mechanism

Mechanisms of homogeneous coordination polymerization differ somewhat, depending on the nature of the metal complex and the monomer. However, the principal features are common to most.

The reaction requires as a first step the attachment of the monomer to the metal of the hydride complex, typically at a free coordinative site as a π-complex of the olefinic double bond. The monomer then inserts itself between the metal and the hydrogen, forming a σ-bond with the metal and vacating the coordinative site it had occupied. That site then accepts another monomer molecule, which subsequently inserts itself between the first monomer molecule and the metal (Cossee mechanism [115,116]; a more complicated pathway has also been suggested [117,118]). This chain of events repeats itself until either a polymer molecule with unsaturated end group splits off after having transferred a hydrogen to the metal and thereby restored the catalyst to its original hydride form, or the chain is intentionally terminated by hydrogenolysis to produce a saturated polymer molecule and restore the hydride catalyst. For ethene polymerization:

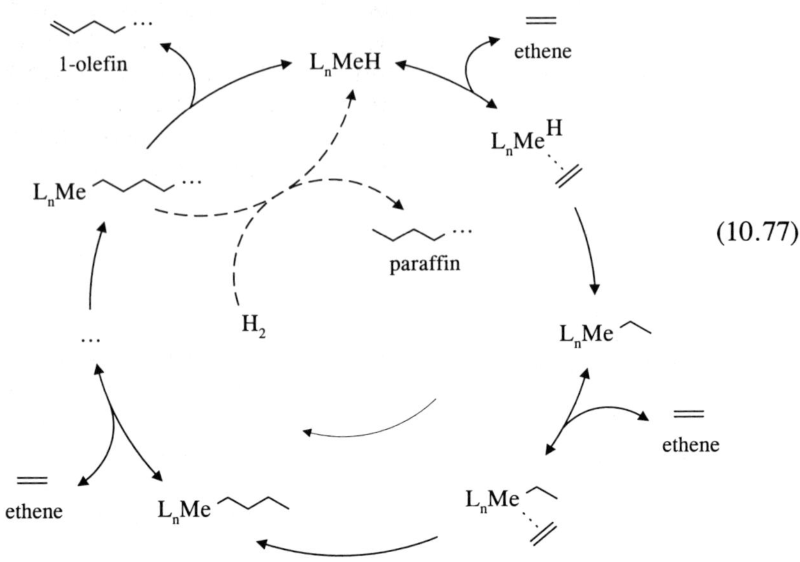

$$(10.77)$$

where the L are ligands such as cyclooctadiene, and the hydrogenolysis pathway is shown dashed.

For the purpose of reaction mathematics and granted quasi-stationary behavior, the two sequential steps of monomer addition and insertion can be consolidated into a single step (see Rule 8.55 in Section 8.6). If this is done, the network can be written

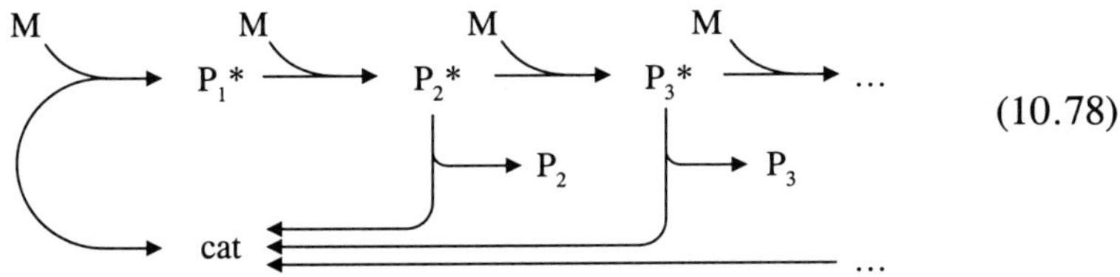

$$(10.78)$$

where M is the monomer, cat is the catalyst, the P_i^* are the metal-complex alkyls, and the P_i are dead polymer molecules, subscript i denoting the respective number of their monomer units.

All growth steps $P_i^* + M \longrightarrow P_{i+1}^*$ are chemically the same except for the length of the polymer chain, and so are all termination steps $P_i^* \longrightarrow P_i + \text{cat}$ (or $P_i^* + H_2 \longrightarrow P_i + \text{cat}$) ($i \geq 2$) in which polymer is split off. The respective rate coefficients, k_p and k_{trm}, can thus be assumed not to vary with chain length i. Note that the addition of the first monomer molecule to the catalyst occurs with a different rate coefficient, to be written k_{01}.

10.5.2. *Molecular-weight distribution and degree of polymerization*

Granted the invariance of the rate coefficients for propagation, k_p, and catalyst split-off, k_{trm}, the net rate of formation of an adduct P_{j+1}^* with $j+1$ monomer units is

$$r_{P_{j+1}^*} = k_p C_{P_j^*} C_M - C_{P_{j+1}^*}(k_p C_M + k_{trm}) \qquad (10.79)$$

Given quasi-stationary behavior of the propagating centers, the Bodenstein approximation $r_{P_{j+1}^*} \cong 0$ can be used. With eqn 10.79 and solved for $C_{P_{j+1}^*}/C_{P_j^*}$ this gives

$$C_{P_{j+1}^*}/C_{P_j^*} = k_p C_M/(k_p C_M + k_{trm}) \qquad (j \geq 1) \qquad (10.80)$$

Equation 10.80 assumes spontaneous termination; for termination by hydrogenolysis, k_{trm} should be replaced by $k'_{trm}p_{H_2}$. [The right-hand side of eqn 10.80 also expresses the probability that an adduct will continue to add monomer rather than split off catalyst and produce a dead polymer molecule.] Since all adducts P_j^* produce polymer P_j with the same rate coefficient k_{trm}, the ratio in eqn 10.80 is also that of the concentrations or mole fractions of the respective polymer molecules P_{j+1} and P_j that are formed:

$$x_{j+1}/x_j = k_p C_M/(k_p C_M + k_{trm}) \qquad (j \geq 1) \qquad (10.81)$$

Moreover, that ratio is independent of the number of monomer units in the polymer molecules, so that eqn 10.81 describes a declining geometrical progression:

Coordination polymerization under stationary or quasi-stationary conditions produces a Schulz-Flory mole-fraction distribution.

The distribution is as in free-radical polymerization with termination by disproportionation or terminating chain transfer (eqn 10.45 in Section 10.3.4) and, with increase of the progression factor as conversion increases, in step-growth polymerization of bifunctional monomers (eqn 10.19 in Section 10.2.3). According to eqn 10.81, the progression factor is

$$g = k_p C_M/(k_p C_M + k_{trm}) \qquad (10.82)$$

and remains constant if the monomer concentration is kept constant.

Example 10.4. Ethene oligomerization in the Shell Higher Olefin Process. The Shell Higher Olefin Process, abbreviated *SHOP*, produces predominantly internal straight-chain olefins in the C_{10} to C_{16} carbon-number range from ethene [107,109,119,120]. Oligomerization of ethene to straight-chain 1-olefins of even carbon numbers is only the first step. It is followed by catalyzed isomerization to internal olefins and finally olefin metathesis that breaks olefins at their double bonds and reassembles the fragments in a random fashion.* Olefins of carbon numbers higher or lower than desired are recycled. Only the oligomerization step is of interest here.

SHOP oligomerization is a three-phase process. The reactor contains a polar liquid catalyst phase, a nonpolar liquid product phase, and ethene gas. Ethene consumed in the catalyst phase is resupplied from the gas phase, and oligomer produced in the catalyst phase separates out to form the product phase. The catalyst is a nickel hydride complex with a phosphine carboxylate group [120,121] in 1,4-butanediol:

Table 10.2. Ethene oligomer formation measured in laboratory batch reactor at 100°C and 80 atm pressure with a SHOP-type catalyst.

product	weight g	mmol	mmol ratio
C_4H_8	3.74	66.79	0.781
C_6H_{12}	4.38	52.14	.783
C_8H_{16}	4.57	40.80	.783
$C_{10}H_{20}$	4.47	31.93	.783
$C_{12}H_{24}$	4.20	25.00	.780
$C_{14}H_{28}$	3.83	19.50	.781
$C_{16}H_{32}$	3.42	15.23	.782
$C_{18}H_{36}$	3.02	11.91	
higher	16.20		

(Ph = organic phosphine). The mechanism is as in 10.77, with spontaneous termination. Table 10.2 shows a typical Schulz-Flory distribution obtained with this process.

Given the Schulz-Flory distribution, the mole and weight fractions of polymer with j monomer units (based on polymer) as a function of that number j are

$$(x_j)_{pol} = g^{j-2} \Big/ \sum_{n=1}^{\infty} g^{n-1} = g^{j-2}(1-g) \qquad (j \geq 2) \qquad (10.83)$$

and

$$(w_j)_{pol} = jg^{j-2} \Big/ \sum_{n=1}^{\infty} (n+1)g^{n-1} = \frac{jg^{j-2}(1-g)^2}{2-g} \qquad (j \geq 2) \qquad (10.84)$$

respectively (the sums converge as shown). The number-average molecular weight is

---

* The metathesis reaction is still occasionally referred to as disproportionation although it bears no resemblance to disproportionation as a chain-growth termination step. Rather, it is analogous to interchange in step-growth polymerization (reaction 10.7 in Section 10.2.2) except that the reactants are broken apart at their double bonds and the fragments reconnected by double bonds.

$$(\bar{N})_{pol} = \sum_{n=1}^{\infty} (n+1)\,g^{n-1} \Big/ \sum_{n=1}^{\infty} g^{n-1} = \frac{2-g}{1-g} \tag{10.85}$$

or, after replacement of g with eqn 10.82:

$$(\bar{N})_{pol} = 2 + k_p C_M / k_{trm} = 2 + \nu \tag{10.86}$$

where ν is the radical chain length (see Section 10.3.2). At sufficiently high degrees of polymerization (g approaching unity), eqns 10.85 and 10.86 reduce to

$$(\bar{N})_{pol} \cong 1/(1-g) \cong \nu \tag{10.87}$$

The degree of polymerization is seen to show the same dependence on the geometric-progression factor as in step-growth polymerization of bifunctional monomers (eqn 10.16) and free-radical polymerization with chain breaking by disproportionation or terminating chain transfer (eqn 10.42).

10.5.3. Polymerization rate

If the condensed network 10.78 is taken at face value, the rate of polymerization (i.e., of monomer consumption) is

$$- r_M = k_{01} C_{cat} C_M + k_p C_{\Sigma P\cdot} C_M - k_{trm} C_{P_1\cdot} \tag{10.88}$$

The exact distribution of metal over free catalyst and the propagating centers is usually not known. A rate equation in terms of total metal, ΣMe, is therefore desired. This turns out to be

$$- r_M = k_p \left[1 + \frac{k_t}{k_p C_M + k_{trm}} \right] \left[\frac{k_{01} C_M}{k_{01} C_M + k_{trm}} \right] C_{\Sigma Me} C_M \tag{10.89}$$

The rate is first order in total metal and, depending on conditions, approximately first order in monomer. Note that eqn 10.89 is in terms of liquid-phase monomer concentration; if the monomer must be supplied from a gas phase, as in SHOP, Langmuir-type absorption may produce a tendency toward saturation kinetics.

Derivation. To replace the catalyst concentration in eqn 10.88, the Bodenstein approximation is applied to that species:

$$r_{cat} = k_{trm} C_{\Sigma P\cdot} - k_{01} C_{cat} C_M \cong 0$$

$$C_{cat} \cong k_{trm} C_{\Sigma P\cdot} / k_{01} C_M \tag{10.90}$$

According to eqn 10.83, the concentration of the first propagating center, $P_1{}^*$, is

$$C_{P_1\cdot} = C_{\Sigma P\cdot} \Big/ \sum_{n=1}^{\infty} g^{n-1} = C_{\Sigma P\cdot} (1-g) \tag{10.91}$$

Lastly, the metal balance is

$$C_{\Sigma Me} = C_{cat} + C_{\Sigma P}. \tag{10.92}$$

Equations 10.90 to 10.92 permit C_{cat} and $C_{P_1^*}$ to be expressed in terms of $C_{\Sigma P^*}$, and then the latter in terms of total metal, $C_{\Sigma Me}$. The result is eqn 10.89.

10.6. Chain-growth copolymerization

The description of chain-growth kinetics in the preceding sections has focused on polymerization of single monomers. However, the great majority of polymers produced on a large scale are copolymers. In fact, our present-day ability to tailor-make polymers of desired mechanical and chemical properties owes a great debt to the progress in the science of copolymerization. To give only two examples: *Butyl rubber* is a copolymer of isobutene and small amounts of isoprene [82], and *Saran* is a copolymer of vinyl chloride and vinylidene chloride [122]. In essence, a second monomer (and maybe even a third) is included to modify polymer properties such as elasticity, tensile strength, etc.

Kinetic aspects of step-growth copolymerization have been examined in Section 10.2.2. The principal features of chain-growth copolymerization are very different, but are alike for all types of chain growth, that is, for free-radical, anionic, cationic, and coordination polymerization.

Many chain-growth copolymerizations include dienes such as divinyl benzene or divinyl adipate that act as crosslinking agents and lead to gel formation. Polymerization kinetics in such cases are complex and are beyond the scope of a book on homogeneous reactions. Here, only binary copolymerization of mono-functional monomers will be examined. For an excellent and extensive treatment that includes copolymerization of more than two monomers as well as crosslinking by bifunctional monomers, the reader is refer to Odian's book [123].

The two principal aspects to be considered here are copolymer composition and polymerization rate. The ways of deducing molecular weights and molecular-weight distributions are essentially the same as in homopolymerization and will not be reiterated.

10.6.1. Polymer composition: reactivity ratios and copolymer equation

A key facet of copolymerization is the possible disparity of reactivities of the monomers. Traditional procedure is to assume, at least as an approximation, that the reactivity of a growing propagating center depends only on the identity of its reactive end unit (i.e., the last monomer added), not on the composition and length of the rest of its chain [124-126] (*first-order Markov* or *terminal* model; see also "shortsightedness" of reaction steps, Section 11.3). Even so and even with only two monomers, four different types of reactions can occur and must be considered: Each of the two monomers can react with an end group of its own kind or with one of the other kind. With two monomers M_A and M_B the possible reactions are:

$$M_A + \ldots -M_A^* \longrightarrow \ldots -M_A M_A^* \qquad \text{rate } k_{AA}C_A C_{A*}$$

$$M_A + \ldots -M_B^* \longrightarrow \ldots -M_B M_A^* \qquad k_{AB}C_A C_{B*}$$

$$M_B + \ldots -M_A^* \longrightarrow \ldots -M_A M_B^* \qquad k_{BA}C_B C_{A*}$$

$$M_B + \ldots -M_B^* \longrightarrow \ldots -M_B M_B^* \qquad k_{BB}C_B C_{B*}$$

where the M_i^* are the reactive end units or propagating centers, be they free-radical, cationic, anionic, or complexed, and the k_{ij} are the respective rate coefficients (first subscript refers to monomer).

The rates at which the monomers are consumed are

$$-r_A = k_{AA}C_A C_{A\cdot} + k_{AB}C_A C_{B\cdot}$$
$$-r_B = k_{BA}C_B C_{A\cdot} + k_{BB}C_B C_{B\cdot}$$

(10.93)

(M in subscripts is suppressed for simplicity).

To find what composition the polymer will have at a given monomer composition, an equation for the ratio of the monomer consumption rates as a function of the concentrations of the monomers is needed. With eqns 10.93 and the Bodenstein approximation of quasi-stationary behavior of either propagating center, M_A^* or M_B^*, one obtains

$$\frac{-r_A}{-r_B} = \frac{r_{A\cdot}}{r_{B\cdot}} = \frac{C_A(\rho_a C_A + C_B)}{C_B(\rho_b C_B + C_A)}$$

(10.94)

where

$$\rho_a \equiv k_{AA}/k_{BA} \qquad \text{and} \qquad \rho_b \equiv k_{BB}/k_{AB}$$

(10.95)

are termed *reactivity ratios*. Equation 10.94, known as the *copolymer equation* (or copolymerization equation) expresses the ratio in which M_A and M_B units are added to the polymer at given monomer concentrations, that is, the *instantaneous* copolymer composition. Alternatively, that composition can be expressed as the mole fraction $y_A \equiv r_{A*}/(r_{A*} + r_{B*})$ of M_A units added (based on added units) as a function of the mole fraction $x_A \equiv C_A/(C_A + C_B)$ (based on total monomer population):

$$y_A = \frac{x_A^2(\rho_a - 1) + x_A}{x_A^2(\rho_a + \rho_b - 2) + 2x_A(1 - \rho_b) + \rho_b}$$

(10.96)

Derivation. The ratio of the monomer consumption rates 10.93 is

$$\frac{-r_A}{-r_B} = \frac{k_{AA}C_A C_{A\cdot} + k_{AB}C_A C_{B\cdot}}{k_{BB}C_B C_{B\cdot} + k_{BA}C_B C_{A\cdot}} = \frac{k_{AA}C_A(C_{A\cdot}/C_{B\cdot}) + k_{AB}C_A}{k_{BB}C_B + k_{BA}C_B(C_{A\cdot}/C_{B\cdot})} \qquad (10.97)$$

The Bodenstein approximation for the end units $-M_A{}^*$ is

$$r_{A\cdot} = k_{AB}C_A C_{B\cdot} - k_{BA}C_B C_{A\cdot} \cong 0$$

and gives

$$C_{A\cdot}/C_{B\cdot} \cong k_{AB}C_A/k_{BA}C_B \qquad (10.98)$$

Replacement of C_{A*}/C_{B*} in eqn 10.97 with this expression and use of eqns 10.95 for the rate coefficient ratios gives eqn 10.94. That equation in terms of mole fractions and with $-r_A/(-r_B) = y_A/y_B = y_A/(1-y_A)$ gives

$$y_A = \frac{\rho_a x_A^2 + x_A x_B}{\rho_a x_A^2 + 2x_A x_B + \rho_b x_B^2}$$

Equation 10.96 is obtained from this with $x_B = 1 - x_A$.

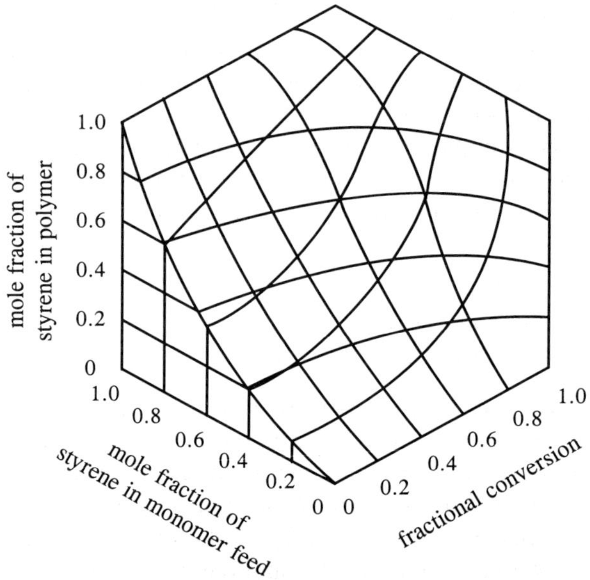

Figure 10.7. Instantaneous copolymer composition in free-radical copolymerization of styrene and 2-vinyl thiophene; mole fraction of styrene in polymer as function of initial mole fraction and fractional conversion of styrene; calculated with reactivity ratios $\rho_a = 0.35$ and $\rho_b = 3.10$ (from Mayo and Walling [127]).

The copolymer equation 10.94 describes the instantaneous copolymer composition at given monomer concentrations. An example is shown in Figure 10.7. In batch polymerization, of course, the concentration ratio of the monomers does not remain constant as conversion progresses, so that the instantaneous copolymer composition varies with time. The prediction of the composition of total polymer at any given instant then requires integration over time [129].

Compilations of reactivity ratios for various pairs of monomers in free-radical polymerization have been provided by Eastmond [130] and Odian [131]. The reactivity ratios for pairs of given monomers can be very different for the different types of chain-growth copolymerization: free-radical, anionic, cationic, and coordination copolymerization. Although the copolymer equation is valid for each of them, the copolymer composition can depend strongly on the mode of initiation (see Figure 10.8).

Several special cases are of interest:

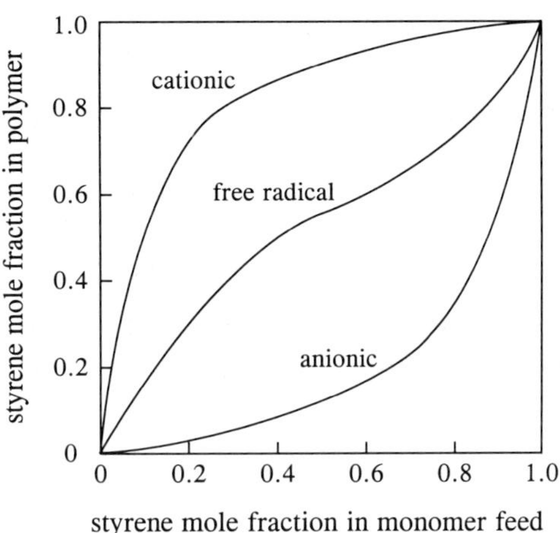

Figure 10.8. Instantaneous copolymer composition in cationic, free-radical, and anionic styrene/methyl methacrylate copolymerization initiated by $SnCl_4$, benzoyl peroxide, and Na in liquid NH_3, respectively (from Pepper [128]).

Case I:	random copolymers (also called ideal or statistical copolymers)	$\rho_a \rho_b = 1$

Here, $k_{AA}/k_{BA} = k_{AB}/k_{BB}$, that is, the probability of adding M_A rather than M_B is the same for both kinds of end groups $-M_A{}^*$ and $-M_B{}^*$. The monomer that is preferentially added becomes enriched in the polymer relative to the monomer mixture. Examples of such behavior are free-radical copolymerizations of styrene and butadiene ($\rho_a \rho_b \cong 1.1$), vinyl chloride and vinyl acetate ($\rho_a \rho_b \cong 0.9$), and vinylidene chloride and vinyl chloride ($\rho_a \rho_b \cong 1.1$) [132].

Mathematically, the dependence of the instantaneous copolymer composition on the monomer composition is analogous to that of liquid-phase on gas-phase compositions in ideal vapor-liquid equilibria. This explains why the term "ideal" was chosen (not to be understood as an optimum or desirable condition).

Case II: alternating copolymers $\rho_a = \rho_b = 0$

In this more interesting case, each monomer adds only to an end unit of the other kind ($k_{AA} = 0$, $k_{BB} = 0$). In the polymer, units M_A and M_B then alternate. Coordination copolymerization of olefins and carbon monoxide, catalyzed by complex hydrides of Pd(II) or Rh(I) in the presence of an alcohol co-solvent to yield polyketo esters, provides an example [133,134]: Olefin and carbon monoxide are added alternatingly, and reaction with alcohol terminates the kinetic chain and restores the catalyst. For ethene as the olefin:

$$n\ H_2C{=}CH_2\ +\ n\ CO\ +\ ROH\ \longrightarrow\ H-(CH_2CH_2CO)_n-OR$$

Case III: block copolymers $\rho_a > 1,\ \rho_b > 1$

In this hypothetical case, the monomers add preferentially to end groups of their own kind ($k_{AA} > k_{AB}$, $k_{BB} > k_{BA}$). As a result, "blocks" $M_AM_AM_A...$ and $M_BM_BM_B...$ of units of the same kind are formed and join to yield a polymer in which they alternate. Such behavior has been reported for some coordination copolymerizations [135,136], but has not been conclusively established.

Block copolymers have great importance for many practical applications, but are more conveniently produced from living polymers (see Section 10.4).

10.6.2. Polymerization rate

While the copolymer equation is universal in that it applies to all kinds of chain-growth copolymerization, an equally universal equation for the polymerization rate cannot be arrived at. For assessing the composition of the copolymer, only the *ratio* of the monomer consumption rates was needed, and that ratio was found to be a unique function of the monomer concentrations and rate coefficients. In contrast, the polymerization rate is composed of the *absolute values* of the monomer consumption rates, and these depend also on the concentrations of the propagating centers and thereby indirectly on the mechanism and rate of termination.

In copolymerization, several different combinations of initiation and termination mechanisms are possible, giving rise to a variety of different poly-merization rate equations. Only two cases will be singled out here: free-radical copolymerization with termination by coupling, and ionic polymerization with termination by chain transfer to a deactivating agent or impurity. For other combinations, the derivation of rate equations follows along the same lines.

Free-radical polymerization. No matter whether the propagating centers are free radicals, anionic, cationic, or coordinated, the propagation rate is equal to the sum of the consumption rates of the two monomers, given by eqns 10.93:

$$r_p = -r_A - r_B \qquad (10.99)$$
$$= k_{AA} C_A C_{A\cdot} + k_{AB} C_A C_{B\cdot} + k_{BA} C_B C_{A\cdot} + k_B C_B C_{B\cdot}$$

In free-radical polymerization with *termination by coupling*, there are three possible termination steps: reaction of end groups $-M_A^*$ with $-M_A^*$, of $-M_A^*$ with $-M_B^*$, and of $-M_B^*$ with $-M_B^*$. Each eliminates two reactive end groups. Leaving the possibility open that all steps contribute significantly, the termination rate is

$$-r_{trm} = 2(k_{tAA} C_{A\cdot}^2 + k_{tAB} C_{A\cdot} C_{B\cdot} + k_{tBB} C_{B\cdot}^2) \qquad (10.100)$$

The initiation rate for free-radical polymerization is

$$r_{init} = 2f k_{init} C_{in} \qquad (10.101)$$

(granted one initiator molecule produces two radicals). With the Bodenstein approximations of quasi-stationary behavior of the individual propagation centers and their total, the rate in terms of the monomer concentrations is found to be [137]:

$$r_p = \frac{(\rho_a C_A^2 + 2 C_A C_B + \rho_b C_B^2)(f k_{init} C_{in})^{1/2}}{\left((k_{tAA}/k_{BA}^2) C_A^2 + (k_{tAB}/k_{AB}k_{BA}) C_A C_B + (k_{tBB}/k_{AB}^2) C_B^2\right)^{1/2}} \qquad (10.102)$$

where ρ_a and ρ_b are the reactivity ratios defined in eqns 10.95 (derivation is given farther below).

Equation 10.102 is rather unwieldy. However, it can often be simplified: Binary terminations as reactions of two free radicals with one another have low activation energies and large rate coefficients that, with few exceptions, are of the same order of magnitude (monomers giving strongly stabilized radicals are poorly suited for polymerization; see also Section 9.4). As a result, in copolymerization:

- *Termination tends to be dominated by the more abundant free radical.*

This is much as in ordinary chain reactions (see Section 9.3). The ratio of the radicals with $-M_A^*$ and $-M_B^*$ end groups is given by eqn 10.98. Accordingly, termination is likely to occur by coupling of $-M_A^*$ end groups if $k_{AB}C_A \gg k_{BA}C_B$, or of $-M_B^*$ end groups if the opposite is true. If termination is by $-M_A^*$ coupling, the second and third denominator terms in eqn 10.102 can be dropped as unimportant; if it is by $-M_B^*$ coupling, the first and second terms can be dropped.

Ionic polymerization. In ionic polymerization with *termination by deactivating chain transfer*, the propagation rate equation is also given by 10.99, but the initiation and termination rates are different. In initiation, each initiator molecule produces only one propagating center:

$$r_{init} = f k_{init} C_{In} \qquad (10.103)$$

Chain transfer can occur from end groups of either or both kinds:

$$-r_{trm} = (k_{trA} \cdot C_{A\cdot} + k_{trB} \cdot C_{B\cdot}) C_{Tr} \qquad (10.104)$$

where Tr is the transfer agent or impurity. With the Bodenstein approximations as for the free-radical case, the propagation rate becomes

$$r_p = \frac{(\rho_a C_A^2 + 2 C_A C_B + \rho_b C_B^2) f k_{init} C_{in}}{((k_{trA} \cdot /k_{BA}) C_A + (k_{trB} \cdot /k_{AB}) C_B) C_{Tr}} \qquad (10.105)$$

where ρ_a and ρ_b are the reactivity ratios defined by eqns 10.95.

Equations 10.102 and 10.105 state the propagation rates. The polymerization rates include in addition the consumption of monomer in initiation. However, this contribution is negligible except in oligomerization.

Derivation of eqns 10.102 and 10.105. For both free-radical and ionic copolymerization, the Bodenstein approximation for the propagating centers $-M_B^*$ yields

$$r_{B\cdot} = k_{BA} C_B C_{A\cdot} - k_{AB} C_A C_{B\cdot} \cong 0$$

$$C_{B\cdot} \cong (k_{BA} C_B / k_{AB} C_A) C_{A\cdot} \qquad (10.106)$$

Using this to replace C_{B*} in eqns 10.99 and 10.100 one obtains

$$r_p = \left[\frac{k_{AA} k_{AB} C_A^2 + 2 k_{AB} k_{BA} C_A C_B + k_{BB} k_{BA} C_B^2}{k_{AB} C_A} \right] C_{A\cdot} \qquad (10.107)$$

and

$$-r_{trm} = 2 \left[\frac{k_{tAA}(k_{AB} C_A)^2 + k_{tAB} k_{AB} k_{BA} C_A C_B + k_{tBB}(k_{BA} C_B)^2}{(k_{AB} C_A)^2} \right] C_{A\cdot}^2 \qquad (10.108)$$

respectively.

The Bodenstein approximation for the total population of propagating centers amounts to equating the initiation and termination rates. For free-radical polymerization this gives, with eqn 10.101 and 10.108 and solved for C_{A*}:

$$C_{A\cdot} = \left[\frac{f k_{init} C_{in}}{k_{tAA}(k_{AB} C_A)^2 + k_{tAB} k_{AB} k_{BA} C_A C_B + k_{tBB}(k_{BA} C_B)^2} \right]^{1/2} k_{AB} C_A$$

Using this to replace C_{A*} in eqn 10.107 one finds

$$r_p = \frac{(k_{AA}k_{AB}C_A^2 + 2k_{AB}k_{BA}C_A C_B + k_{BB}k_{BA}C_B^2)\,(fk_{init}C_{in})^{1/2}}{\left(k_{tAA}(k_{AB}C_A)^2 + k_{tAB}k_{AB}k_{BA}C_A C_B + k_{tBB}(k_{BA}C_B)^2\right)^{1/2}} \qquad (10.109)$$

Dividing numerator and denominator by $k_{AB}k_{BA}$ and replacing the ratios k_{AA}/k_{BA} and k_{BB}/k_{AB} by the respective reactivity ratios with eqns 10.95 one obtains eqn 10.102.

For ionic polymerization, equating the initiation and termination rates (eqns 10.103 and 10.104), replacing C_{B*} with eqn 10.106, and solving for C_{A*}, one obtains

$$C_{A\cdot} = \frac{(k_{AB}C_A)\,(k_{init}C_{in})}{(k_{trA}k_{AB}C_A + k_{trB}k_{BA}C_B)\,C_{Tr}}$$

Replacement of C_{A*} in eqn 10.107 by this expression, division of numerator and denominator with $k_{AB}k_{BA}$, and introduction of the reactivity ratios yields eqn 10.105.

Summary

A distinction can be made between condensation and addition polymerization, depending on whether or not a small molecule such as water or hydrogen halide is cast off when monomers link up. With respect to kinetics, a more relevant distinction is between step growth and chain growth. In step-growth polymerization, molecules link up with one another by reaction of their functional end groups, and that is the only reaction occurring. Molecular weight increases with progressing conversion. In chain-growth polymerization, initiation is required to produce chain carriers or reactive centers that then add monomer molecules successively until some event terminates the kinetic chain or monomer is used up. The number of polymer molecules increases with progressing conversion, the molecular weight as a rule remains constant. Free-radical, anionic, cationic, and coordination polymerization proceed with chain-growth mechanisms.

Step growth is essentially a second-order reaction of the functional groups. The number-average molecular weight is related in a simple fashion to the fractional conversion of functional groups by the Carothers equation. That equation can also be used to estimate the gel point (state of conversion at which crosslinks begin to form) in polymerization of monomers with more than two functional groups per molecule. If the monomers are bifunctional and statistical growth can be assumed, the mole fractions of successive polymer molecules (with increasing number of monomer units) are in a declining geometrical progression. This is called a Schulz-Flory distribution.

Free-radical polymerization requires initiation to produce free radicals that link up with monomer molecules to produce reactive centers. Additional monomer molecules are then added successively at these centers. In this way, a small family of polymer radicals acts as an assembly line to produce "dead" polymer. The most common termination mechanisms are reactions of two polymer radicals with one another, either by coupling to yield one larger dead polymer molecule or, more rarely, by disproportionation to convert two radicals into a saturated and an unsaturated dead polymer molecule. With either of these terminations, the polymerization rate is first order in monomer and of order one half in initiator. Chain transfer to monomer, polymer, or solvent can also occur. Such chain breaking stops the growth of the polymer radical and may or may not terminate the kinetic chain, which might continue on another molecule. If transfer terminates the kinetic chain, the polymerization rate is first order rather than half order in initiator. In bulk polymerization, the increase in viscosity with conversion reduces the rate coefficients. Termination by coupling or disproportionation, involving two polymeric radicals, is more strongly affected than propagation. This causes self-acceleration (Trommsdorff effect) and, under certain conditions, can result in a runaway. Chain breaking exclusively by disproportionation or chain transfer produces a Schulz-Flory molecular-weight distribution. Chain breaking predominantly by coupling produces a higher degree of polymerization and a narrower, Poisson-type molecular-weight distribution.

Compounds capable of forming carbanions stabilized by delocalization of the negative charge can be made to undergo anionic polymerization. A key feature distinguishing anionic (and cationic) from free-radical polymerization is that binary termination cannot occur because ionic charges of same sign repel one another. Other termination mechanisms can be suppressed, and the polymer then still contains propagating centers when all monomer is used up ("living polymers"). Typical initiators are alkali metals, their alkyls, and metal amides. Rate behavior, reaction orders, and molecular-weight distributions depend on conditions. In particular, the rate can be reduced by addition of a salt with the same cation as that of the initiator. As a rule, initiation is fast compared with propagation, so that the propagating centers start growing at the same time and add monomer at the same rate. This makes it possible to produce polymers of specified molecular weights and narrow molecular-weight distributions by addition of a deactivating agent when polymerization has progressed as far as desired.

Cationic polymerization is similar to anionic polymerization in that binary termination by recombination or disproportionation cannot occur. The most common initiators are Brønsted or Lewis acids and iodine. A plethora of possible side reactions make it difficult to attain high molecular weights or prepare living polymers. Also, theoretical predictions of rates, molecular weights, and molecular-weight distributions are in general not reliable.

In coordination polymerization, monomer forms an adduct with a transition-metal complex, and further monomer is then successively inserted between metal and carbon. Termination occurs when the metal complex splits off from the polymer or the chain is broken intentionally by hydrogenolysis. Since the initiator is restored to its original form, the process is catalytic. The most important industrial processes are Ziegler-Natta polymerizations of α-olefins and employ heterogeneous (solid) catalysts. Most homogeneous catalysts for coordination polymerization are hydride complexes of transition metals. An important example is the Shell Higher Olefin Process (SHOP) for oligomerization of ethene with a complex nickel catalyst. The molecular-weight distribution is a Schulz-Flory distribution. The rate is first order in the catalyst metal. Saturation kinetics may result from Langmuir-type absorption of gaseous monomer by the liquid catalyst phase.

In chain-growth copolymerization, the composition of the polymer depends on the concentrations and relative reactivities of the monomers. The relative reactivities can be drastically different in free-radical, ionic, and coordination polymerization. Three special cases are random (also called statistical or ideal), alternating, and block copolymerization. In random copolymerization, the preference of adding monomer M_A rather than M_B is the same for polymers with reactive end group M_A as for those with reactive end group M_B; the sequence of M_A and M_B units in the polymer then is random. In alternating copolymerization, each monomer adds preferentially to reactive end groups of the other kind; in the product, M_A and M_B units then alternate. In rarely observed block copolymerization, each monomer adds preferentially to end groups of its own kind; the product then consists of alternating long "blocks" of monomer units of the same kind. However, a more practical method of producing block copolymers is via living polymers.

Rate behavior in chain-growth copolymerization is complex. The presence of reactive end groups of different types and different reactivities makes for a profusion of possible propagation and termination steps. While copolymer composition is essentially dictated by the relative amounts and relative reactivities of the monomers, the rate depends in addition on the population of propagating centers and thereby on the termination mechanism.

Examples include control of molecular weight in step-growth polymerization, number-average degree of polymerization in step-growth polymerization of nonstoichiometric monomer mixtures, free-radical and anionic polymerizations of styrene, and ethene oligomerization to linear 1-olefins in the Shell Higher Olefins Process.

References

General references

G1. G. C. Eastmond, *The kinetics of free radical polymerization of vinyl monomers in homogeneous solutions*, in *Comprehensive chemical kinetics*, Vol. 14a, C. H. Bamford and C. F. H. Tipper, eds., Elsevier, Amsterdam, 1967, ISBN 044441486X.

G2. P. J. Flory, *Principles of polymer chemistry*, Cornell University Press, Ithaca, 1953.

G3. P. C. Hiemenz, *Polymer chemistry: the basic concepts*, Dekker, 1984, ISBN 082477082X.

G4. C. D. Holland and R. G. Anthony, *Fundamentals of chemical reaction engineering*, Prentice Hall, Englewood Cliffs, 2nd ed., 1989, ISBN 0133356396, Chapter 10.

G5. J. P. Kennedy and E. Maréchal, *Carbocation polymerization*, Wiley, New York, 1982, ISBN 0471017876.

G6. G. Moad and D. H. Solomon, *The chemistry of free radical polymerization*, Pergamon, Oxford, 1995, ISBN 0080420788.

G7. M. Morton, *Anionic polymerization: principles and practice*, Academic Press, New York, 1983, ISBN 0125080808.

G8. G. Odian, *Principles of polymerization*, Wiley, New York, 3rd ed., 1991, ISBN 0471610208.

G9.   A. Rudin, *The elements of polymer science and engineering*, Academic Press, San Diego, 2nd ed., 1999, ISBN 0126016852.

G10.   S. R. Sandler and W. Karo, *Polymer syntheses*, Academic Press, Boston, 2nd ed., Vol I-III, 1992-1996, ISBN 0126185115, 0126185123, 0126185131.

G11.   M. Szwarc and M. van Beylen, *Ionic polymerization and living polymers*, Chapman & Hall, New York, 1993, ISBN 0412036614.

G12.   M. Szwarc, *Ionic polymerization fundamentals*, Hanser, Munich, 1996, ISBN 3446185062.

Specific references

1.   Flory (ref. G2), Section II-2.

2.   W. H. Carothers, *J. Am. Chem. Soc.*, **51** (1929) 2548.

3.   F. G. Helfferich and P. E. Savage, *Reaction kinetics for the practical engineer*, Course #195, AIChE Educational Services, New York, 7th ed., 1999, Section 6.10.

4.   L. H. Baekeland, *J. Ind. Eng. Chem.*, **1** (1909) 149; **6** (1913) 506.

5.   J. N. Weber, *Polyamides*, in *Kirk-Othmer, Encyclopedia of chemical technology*, 4th ed., J. I. Kroschwitz and M. Howe-Grant, eds., Wiley, New York, Vol. 19, 1996, ISBN 0471526886, p. 472.

6.   Sandler and Karo (ref. G10), Vol. II, Chapter 2, Section 2.

7.   P. W. Kopf, *Phenolic resins*, in *Kirk-Othmer, Encyclopedia of chemical technology*, 4th ed., J. I. Kroschwitz and M. Howe-Grant, eds., Wiley, New York, Vol. 18, 1996, ISBN 471526878, p. 6037.22.

8.   Flory (ref. G2), Section III-7.

9.   E. G. Lovering and K. J. Laidler, *Can. J. Chem.*, **40** (1962) 31.

10.   M. Kronstadt, P. L. Dubin, and J. A. Tyburczy, *Macromolecules*, **11** (1978) 37.

11.   Odian (ref. G8) p. 57.

12.   R. W. Missen, C. A. Mims, and B. A. Saville, *Introduction to chemical reaction engineering and kinetics*, Wiley, New York, 1999, ISBN 0471163392, Section 7.3.2.

13.   J. W. Moore and R. G. Pearson, *Kinetics and mechanism: a study of homogeneous chemical reactions*, Wiley, New York, 3rd ed., 1981, ISBN 0471035580, p. 23.

14.   S. M. Walas, *Reaction kinetics*, in *Perry's chemical engineers' handbook*, 7th ed., D. W. Green, and J. O. Maloney, eds., McGraw-Hill, New York, 1997, ISBN 0070498415, Table 7.4.

15.   Flory (ref. G2), Section III-1.

16.   Odian (ref. G8), Section 2-5.

17.   M. Stoll, A. Rouvé, and G. Stoll-Comte, *Helv. Chim. Acta*, **17**(1934) 1289.

18.   W. H. Carothers, *Trans. Faraday Soc.*, **32** (1936) 39.

19.   S. H. Pinner, *J. Polymer Sci.*, **21** (1956) 153.

20.   P. J. Flory, *J. Am. Chem. Soc.*, **63** (1941) 3083, 3091, and 3096.

21. Flory (ref. G2), Section IX-1.
22. W. H. Stockmayer, *J. Chem. Phys.*, **11** (1943) 45.
23. W. H. Stockmayer, *J. Polymer Sci.*, **9** (1952) 69; **11** (1953) 424.
24. Hiemenz (ref. G3), Section 5.8.
25. R. H. Kienle and F. E. Petke, *J. Am. Chem. Soc.*, **62** (1940) 1053; **63** (1941) 481.
26. Y.-K. Leung and B. E. Eichinger, in *Characterization of highly crosslinked polymers,* ACS Symp. Ser., **243** (1984) 21.
27. J. P. Flory, *Chem. Rev.*, **39** (1946) 137.
28. Flory (ref. G2), Section VIII-1.
29. G. V. Schulz, *Z. physik. Chem.*, **B 30** (1935) 379.
30. P. J. Flory, *J. Am. Chem. Soc.*, **62** (1940) 1561.
31. Sandler and Karo (ref. G10), Vol. I, Chapter 1, Section 2.
32. D. B. Priddy, *Styrene plastics, in Kirk-Othmer, Encyclopedia of chemical technology*, 4th ed., J. I. Kroschwitz and M. Howe-Grant, eds., Wiley, New York, Vol. 22, 1996, ISBN 0471526916, p. 1034.
33. Moad and Solomon (ref. G5), p. 92.
34. Eastmond (ref. G1), pp. 64-65, Chapter 1.
35. G. Moad and D. H. Solomon, in *Comprehensive polymer science*, G. C. Eastmond, A. Ledwith, S. Russo, and P. Sigwalt, eds., Vol. 3, Pergamon, Oxford, 1989, ISBN 0080325157, p. 147.
36. T. Sugimura and Y. Minoura, *J. Polymer Sci., A-1*, **4** (1966) 2735.
37. G. V. Schulz and F. Blaschke, *Z. physik. Chem.* (Leipzig), **B 51** (1942) 75.
38. Kennedy and Maréchal (ref. G5), pp. 193-194.
39. Rudin (ref. G9), Section 6.8.2.
40. P. D. Bartlett and R. Altschul, *J. Am. Chem. Soc.*, **67** (1946) 816.
41. Moad and Solomon (ref. G6), Section 5.3.3.4.
42. Sandler and Karo (ref. G10), Vol. III, Chapter 8, Section 2.
43. Moad and Solomon (ref. G6), Section 5.3.4.
44. Odian (ref. G8), Section 3-6d.
45. M. J. Roedel, *J. Am. Chem. Soc.*, **75** (1953) 6110.
46. Odian (ref. G8), pp. 257-258.
47. Rudin (ref. G9), p. 217.
48. C. M. Starks, *Free radical telomerization*, Academic Press, New York, 1974, ISBN 0126636508.
49. T. Corner, *Adv. Polymer Sci.*, **62** (1984) 95.
50. W. Heitz, in *Telechelic polymers: synthesis and applications*, E. J. Goethals, ed., CRC Press, Boca Raton, 1989, ISBN 0849367646, p. 61.
51. B. Boutevin, *Adv. Polymer Sci.*, **94** (1990) 69.
52. Eastmond (ref. G1), Chapter 3.

53. *Landolt-Börnstein, New Series, Radical reaction rates in liquids*, H. Fischer, ed., Springer, Berlin, Part II, Vol. 13a, 1984, ISBN 0387126074.

54. Moad and Solomon (ref. G6), pp. 211-214.

55. G. V. Schulz and G. Harborth, *Makromol. Chem.*, **1** (1948) 106.

56. E. Trommsdorff, H. Köhle, and P. Lagally, *Makromol. Chem.*, **1** (1948) 169.

57. G. V. Schulz, *Z. physik. Chem.* (Frankfurt), **8** (1956) 290.

58. Odian (ref. G8), Section 3-10a.

59. M. Kučera, *Mechanism and kinetics of addition polymerizations*, in *Comprehensive chemical kinetics*, Vol. 31, H. G. Compton, ed., Elsevier, Amsterdam, 1992, ISBN 0444987959, Chapter 3, Section 1.3.

60. Odian (ref. G8), Section 3-4c.

61. Odian (ref. G8), Section 3-5a.

62. Rudin (ref. G9), Section 6.6.

63. Flory (ref. G2), pp. 334-336.

64. E. Tschunkur and W. Bock, *Ger. Pat. 532,456*, 1929; *570,980*, 1933 (to I. G. Farbenindustrie).

65. K. Ziegler, *Angew. Chem.*, **49** (1936) 499.

66. M. Szwarc, M. Levy, and R. Milkovich, *J. Am. Chem. Soc.*, **78** (1956) 2656.

67. M. Szwarc, *Carbanions, living polymers, and electron transfer processes*, Interscience, New York, 1968, ISBN 0470843055.

68. D. C. Allport, *Block copolymers*, Elsevier, Amsterdam, 1991, ISBN 0853345570.

69. Morton (ref. G7), Chapter 9.

70. H. L. Hsieh, *J. Polymer Sci.*, **A 3** (1965) 163.

71. K. F. O'Driskoll, E. N. Ricchezza, and J. E. Clark, *J. Polymer. Sci.*, **A 3** (1965) 3241.

72. Sandler and Karo (ref. G10), Vol. I, p.36.

73. W. C. E. Higginson and N. S. Wooding, *J. Chem. Soc.*, **1952**, 760 and 1178.

74. C. S. Marvel, W. J. Bailey, and G. E. Inskeep, *J. Polymer Sci.*, **1** (1946) 275.

75. N. G. Gaylord and S. S. Dixit, *J. Polymer Sci. Macromol. Rev.*, **8** (1974) 51.

76. M. Szwarc, in *Ions and ion pairs in organic reactions*, Vol. 2, M. Szwarc, ed., Wiley, New York, 1974, ISBN 0471843083, Chapter 4.

77. S. Winstein, E. Clippinger, A. H. Fainberg, and G. C. Robinson, *J. Am. Chem. Soc.*, **76** (1954) 2597.

78. M. Szwarc, in *Ions and ion pairs in organic reactions*, M. Szwarc, ed., Wiley, New York, 1972, ISBN 0471843075, Chapter 1.

79. B. J. Schmitt and G. V. Schulz, *Eur. Polymer J.*, **11** (1975) 119.

80. Morton (ref. G7), Section 8.3.

81. Hiemenz (ref. G3), pp. 407-410.

82. E. Kresge and H.-C. Wang, *Butyl rubber*, in *Kirk-Othmer, Encyclopedia of chemical technology*, 4th ed., J. I. Kroschwitz and M. Howe-Grant, eds., Wiley, New York, Vol. 18, 1996, ISBN 471526878, p. 603.

83. J. P. Kennedy and R. G. Squires, *Polymer*, **6** (1965) 579.

84. Kennedy and Maréchal (ref. G5), Section 2.2.

85. M. Chmelir and M. Marek, *Coll. Czech. Chem. Comm.*, **32** (1967) 3047.

86. A. Gandini and H. Cheradame, *Adv. Polymer Sci.*, **34/35** (1980) 1.

87. P. Sigwalt, *Makromol. Chem.*, **175** (1974) 1017.

88. G. Sauvet, J. P. Vairon, and P. Sigwalt, *J. Polymer Sci., Polym. Chem. Ed.*, **16** (1978) 3047.

89. G. Sauvet and P. Sigwalt, *Carbocation polymerization: general aspects and initiation*, in *Comprehensive polymer science*, Vol. 3, G. C. Eastmond, A. Ledwith, S. Russo, and P. Sigwalt, eds., Pergamon, Oxford, 1989, ISBN 0080325157, Chapter 39.

90. J. V. Crivello, *Ann. Rev. Mat. Sci.*, **13** (1983) 173.

91. V. T. Stannett, J. Silverman, and J. L. Garnett, *Polymerization by high-energy radiation*, in *Comprehensive polymer science*, Vol. 4, G. C. Eastmond, A. Ledwith, S. Russo, and P. Sigwalt, eds., Pergamon, London, 1989, ISBN 0080325157, p.317.

92. O. F. Olaj, *Makromol. Chem., Macromol. Symp.*, **8** (1987) 235.

93. Kennedy and Maréchal (ref. G5), Section 4.3.

94. M. Biswas and P. Kamannarayana, *J. Polymer Sci., Polym. Chem. Ed.*, **14** (1976) 2071.

95. A. R. Mathieson, in *The chemistry of cationic polymerization*, P. H. Plesch, ed., Macmillan, New York, 1963, ISBN 0080102891, Chapter 6.

96. G. Heublein, *Zum Ablauf ionischer Polymerisationsreaktionen*, Akademie Verlag, Berlin, p. 125.

97. Kennedy and Maréchal (ref. G5), p. 220.

98. Kennedy and Maréchal (ref. G5), Section 3.1.

99. D. C. Pepper and P. J. Reilly, *Proc. Roy. Soc.*, **A 291** (1966) 41.

100. P. H. Plesch and A. Gandini, *The chemistry of polymerization processes*, Monograph No. 20, Society of Chemical Industry, London, 1966.

101. D. J. Dunn, *The cationic polymerization of vinyl monomers*, in *Developments in polymerization*, Vol. 1, R. N. Haward, ed., Appl. Sci. Publishers, London, 1979, ISBN 0853348227, Chapter 2.

102. P. H. Plesch, *Makromol. Chem., Macromol. Symp.*, **13/14** (1988) 375 and 393.

103. K. A. Matyjaszewski, *Makromol. Chem., Macromol. Symp.*, **13/14** (1988) 389.

104. R. Faust, A. Fehervari, and J. P. Kennedy, *J. Macromol. Sci., Chem.*, **A 18** (1982-83) 1209.

105. J. Puskás, G. Kaszás, J. P. Kennedy, T. Kelen, and F. Tüdös, *J. Macromol. Sci., Chem.*, **A 18** (1982-83) 1229 and 1263.

106. M. Sawamoto and J. P. Kennedy, *J. Macromol. Sci., Chem.*, **A 18** (1982-83) 1275.

107. J. P. Collman, L. S. Hegedus, J. R. Norton, and R. G. Finke, *Principles and applications of organotransition metal chemistry*, University Science Books, Mill Valley, 2nd ed., 1987, ISBN 0935702512, Chapter 11.

108. Odian (ref. G8), Chapter 8.

109. G. W. Parshall and S. D. Ittel, *Homogeneous catalysis: the application and chemistry of catalysis by soluble transition metal complexes*, Wiley, New York, 1992, ISBN 0471538299, Chapter 4.

110. Sandler and Karo (ref. G10), Vol. I, Chapter 1, Section 5.

111. Y. V. Kissin, *High density polyethylene*, in *Kirk-Othmer, Encyclopedia of chemical technology*, 4th ed., J. I. Kroschwitz and M. Howe-Grant, eds., Wiley, New York, Vol. 17, 1996, ISBN 047152686X, p. 724.

112. K. Ziegler, E. Holzkamp, H. Breil, and H. Martin, *Angew. Chem.*, **67** (1955) 541.

113. G. Natta, *J. Polymer Sci.*, **16** (1955) 143 (in French).

114. G. Natta, *Chim. é Ind.* (Milan), **37** (1955) 888.

115. P. Cossee, *J. Catal.*, **3** (1964) 80.

116. E. J. Arlman and P. Cossee, *J. Catal.*, **3** (1964) 99.

117. K. J. Ivin, J. J. Rooney, C. D. Stewart, M. L. H. Green, and J. Mahtab, *J. Chem. Soc., Chem. Comm.*, **1978**, 604.

118. M. L. H. Green, *Pure Appl. Chem.*, **100** (1978) 2079.

119. E. R. Freitas and C. R. Gum, *Chem. Eng. Progr.*, **75**(1) (1979) 73.

120. M. Peuckert and W. Keim, *Organometallics*, **2** (1983) 594.

121. W. Keim, F. H. Kowaldt, R. Goddard, and C. Krüger, *Angew. Chem. Int. Ed. English*, **17** (1978) 466.

122. R. A. Wessling, D. B. Gibbs, P. T. DeLassus, B. E. Obi, and B. A. Howell, *Vinylidene monomer and polymers*, in *Kirk-Othmer, Encyclopedia of chemical technology*, 4th ed., J. I. Kroschwitz and M. Howe-Grant, eds., Wiley, New York, Vol. 24, 1997, ISBN 0471526932, p. 882.

123. Odian (ref. G8), Section 6-6.

124. T. Alfrey, Jr., and G. Goldfinger, *J. Chem. Phys.*, **12** (1944) 205 and 332; **14** (1946) 115.

125. F. R. Mayo and F. M. Lewis, *J. Am. Chem. Soc.*, **66** (1944) 1594.

126. F. T. Wall, *J. Am. Chem. Soc.*, **66** (1944) 2050.

127. F. R. Mayo and C. Walling, *Chem. Revs.*, **46** (1950) 191.

128. D. C. Pepper, *Quart. Revs.* (London), **8** (1954) 88.

129. I. Skeist, *J. Am. Chem. Soc.*, **68** (1946) 1781.

130. Eastmond and E. G. Smith (ref. G1), Appendix to Chapter 4.

131. Odian (ref. G8), Table 6-2.

132. Rudin (ref. G9), pp. 247-248.

133. T. W. Lai and A. Sen, *Organometallics*, **3** (1984) 866.

134. A. Sen and J. S. Brumbaugh, *J. Organomet. Chem.*, **279** (1985) C 5.

135. P. Prabhu, A. Schindler, M. H. Theil, and R. D. Gilbert, *J. Polymer Sci., Polym. Lett. Ed.*, **18** (1980) 389.

136. R. W. Gumbs, S. Penczek, J. Jagur-Grodzinski, and M Szwarc, *Macromolecules*, **2** (1969) 77.

137. C. Walling, *J. Am. Chem. Soc.*, **71** (1949) 1930.

Chapter 11

Mathematical Modeling

Mathematical models play an essential role in process development. How best to construct a kinetic model depends on the process development strategy that has been chosen. The present chapter discusses such strategies and suggests approaches to mathematical modeling suited to them.

11.1. Strategies of process development

Evolutionary approach (left-hand column in Figure 11.1). Until the 1950s, practically all chemical processes were developed with evolutionary methods, and many still are: The chemist, having discovered a new and potentially useful reaction, replicates his bench-scale experiments in a larger vessel. If there is commercial promise, engineers then take over and construct a still larger "semi-technical unit." A small pilot plant might follow, and eventually a fair-sized pilot plant that provides the operating experience on which the design of the full-sized plant can be based. In a nutshell, scale-up is by a number of small steps. In each of these, the vagaries of scale-up are apt to cause some problems or failures that must be remedied by tinkering, but the scale-up factors are small enough that no serious risk is incurred.

This approach is reliable and almost always successful. Its disadvantage is that, for a large-scale process, it takes a lot of time and manpower, both chronically in short supply. A knowledge of mechanisms is not required and even serves no useful purpose other than providing guidance to the research chemist or trouble-shooting engineer.

Empirical approach (center column in Figure 11.1). About halfway through the twentieth century, competitive pressures in chemical industry increased to the point that a shortening of development time was accorded a high priority. At the same time, computers were becoming more powerful and more readily available and statistics gained in popularity. This combination instigated an empirical approach to process development that relies heavily on statistics. Typically, after a chemist has done his job, engineers take an educated guess at probable optimum design and

Figure 11.1. Three strategies of process development (from Helfferich and Savage [1]).

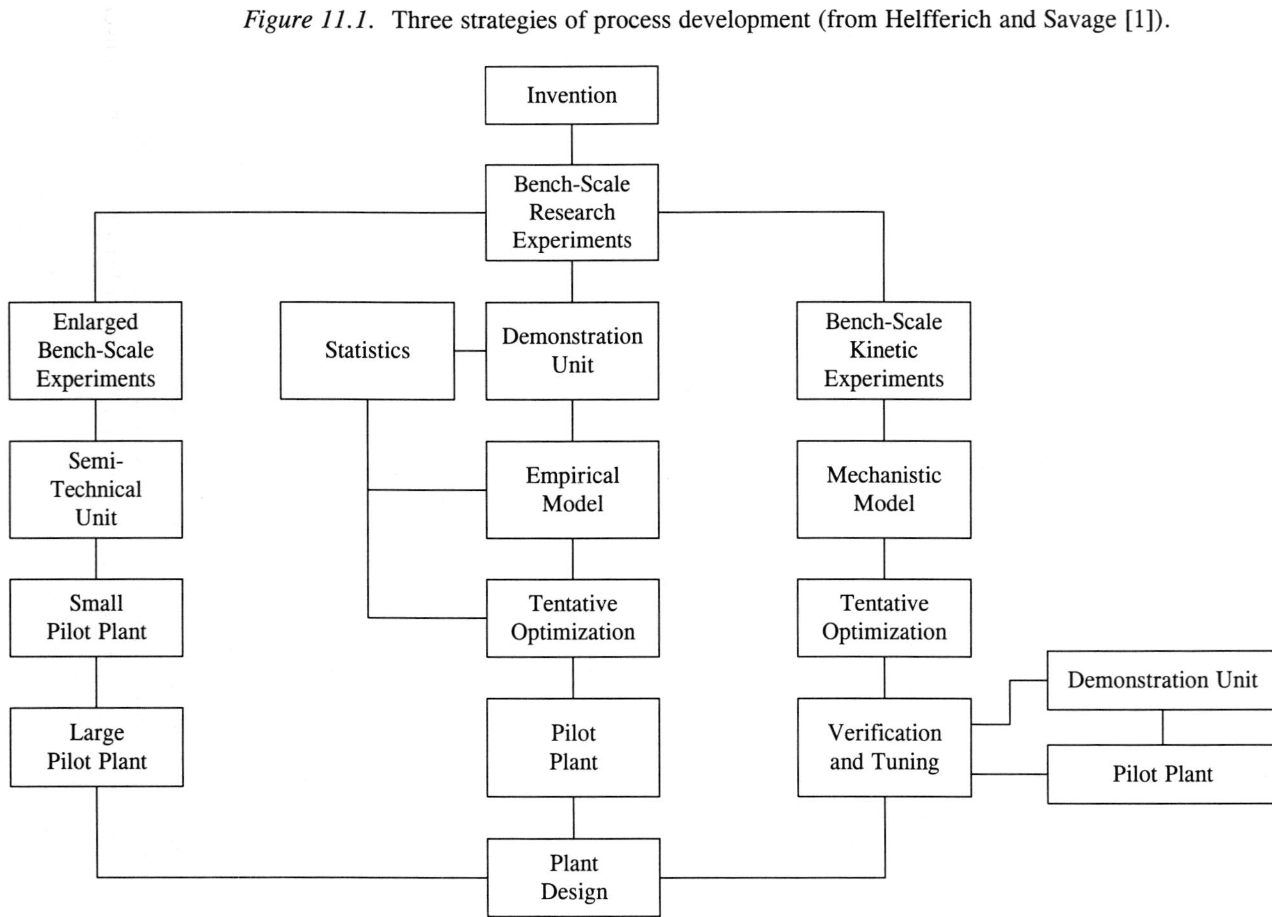

operating conditions for the future plant and build an integrated *demonstration unit* —really a miniature pilot plant—that faithfully models the anticipated plant with all its reactors and separation trains, recycle loops, etc., on a laboratory scale. This unit is then used to explore a broad range of operating conditions by varying the operating parameters with greatest economy of effort, guided by theory of statistical design of experiments [2,3]. When sufficient data have been collected, a computer is used to construct a multidimensional operating surface and search on it for an optimum. A pilot plant is then built and operated within a narrower range of conditions in the vicinity of the supposed optimum, for tuning the design of the full-scale plant. In this approach, the chemistry is essentially a "black box." No effort is made to elucidate mechanisms. Rather, empirical equations are fitted to observed reactor dynamics.

This approach is quite workable and may be the best if the process is not complicated and the scale not large. However, with, say, six or seven or more design parameters, hundreds of runs are needed for a complete statistical design. Each run may take two to three days to line the demonstration unit out to a steady state. Development time and consumption of man-time and chemicals then become prohibitive, so that short-cuts must be taken, a risky procedure if the scale-up factor is large. Moreover, information about the chemistry, thermodynamics, quantum mechanics, etc., and observations of transient behavior upon start-up, shut-down, and switch to different operating conditions have no input into the mathematical optimization. Whenever he makes no use of information available for free, the engineer does not work at highest efficiency. All this militates against the empirical approach to large and complex processes.

Fundamental approach (right-hand column in Figure 11.1). Concurrent with the empirical approach just described, a quite different philosophy was developed in the 1960s, pioneered chiefly by Mobil Oil in petroleum processing and Shell in industrial chemicals. The basic idea is to establish the true mathematics of the chemical reactions by elucidation of mechanisms in short-duration bench-scale experiments in order to make possible a direct and safe scale-up to the full-sized plant. Demonstration unit and pilot plant are relegated to assessing long-term effects such as catalyst life and corrosion, fine-tuning the design, providing proof of operability, piloting process control, and producing samples for customers, but are freed from the task of scanning wide ranges of operating conditions. The shining goal of direct scale-up from the laboratory bench can rarely be attained. Nevertheless, the approach has proved it can come close to the ideal of process development: to do the best possible job in the shortest possible time with least possible manpower at least possible expense.

The fundamental approach is at its best if the scale is large, the process can be expected to have a long life with additional plants being built as market share is conquered, and the chemistry is neither trivial nor overwhelmingly complex, conditions often given especially in industrial chemicals production. In such situations the fundamental approach requires less equipment, manpower, and chemicals than does the empirical approach. It also has the advantage of providing more insight into details of chemical mechanisms, a knowledge often valuable for research as well as for development of future processes based on the same or a similar chemistry. A drawback is that the approach is also the most demanding on the expertise of the development team. This problem is aggravated at least at the present time by a lack of thorough schooling in chemical kinetics (as distinct from reactor design) in our current chemical engineering curriculum. Moreover, if the scale is small, the chemistry is very complex, timing is more important than efficiency, and the lifetime of the process is uncertain, as is often the case in biotechnology, an effort to establish mechanisms in detail may not pay off.

Addressing fundamental kinetics, this book is chiefly intended as an aid to practitioners of the fundamental approach. In part this is because little can be said in generally applicable terms about the other approaches as they largely rely on experience with the specific chemistry at hand and on intuition—not to add folklore—as well as the well-documented principles of statistical design. Balanced coverage requires, however, that the limitations of the fundamental approach be pointed out and alternatives be mentioned.

Pursuit of strategies.   The brief outlines above may serve to characterize the possible options in process development. They describe clear-cut, "pure" strategies, but, of course, modifications are possible. For efficient development it is important to decide as early as possible on the merits of the case which tack to take. If the chosen approach is evolutionary or empirical, expected best operating conditions for the eventual plant should dominate the design of the experimental program. In contrast, if the fundamental approach is taken, the clear aim should be to establish networks and mechanisms beyond reasonable doubt with the fewest possible experiments, even if their conditions are far remote from those likely to be employed in the plant. Only when networks and mechanisms are believed to be in hand, should predictions for expected plant conditions be made and verified. An attempt to look at plant conditions while pursuing the fundamental approach is apt to syphon off valuable manpower and time from the principal effort:

Don't mix your strategies !

To be sure, this dictum is not absolute. Complementary experiments on an as-time-permits basis might settle pressing questions or give valuable leads. However, the delay of the principal task they almost always entail should be weighed against the expected benefits.

11.2. Effective mathematical modeling

How best to approach mathematical modeling of kinetics depends on what strategy of process development was chosen.

The evolutionary strategy, which essentially relies on tinkering upon stepwise scale-up, has little use for kinetic modeling. For the empirical strategy, empirical equations are sought that can fit observation, a task for which, as a rule, mathematicians are better suited than engineers, and whose details have no proper place in this book. As far as the chemistry of the process is concerned, the most common procedure is to start out with power-law rate equations. Here, fundamental kinetics can offer one piece of advice:

If a power-law rate equation requires fractional exponents, one-plus equations with integer exponents should be tried instead. If chain mechanisms or pre-dissociation may be involved, one-plus equations with exponents that are integer multiples of one half should also be tried.

This is because, as has been seen, such one-plus equations are more likely to reflect actual mechanisms and so to provide approximate fits over wider ranges of conditions than do power laws. Of course, there is no guarantee that one-plus equations will do better. In any event, whether power law or one-plus, the equations chosen must be fine-tuned and, in the end, will rarely resemble their simple initial forms.

Other than this, fundamental kinetics has little, if anything, to offer in an evolutionary or empirical process development. In fact, an attempt to introduce it into what aims to be the best possible empiricism might only cause distracting complications. The balance of the present section will therefore be restricted to fundamental kinetic modeling based on the concepts and procedures developed in the preceding chapters.

11.2.1. Complete fundamental modeling with Bodenstein approximation

A prerequisite for fundamental mathematical modeling is that the reaction network or networks have been established. This will be taken for granted here (for network elucidation, see Chapter 7).

Software for direct, "brute-force" solution of the rate equations is available [4-9] and can be used if the network consists of only a few elementary steps. In practice, however, effective fundamental modeling usually calls for a reduction in the number of simultaneous rate equations and their coefficients. As Chapter 6 has shown, a systematic application of the Bodenstein approximation to all trace-level intermediates can achieve this, at least unless the network is largely non-simple.

The approximations of a rate-controlling step, quasi-equilibrium steps, and long chains in chain reactions and the concept of relative abundance of catalyst-containing species in catalysis or propagating centers in ionic polymerization can often be used for additional simplification (see Sections 4.1, 4.2, 8.5, 9.3, and 10.4.1). A procedure suited in many cases consists essentially of the following steps [10]:

(1)   *Identify any non-trace intermediates and steps with two or more molecules of intermediates as reactants*;

(2)   *cut the network into piecewise simple portions at such intermediates and steps* (see Sections 6.5 and 7.3.3);

(3)   *establish all independent stoichiometric constraints and yield-ratio equations* (see Section 6.4.3) for possible replacement of rate equations;

(4)   *reduce the network or its piecewise simple portions to a form or forms with only pseudo-single, pseudo first-order steps between adjacent nodes and between nodes and adjacent end members* (see Section 6.4);

(5)   *compile the rate equations in terms of Λ coefficients* (segment coefficients) *for all end members of the network or its piecewise simple portions, except those which can be replaced by stoichiometric constraints* (see Sections 6.3 for pathways and 6.4.2 for networks);

(6)   *use yield-ratio equations to replace rate equations where the former are of simpler form than the latter*;

(7)   *establish equations for all Λ coefficients in terms of λ coefficients* (pseudo-first order coefficients);

(8)   *replace all λ coefficients by k coefficients* (true coefficients), *multiplied by co-reactant or co-product concentrations where appropriate* (see Section 6.2);

(9)   *for single pathways and small networks or network portions, reduce the rate equation or equations to their most convenient one-plus forms with lumped phenomenological coefficients* (see Section 7.2.1); *for large networks or network portions, do so with the equations for the Λ coefficients.*

In networks with loops or multiple catalytic cycles, the use of $\mathscr{L}$ (loop) or Γ (collective) instead of Λ coefficients may provide further simplification (see Sections 6.4 and 8.8). A detailed example will illustrate the procedure.

Example 11.1. Hydroformylation of cyclohexene with phosphine-substituted cobalt hydrocarbonyl catalyst. The most probable network of cyclohexene hydroformylation catalyzed by a phosphine-substituted cobalt hydrocarbonyl is shown on the facing page. $HCo(CO)_3Ph$ (cat) is in equilibrium with the CO-deficient $HCo(CO)_2Ph$ (cat') and CO. For greater generality, quasi-equilibrium of these species with the π-complex, X_1, is not assumed. Actual hydroformylation olefin $\longrightarrow$ aldehyde proceeds via a Heck-Breslow pathway (cycle 6.9 that includes the trihydride, X_2) but without

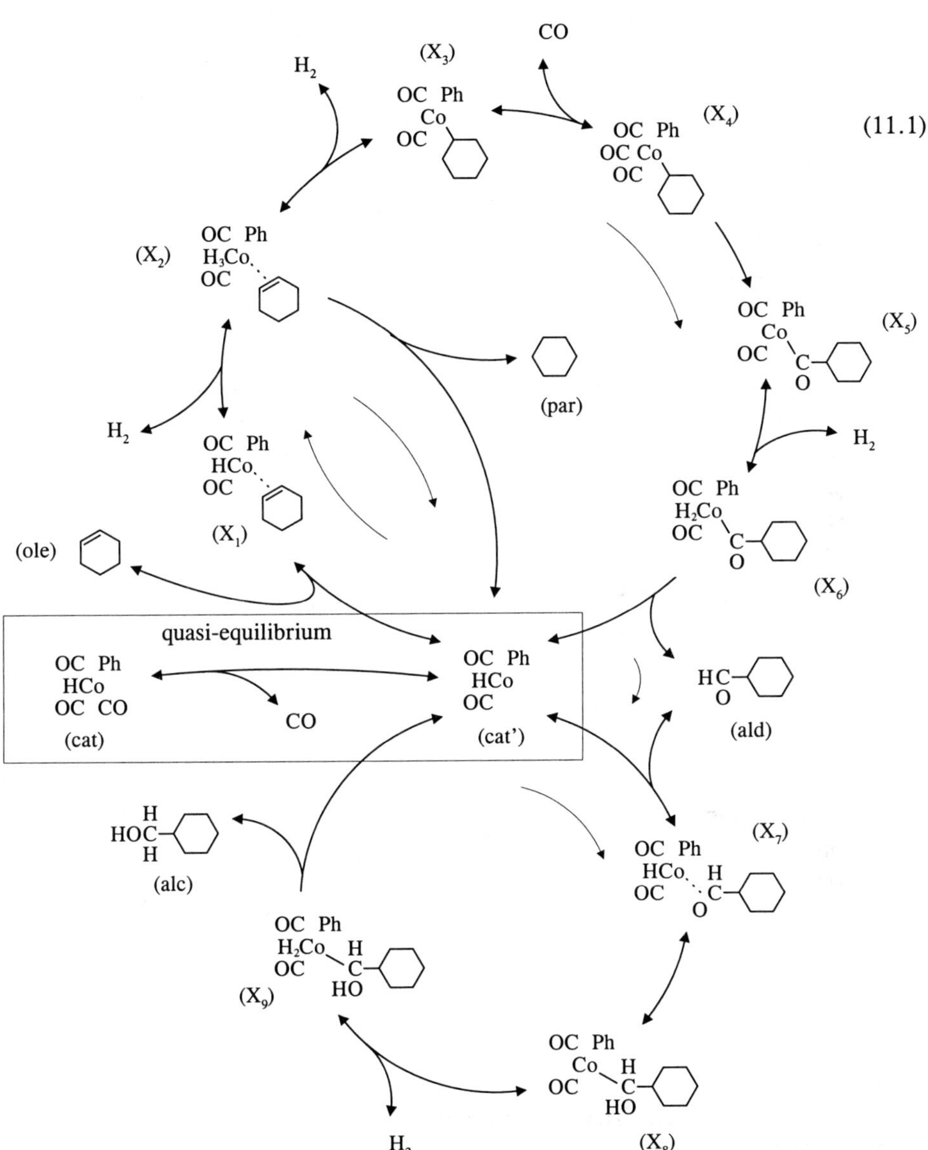
CO
H2
(X3)
OC Ph
Co
OC
OC Ph
OC Co
OC
(X4)
(11.1)
OC Ph
H3Co
OC
(X2)
OC Ph
Co
OC C
O
(X5)
H2
OC Ph
HCo
OC
(X1)
(par)
OC Ph
H2Co
OC C
O
(X6)
(ole)
quasi-equilibrium
HC
O
(ald)
OC Ph
HCo
OC CO
(cat)
OC Ph
HCo
OC
(cat')
CO
OC Ph
HCo H
OC C
O
(X7)
H
HOC
H
(alc)
OC Ph
H2Co H
OC C
HO
(X9)
OC Ph
Co H
OC C
HO
(X8)
H2

the tetracarbonyl acyl (Y in 6.9), whose concentration remains insignificant at the lower pressures used with the phosphine-substituted catalyst. Paraffin by-product is presumed to be formed from the trihydride (see Example 7.5, pathway III), and aldehyde hydrogenation is via pathway 7.28 of Example 7.4. The steps $X_2 \longrightarrow$ paraffin + cat', $X_4 \longrightarrow X_5$, $X_6 \longrightarrow$ ald + cat', and $X_9 \longrightarrow$ alc + cat' are irreversible. Equilibrium with other catalyst species (Example 8.3) is not accounted for.

This example is to illustrate procedure, not demonstrate the power of the method, therefore the choice of an olefin with relatively simple kinetics: Cyclohexene exists only as internal *cis*-olefin, so that olefin isomerization and production of isomeric aldehydes and alcohols need not be considered, and the aldehyde formed carries its $-$CHO group on a secondary carbon atom, so that aldol condensation remains insignificant (these complications will be included in the next example).

Break-up into piecewise simple portions and their reduction. The network 11.1 is "simple" except that aldehyde, an intermediate, builds up to higher than trace concentrations. Thus it can be cut at the aldehyde into two piecewise simple portions which share aldehyde and the ligand-deficient catalyst, cat'. After reduction the portions can be written

$$\text{ole} \leftrightarrow X_2 \longrightarrow \text{ald} \qquad\qquad \text{and} \qquad\qquad \text{ald} \longrightarrow \text{alc}$$
$$\downarrow$$
$$\text{par}$$

with five end members and five Λ coefficients (ole and ald rather than cat' are written as starting species, for better distinction between the two portions.)

Stoichiometric constraints and yield ratio equation. The stoichiometric constraints are

conservation of skeletal carbon: $-r_{\text{ole}} \;=\; r_{\text{par}} + r_{\text{ald}} + r_{\text{alc}}$ (11.2)

conservation of hydrogen: $-r_{\text{H}_2} \;=\; r_{\text{par}} + r_{\text{ald}} + 2\,r_{\text{alc}}$ (11.3)

conservation of oxygen: $-r_{\text{CO}} \;=\; r_{\text{ald}} + r_{\text{alc}}$ (11.4)

and obviate compilation of the rate equations for olefin, H_2, and CO. The yield ratio paraffin-to-hydroformylation products is

$$\frac{r_{\text{par}}}{r_{\text{ald}} + r_{\text{alc}}} \;=\; \frac{\Lambda_{2,\text{par}}}{\Lambda_{2,\text{ald}}} \qquad\qquad (11.5)$$

and can be used instead of the rate equation for paraffin.

Rate equations in terms of Λ coefficients. With the rate equations for olefin, paraffin, H_2, and CO replaced by the stoichiometric constraints and the yield ratio of paraffin to hydroformylation products, the rate equations for aldehyde and alcohol remain to be established. In terms of Λ coefficients these are:

$$r_{ald} = \left[\frac{\Lambda_{02}\Lambda_{2,ald}}{\Lambda_{20} + \Lambda_{2,par} + \Lambda_{2,ald}} - \Lambda_{0,alc}C_{ald} \right] C_{cat'}$$

$$r_{alc} = \Lambda_{0,alc}C_{cat'}$$

(first term in equation for r_{ald} is obtained with eqn 6.19).

These equations are in terms of the CO-deficient catalyst, cat'. In view of the equilibrium cat $\longleftrightarrow$ cat' + CO:

$$C_{cat'} = K_{cat}C_{cat}/p_{CO}$$

(K_{cat} = dissociation constant of catalyst) the rates can be written instead:

$$r_{ald} = \left[\frac{\Lambda'_{02}\Lambda_{2,ald}}{\Lambda_{20} + \Lambda_{2,par} + \Lambda_{2,ald}} - \Lambda'_{0,alc}C_{ald} \right] \frac{C_{cat}}{p_{CO}} \tag{11.6}$$

$$r_{alc} = \Lambda'_{0,alc}C_{cat} /p_{CO} \tag{11.7}$$

where

$$\Lambda'_{02} \equiv \Lambda_{02}K_{cat} \qquad \text{and} \qquad \Lambda'_{0,alc} \equiv \Lambda_{0,alc}K_{cat}$$

Λ *coefficients and their one-plus forms.* The derivation of equations for the Λ coefficients can be simplified with rules from Section 7.3.1 by shortening and consolidating the pathways $X_2 \longrightarrow$ ald and ald $\longrightarrow$ alc: Since the step $X_4 \longrightarrow X_5$ is irreversible, the pathway portion $X_5 \longrightarrow$ ald does not affect rate behavior (Rule 7.12), and the step sequences $X_3 + CO \longrightarrow X_4 \longrightarrow X_5$, ald + cat' $\longrightarrow X_7 \longrightarrow X_8$, and $X_8 + H_2 \longrightarrow X_9 \longrightarrow$ alc + cat' can be consolidated into single steps $X_3 + CO \longrightarrow X_5$, ald + cat' $\longrightarrow X_8$, and $X_8 + H_2 \longrightarrow$ alc + cat', respectively (Rule 7.24). With these simplifications, the general procedure for simple pathways or segments (Section 6.3) gives

$$\Lambda'_{02} = \frac{\lambda_{01}\lambda_{12}K_{cat}}{\lambda_{12} + \lambda_{10}} = \frac{k_{01}k_{12}K_{cat}C_{ole}p_{H_2}}{k_{12}p_{H_2} + k_{10}} = \frac{k_a C_{ole}p_{H_2}}{1 + k_b p_{H_2}} \tag{11.8}$$

$$\Lambda_{20} = \frac{\lambda_{10}\lambda_{21}}{\lambda_{12} + \lambda_{10}} = \frac{k_{10}k_{21}}{k_{12}p_{H_2} + k_{10}} = \frac{k_{21}}{1 + k_b p_{H_2}} \tag{11.9}$$

$$\Lambda_{2,par} = \lambda_{2,par} = k_{2,par} \tag{11.10}$$

$$\Lambda_{2,ald} = \frac{\lambda_{23}\lambda_{35}}{\lambda_{35} + \lambda_{32}} = \frac{k_{23}k_{35}p_{CO}}{k_{35}p_{CO} + k_{32}} = \frac{k_c p_{CO}}{1 + k_d p_{CO}} \tag{11.11}$$

$$\Lambda'_{0,alc} = \frac{\lambda_{08}\lambda_{8,alc}K_{cat}}{\lambda_{8,alc} + \lambda_{80}} = \frac{k_{08}k_{8,alc}K_{cat}C_{ald}p_{H_2}}{k_{8,alc}p_{H_2} + k_{80}} = \frac{k_e C_{ald}p_{H_2}}{1 + k_f p_{H_2}} \tag{11.12}$$

The phenomenological coefficients k_a to k_f in the one-plus forms are related to the coefficients of the individual steps and the catalyst dissociation constant by

$$k_a \equiv k_{01}k_{12}K_{cat}/k_{10}, \qquad k_b \equiv k_{12}/k_{10}, \qquad k_c \equiv k_{23}k_{35}/k_{32},$$

$$k_d \equiv k_{35}/k_{32}, \qquad k_e \equiv k_{08}k_{8,alc}K_{cat}/k_{80}, \qquad k_f \equiv k_{8,alc}/k_{80}$$

Model equations. The mathematical model requires eight concentration-independent coefficients: k_a to k_f, $k_{2,par}$, and k_{21}. From these it calculates the five Λ coefficients with eqns 11.8 to 11.12; from these, the rates of aldehyde and alcohol with eqns 11.6 and 11.7; and finally the rates of olefin, paraffin, H_2, and CO with eqns 11.2, 11.5, 11.10, 11.11, 11.3, and 11.4, respectively. Alternatively, eqns 11.8 to 11.12 can be used to replace the Λ coefficients in eqns 11.6 and 11.7 in order to obtain explicit rate equations for aldehyde and alcohol in terms of the phenomenological coefficients. However, the resulting rate equations are more cumbersome.

This example has shown how the procedures developed in earlier chapters can be used effectively for modeling. The reaction system has seventeen participants: olefin, paraffin, aldehyde, alcohol, H_2, CO, $HCo(CO)_3Ph$, $HCo(CO)_2Ph$, and nine intermediates. "Brute force" modeling would require one rate equation for each, four of which could be replaced by stoichiometric constraints (in addition to the constraints 11.2 to 11.4, the brute-force model can use that of conservation of cobalt). Such a model would have 22 rate coefficients (arrowheads in network 11.1, not counting those to and from co-reactants and co-products), whose values and activation energies would have to be determined. This has been reduced to two rate equations and nine simple algebraic relationships (stoichiometric constraints, yield ration equations, and equations for the Λ coefficients) with eight coefficients. Most impressive here is the reduction from thirteen to two rate equations because these may be differential equations.

For this example and seven others from this book, Table 11.1 illustrates the reduction of complexity achieved, showing a comparison of the numbers of rate and other equations and their coefficients of reduced and "brute force" models. The latter are understood to consist of the rate equations for all participants except those that can be replaced by stoichiometric constraints, and the constraints used in this fashion. The greatest reductions are where it counts most: in the possibly differential rate equations. Also important is the reduction in the number of coefficients. This is because the problem with brute-force modeling today is not so much the demands of the actual calculations, but the experimental work required to obtain values for all the coefficients and their activation energies.

The reduced models in Table 11.1 rely on the validity of the Bodenstein approximation for all intermediates except the aldehyde in hydroformylation, but are otherwise free of assumptions. In every case, equations that are as simple or even simpler have long been derived, but only with much more restrictive assumptions, most commonly that of a single rate-controlling step and quasi-equilibrium everywhere else. Of course, such equations should be used in preference if their assumptions can be substantiated.

Table 11.1. Reduction of complexity: comparison with "brute force" models.

reaction	model	number of			
		rate equations*	stoichiom. constraints	yield ratios	coeffi- cients
nitration of aromatics (Example 6.1)	brute force	5	4		6
	reduced	1			3
aldol condensation (Example 8.2)	brute force	5	2		5
	reduced	1			3
aldehyde hydrogenation (Example 7.4)	brute force	9	4		5
	reduced	1			2
"oxo" reaction (Example 6.2)	brute force	8	4		16
	reduced	1			2
hydrocyanation (Example 8.7)	brute force	7	4		12
	reduced	1			4
Wilkinson reaction (Example 8.9)	brute force	6	4		10
	reduced	1			8
hydroformylation (Example 11.1)	brute force	13	4		22
	reduced	2	3	1	8**

* Does not include equations that can be replaced by stoichiometric constraints or yield ratios.
** If quasi-equilibrium ole + cat' $\longrightarrow$ X$_1$ can be assumed as in network 6.9, k_b is negligible and only seven coefficients are needed.

11.2.2. *Streamlining for large networks*

Industrial practice often confronts the development engineer with networks that are considerably more complicated than that of cyclohexene hydroformylation in the example above. Additional simplifications may then be desirable or necessary in order to arrive at a model that remains manageable in the highly iterative applications called for in reactor design and optimization and possibly on-line process control. A useful and usually successful way of achieving such streamlining is to place all network nodes at end members or non-trace intermediates, ignoring the fact that some of them may be at trace-level intermediates [10].

The segments in a streamlined model might not correspond to those in the actual network with trace-level node intermediates. This raises the question what to use as Λ coefficients. The choice is somewhat arbitrary. A good starting point is to trace, in the actual network, the most direct paths between what are the node and end members in the streamlined network, and to establish the Λ coefficients for these paths regardless of any branches along them. The following example will illustrate this procedure.

Example 11.2. Streamlined network for hydroformylation of n-heptene catalyzed by phosphine-substituted cobalt hydrocarbonyl. In hydroformylation of straight-chain olefins with a phosphine-substituted cobalt hydrocarbonyl catalyst, the model must account for three complications that are absent with cyclohexene: isomerization by migration of the double bond along the hydrocarbon chain, formation of isomeric aldehydes and alcohols, and condensation of the straight-chain aldehyde to "heavy ends" (chiefly an alcohol of twice the carbon number, such as 2-ethylhexanol from propene via *n*-butanal and a C_8 aldol). A streamlined network for *n*-heptene is:

$$
\begin{array}{ccc}
 & (P_0)\ \text{2-hexyldecanol} & \\
 & \uparrow & \\
n\text{-octanal }(K_1) & \longrightarrow & n\text{-octanol }(P_1) \\
\nearrow & & \\
\text{1-heptene }(A_1) & \searrow & \\
 & \text{2-methylheptanal }(K_2) \longrightarrow \text{2-methylheptanol }(P_2) \\
\nearrow & & \\
n\text{-heptane }(Q) \longleftarrow \text{2-heptene }(A_2) & & \\
 & \searrow \text{2-ethylhexanal }(K_3) \longrightarrow \text{2-ethylhexanol }(P_3) \\
\nearrow & & \\
\text{3-heptene }(A_3) & \searrow & \\
 & \text{2-propylpentanal }(K_4) \longrightarrow \text{2-propylpentanol }(P_4) \\
\end{array}
\tag{11.13}
$$

(catalyst and co-reactants not shown to avoid clutter).

The aldehydes arise from the olefin isomers by addition of CO to a carbon atom on either side of the double bond in coupled parallel steps (see Example 5.3 in Section 5.3) and are hydrogenated to the respective alcohol isomers. Fortunately, only the straight-chain aldehyde condenses to a significant extent. In comparison, the detailed network consist of 111 steps.

A derivation of the rate equation for 2-propylpentanal (K_4) may serve to illustrate the resulting streamlined mathematics. The rate is given by

$$r_{K_4} = \Lambda_{A_3,K_4} C_{A_3} - \Lambda_{K_4 P_4} C_{K_4} \tag{11.14}$$

The shortened and consolidated, direct pathways $A_3 \longrightarrow K_4$ and $K_4 \longrightarrow P_4$ are

$$\text{cat} \rightleftharpoons \text{cat'} \xrightarrow{\text{ole}} \text{cat'} \rightleftharpoons X_1 \xrightarrow{H_2} X_1 \rightleftharpoons X_3 \xrightarrow{CO} X_5 \cdots$$
$$(\text{CO}) \qquad\qquad (\text{ald}) \qquad\qquad (H_2)$$

$$\text{cat} \rightleftharpoons \text{cat'} \rightleftharpoons X_8 \xrightarrow{H_2} \text{alc} + \text{cat'}$$
$$(\text{CO})$$

with Λ coefficients of the forms

$$\Lambda_{A_3,K_4} = \frac{k_a C_{A_3}}{1 + k_b p_{CO} + k_c p_{CO}/p_{H_2}}, \qquad \Lambda_{K_4,P_4} = \frac{k_d C_{K_4}}{p_{CO}(1 + k_e p_{H_2})} \tag{11.15}$$

respectively (in both equations, the first denominator term before reduction to one-plus form has been omitted because of equilibrium in the first step of the pathway). A comparison shows that eqn 11.14 with eqns 11.15 gives a much simpler rate equation than does eqn 11.6 with eqns 11.8 to 11.11 for an analogous step sequence. There is no simplification in the rate equation for aldehyde $\longrightarrow$ alcohol because that pathway has no branch in the detailed network.

While the reduced models discussed previously invoke only the Bodenstein approximation for trace-level intermediates, the additional streamlining is apt to introduce some errors, and these are hard to estimate beforehand. A safe way to proceed is to compile both a "research model" based on the detailed network, and a streamlined "process model." Apart from its use in evaluation of bench-scale experiments, the research model can serve to assess, by comparison, the nature, direction, and magnitude of the error of the process model under operating conditions of the plant. If found satisfactory, the process model can then be used for reactor design and optimization, possibly after some tuning.

Provided steric considerations and experience strongly suggest its topology, a streamlined model may also come into play as working hypothesis before the results of kinetic studies are in, or as a poor-man's mechanistic model if elucidation of kinetics in detail is not possible. In such applications, the concentration dependence of the Λ coefficients must be determined empirically.

11.2.3. *Determination of coefficients*

The discussion up to this point has addressed the compilation of a mathematical model, that is, how best and most concisely to express a network and mechanism in terms of equations. A second and no less important task is to obtain the best numerical values of the model's coefficients.

In general terms, the determination of coefficients in power-law and one-plus rate equations has been described in Sections 3.3 and 7.2, respectively. Here, some detail will be added.

There are two ways of arriving at values of coefficients. The conventional, old-fashioned, graphical procedure seeks plots that give straight lines and determines coefficient values from their slopes and intercepts. Examples have been seen in earlier chapters. Instead, linear- or nonlinear-regression computer software can be used for "optimization," that is, finding the set of coefficient values that fits the experimental data with least error [11-15]. Both methods have their pros and cons, and each has its place in kinetic modeling.

The main shortcoming of the graphical method is that it relies on "eye-balling," a subjective activity, and is therefore unlikely to yield values that provide an optimum fit. It has the advantage of presenting a picture to look at, making it easier to spot any untoward behavior. On the other hand, statistical regression, if properly used, turns out best values. However, unless the program is equipped with a plotting routine, it is largely a black-box affair, and that makes it difficult in complicated cases to discern where these values really come from. For example, a graph may show a fit to be excellent except for one point or a pair of points that are far off, raising a red flag that an experimental error may have occurred and prompting a check; in contrast, a primitive regression program accepts the point or points at face value and has them bias the entire set of values it returns as "best."

With statistics there are still other pitfalls to beware of. First, primitive regression programs seek the best correlation without screening for systematic errors. Given the choice between two models, they will select that with lesser error even if its error is systematic and that of the rejected model is random scatter attributable to experimental inaccuracy. Second, in complicated situations the program may converge onto a false, local optimum unless started with an approximately correct guess. Lastly, there is the psychological danger of being overly impressed by a good correlation coefficient, misinterpreting it as evidence of correctness of the model.* Statistics is a very fine tool, but no substitute for human judgment.

---

* A story, possibly apocryphal and admittedly from old times, tells of a research team at Dow Chemical's Midland facility submitting to their brethren in Applied Mathematics a bogus model and a heap of random numbers alleged to be experimental results, and receiving within a day the reply that statistical optimization of the coefficients gave an excellent correlation, proving the model to be correct. And there is no end of jokes about statisticians misapplying the tenets of their profession to situations in their everyday lives.

In the early stages of model development, the engineer should keep his eyes open for any symptoms that might point to effects overlooked or not yet accounted for. Here, graphs are generally preferable to abstract statistics. Once the equations of a model have been established, statistical computer programs are invaluable for obtaining the objectively best set of coefficients. Even here, however, old-fashioned graphics still has an input in that it can provide a good initial guess for the regression, thereby greatly reducing the chance of convergence to a false optimum.

11.3. "Shortsightedness" of elementary reaction steps

A useful principle that often holds and can be applied to ease the work load of development is:

> The reactivity of a group depends strongly on the local configuration, but little or not at all on the size and shape of the reactant molecule more than two atoms away.

For example, the reactivities of primary and secondary hydroxyl groups of alcohols differ greatly, but neither depends much on the lengths and configurations of the hydrocarbon chains to which they are attached, granted the chains are longer than three carbon atoms. One might say, the elementary reaction at a group usually is "shortsighted." There are exceptions, especially in structures with steric hindrance or resonance, the most notable being the effect of conjugation in molecules with multiple double bonds or aromatic rings. Therefore, the principle of shortsightedness must be tested in each case before it can be relied upon.

Where applicable, the shortsightedness principle can significantly simplify quantitative modeling, especially in networks with coupled parallel steps. Examples are olefin reactions that involve double-bond migration in parallel to conversion to products, as in homogeneous catalytic hydrogenation, hydroformylation, hydrocyanation, and hydrohalogenation [16].

Consider as a prototype the network 11.13 of *n*-heptene hydroformylation, keeping in mind that the arrows represent multistep pathways and that the reactions of higher straight-chain olefins involve still more parallel pathways of internal olefin isomers to aldehyde isomers and on to alcohol isomers. In such networks, all but one of the aldehyde-to-alcohol conversions involve the reaction of an aldehyde group on a secondary carbon atom, so that all these pathways can be assumed to involve essentially the same rate coefficients of their steps. Only the conversion of the straight-chain aldehyde (*n*-octanal to *n*-octanol in network 11.13) must be expected to occur with somewhat different rate coefficients. Likewise, all conversions of the internal olefin isomers to aldehydes and paraffin, except that of the 2-olefin, can be assumed to occur with practically equal rate coefficients. If the network is large, such equalities of coefficients can greatly reduce the number of coefficient values that must be determined experimentally.

Apart from its usefulness in the construction of mathematical models, the shortsightedness principle packs notable predictive power. As an example, an olefinic double bond can exist in only six configurations:

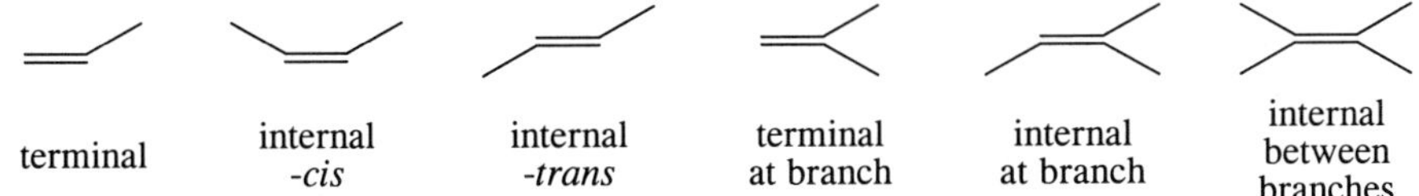

Usually, the reactivity is highest for the terminal position and lowest for the positions at branches. Once the reactivities with regard to a specific reaction have been determined for all six positions, the reaction behavior of all types of mono-olefins except those with strained rings or bulky substituents can be predicted with reasonable confidence. The reactivities of three structurally differed hexene isomers in hydroformylation catalyzed by phosphine-substituted cobalt hydrocarbonyls may serve as an example [16]:

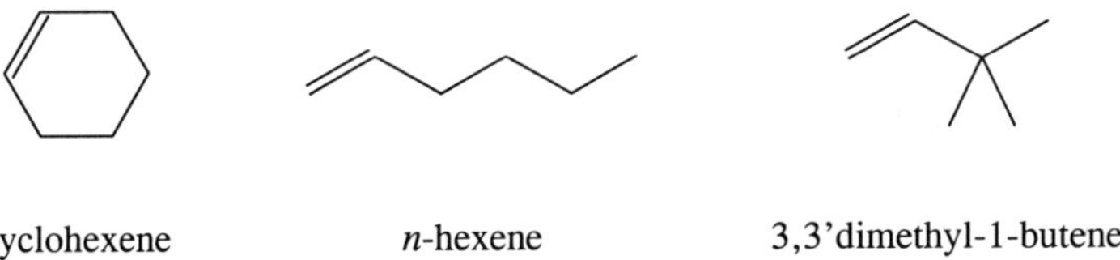

Cyclohexene exists only as internal *cis*-olefin and is moderately reactive. In contrast, *n*-hexene, regardless of whether charged as 1- or internal olefin or a mixture of these, is quickly isomerized to a near-equilibrium mixture containing some 5 to 10% of the isomer with terminal double bond, whose reactivity is about two orders of magnitude higher than those with internal ones. Accordingly, *n*-hexene is more reactive than cyclohexene with only internal double-bond positions. Lastly, neo-hexene (3,3'-dimethyl-1-butene) has its double bond locked in the terminal position —no double bond can exist adjacent to a quaternary carbon atom—and so should have the highest reactivity if not sterically hindered. (This is an unsubstantiated prediction, as the hydroformylation reactivity of that olefin seems not to have been studied to date.)

Once its validity has been confirmed, the shortsightedness principle can also be used in exploratory evaluations and process development for prediction of reactivities within homologous series. Keep in mind that equal reactivity of a group on molecules of different sizes does not necessarily entail equal overall reaction rates in a practical context. Again taking hydroformylation of straight-chain olefins as an example: Ethene and propene exist only as 1-olefins and so are highly and about equally reactive. With each carbon atom added to the chain, the percentage of 1-olefin in the quickly established near-equilibrium isomer mixture decreases, and so does the overall reactivity. If the catalyst is a phosphine-substituted cobalt hydrocarbonyl, the overall rate at near-equilibrium of isomerization decreases more than a hundred-fold from ethene and propene to *n*-octadecene. A fundamental model allows the reaction behavior of all olefins of the series to be predicted with reasonable accuracy once the individual rate coefficients for reaction of the terminal and internal double-bond positions have been determined by experiments with only one member of the series.

A historical footnote may illustrate this point. In the 1960s, Shell had developed and put into operation hydroformylation processes for *n*-butanol from propene and for detergent-range alcohols (C_{10} to C_{16}) from higher olefins. A new process for plasticizer-range alcohols (C_5 to C_9) was then developed, based entirely on the predictions of a fundamental model and only a few confirmatory bench-scale kinetic experiments with one olefin in that range. Development and plant construction were completed in record time and design capacity was reached almost immediately, even though the plant processed some olefins with which no experiments at all had been done. [However, the plant was shut down and cannibalized in short order: The product tanks were full and customers failed to materialize. A brilliant technical success, but a sad economic failure. Such is life!]

11.4. Model validation

The goal of fundamental mathematical modeling is to make a risk-free scale up by large factors possible, ideally even a scale-up from the laboratory bench to the full-sized plant. For this purpose, the model cannot be based on mere plausibility, it must be established beyond reasonable doubt. Plausible models will do as working hypotheses in research, in exploratory evaluations, and in the early stages of development, but not as a basis for plant design. One of the most important lessons the young development engineer must learn is to distinguish sharply between plausible hypotheses and well-established fact, in his own mind as well as in his presentations to his management.

It is true that in chemical kinetics one can disprove, but never definitely prove. A theory or model is conclusively disproved by experimental evidence to the contrary, but compatibility with such evidence is no proof because alternative explanations are always possible and new experiments might support those.*

---

* Chad Tolman [17] quotes Oxford's Peter Atkins as saying, "a proof in a chemical reaction is less like a mathematical proof and more like a proof in a court of law," and adds, "as in the court, the reputation and rhetoric of the lawyer ... may swing opinions regarding correctness of the mechanism."

Nevertheless, a very high degree of reliability can certainly be attained. Moreover, what matters for a design is only the accuracy (within the required limits) of the model's equations, not the correctness of all assumptions that have led to them.

While a design engineer may do fine with just two, his model needs three legs to stand on:

A model for scale-up, design, and optimization must:
- conform with all experimental evidence,
- be explainable by an eminently plausible mechanism, and
- have proved its predictive power.

No model should be accepted that does not meet all three of these criteria.

To be sure, the first criterion, of conforming with all experimental results, is hardly ever met in practice. Where deviations show up, the design of the experiments, their evaluation, and the reproducibility of the results must be checked. Often, the disagreement can be traced to an artifact, an equipment malfunction, a calculation error, an incorrect entry in a lab record book, or some other extraneous source, and can then be safely disregarded. However, if a divergent result is confirmed, the proposed model is inadequate and must be refined or entirely rejected.

Regarding the second criterion, the proposed mechanism must withstand scrutiny from all angles such as stereochemistry, thermodynamics, molecular-orbital theory, experience with analogous chemistry, etc. Such matters have been discussed in the context of network elucidation in Section 7.4.

The third criterion, a good track record of correct predictions, is vital because any set of experimental observations can be explained in different ways, each of which is apt to give different projections of behavior under other conditions. The most conclusive support for a model comes from experimental confirmation of counterintuitive predictions. If a prediction is rejected as absurd because it runs counter to common sense, and experiments then confirm it, the engineer can finally have reasonable confidence in the reliability of his model. The seasoned fundamental modeler will go to great lengths to seek out such counterintuitive features and take devious delight in seeing them confirmed. A case in point is the (correct) prediction of the fundamental model of hydroformylation with base-stabilized, phosphine-substituted cobalt hydrocarbonyl catalysts that both rate and yield are higher at lower pressure although gas is consumed and volume therefore shrinks. The higher rate appears to contradict the Le Châtelier principle [it does not, because that principle applies to equilibria, not to rates], and the higher yield at lesser effort and lower cost offends the third law of engineering: no free lunch. The higher rate stems from a better utilization of cobalt as $HCo(CO)_3Ph$ catalyst thanks to a higher phosphine-to-CO ratio at lower pressure (see Example 8.3 in Section 8.2.2); the better yield at lower pressure results from the increase in the aldehyde hydrogenation rate (see Example 7.3 in Section 7.2.2) without increase in the competing rate of aldehyde condensation to heavy ends (see Example 8.2.1).

Summary

What kind of mathematical modeling is best for process development depends on the development strategy chosen. Evolutionary development by stepwise scale-up by small factors essentially relies on tinkering rather than modeling. Empirical development with statistical methods seeks empirical equations to fit observed reactor dynamics. There is little input for reaction-kinetic modeling except that, for first trials, one-plus equations are preferable to power-law equations with fractional exponents. Fundamental development based on fully elucidated networks and mechanisms, addressed in this book, is best suited for large processes with long life expectancy. The experimental program to support evolutionary or empirical development should be dictated by anticipated plant conditions. The program for a fundamental development should aim for the most efficient way of elucidating networks, even if that calls for experiments under conditions far from those expected for the plant.

In a typical industrial process, the key problem in fundamental modeling of an established network is to reduce the number of rate equations and coefficients. The issue is not so much one of computation time, but of reduction of the experimental effort needed to obtain values for the coefficients and their temperature dependences. The most powerful tool for this purpose is the Bodenstein approximation. A detailed example describes step by step how a model with a minimum of equations and coefficients can be established.

For very large networks, a detailed fundamental model may be too cumbersome for highly iterative use in reactor design or optimization. An option then is to use a streamlined model in whose network all branches are placed at non-trace intermediates. This introduces an error that can be assessed by comparison with the detailed model.

Numerical values of the coefficients in a model can be determined graphically or by regression with statistical computer programs. Graphical methods are more likely to show up symptoms that indicate effects not yet accounted for, and are therefore preferable in the early stages of modeling. They can also help to provide good initial guesses for regression by computer. A final best set of coefficients should be obtained by regression.

An often helpful principle is that of "shortsightedness" of reaction steps: The reactivity of a group on a molecule usually does not depend on the size and shape of the molecule more than two atoms away. Pathways involving like chemistry, such as conversion of aldehyde groups on secondary carbon atoms of molecules with different substituents, can thus be assumed to involve closely similar rate coefficients. In large networks this can significantly reduce the number of coefficient values to be determined. The principle can also provide predictions of still unknown rate behavior, particularly in homologous series. It is not valid without exceptions, and must therefore be tested before use.

  To be acceptable for scale-up and reactor design, a model must meet three criteria: It must conform with all experimental evidence, be explainable by an eminently plausible mechanism, and have established a good track record of correct predictions, preferably counterintuitive ones.

  The establishment of the model equations of a detailed and a streamlined network and the shortsightedness principle are illustrated with cases of olefin hydroformylation.

References

1.   F. G. Helfferich and P. E. Savage, *Reaction kinetics for the practical engineer*, Course #195, AIChE Educational Services, New York, 7th ed., 1999, Chapter 1.

2.   C. D. Hendrix, *Chem. Tech.*, **9**(3) (1979) 167.

3.   M. J. Anderson and P. J. Whitcomb, *Chem. Eng. Progr.*, **92**(12) (1996) 51.

4.   F. J. Weigert, *Comp. Chem.*, **11** (1987) 273.

5.   R. J. McKinney and F. J. Weigert, GEAR and GIT, Project SERAPHIM disks PC3003A and PC3003B, 1987.

6.   ACUCHEM and ACUPLOT programs, NIST Chemical Kinetics Data Center, Gaithersburg, MD, 1988 (for isothermal constant-volkume batch only).

7.   W. Braun, J. T. Herron, and D. K. Kahaner, *Internat. J. Chem. Kinetics*, **20** (1988) 51.

8.   CHEMKIN (FORTRAN program), Sandia National Laboratory, Livermore, CA, 1990.

9.   R. J. Kee, F. M. Rupley, and J. A. Miller, Sandia Report SAND 889-8009, UC-401, 1990.

10.   Helfferich and Savage (ref. 1), Section 8.2.

11.   N. R. Draper and H. Smith, *Applied regression analysis*, Wiley, New York, 3rd ed., 1998, ISBN 0471170828.

12.   R. D. Cook and S. Weisberg, *Applied regression including computing and graphics*, Wiley, New York, 1999, ISBN 047131711X.

13.   S. Chatterjee, A. S. Hadi, and B. Price, *Regression analysis by example*, Wiley, New York, 3rd ed., 2000, ISBN 471319465.

14.   B. S. Gottfried, *Spreadsheet tools for engineers: EXCEL 2000 Version*, McGraw-Hill, New York, 2000, *Solver* program.

15.   SYSTAT (from Systat, Inc., Evanston IL).

16.   Helfferich and Savage (ref. 1), Chapter 10.

17.   C. A. Tolman and J. W. Faller, in *Homogeneous catalysis with metal phosphine complexes*, L. H. Pignolet, ed., Plenum, New York, 1983, ISBN 030641211X, p. 22.

Chapter 12

Unusual Thermal and Mass-Transfer Effects

Thermal and mass-transfer effects are matters of reaction engineering and reactor design, not of kinetics as understood here. For standard situations—that is, reactions with positive activation energies and rate equations with reaction orders that are positive or zero for reactants and negative or zero for products—current texts on reaction engineering provide excellent treatments [1-10]. In some instances, however, the special nature of a multistep reaction may result in unusual and quite different behavior. Only such cases will be examined here.

12.1. Anomalous temperature dependence

As discussed in Section 2.3, a single reaction step obeys a power-law rate equation whose algebraic form is independent of temperature, with a rate coefficient whose temperature dependence in general is given by the Arrhenius equation

$$k = A \exp(-E_\mathrm{a}/RT) \tag{1.3}$$

The activation energy, E_a, is positive and in good approximation constant. Accordingly, the logarithm of the rate coefficient varies linearly with reciprocal absolute temperature:

$$\frac{\mathrm{d}\ln k}{\mathrm{d}(1/T)} = -\frac{E_\mathrm{a}}{R} \tag{2.2}$$

and the rate increases with increasing temperature. However, this is not always true. For multistep reactions, the following anomalies are occasionally observed:

- *The activation energy may be negative, so that the rate decreases with increasing temperature;*

- *the algebraic form of the rate equation may change with a change in temperature;* and

- *although the form of the rate equation remains the same, the apparent activation energy may change to a different value with a change in temperature.*

A negative but approximately temperature-independent activation energy is the more frequent anomaly and will be examined first. The other two anomalies result from a change in rate control to a different step or pathway with change in temperature. They will be discussed jointly in Section 12.1.2. Although the latter anomalies are rare, the fact that they may occur makes it risky to extrapolate with the Arrhenius equation into temperature regions not covered by experiments.

At very low temperatues, even some single reaction steps involving free radicals behave abnormally: They do not obey the Arrhenius equation, and their rates increase if the temperature is lowered. Examples include reactions of CN · with O_2, ethene, ethyne, and ammonia and of OH · with O:, butenes, and HBr, among others. Typically, $\ln k$ decreases linearly with $\ln T$.

Normal Arrhenius behavior arises from an activation barrier which the reaction must surmount. In many exothermic reactions of free radicals with one another or with small molecules, there is no such barrier to speak of (see Section 9.4), and a hindrance from other effects, say, molecular rotation, may decrease as the temperature is lowered. This interesting anomaly has recently been reviewed by Sims [11].

Such behavior is exceptional and appears to be restricted to reactions of free radicals at low temperatures. The rules for temperature dependence of multistep reactions in the following assume that no such non-Arrhenius steps are involved.

12.1.1. *Negative apparent activation energy*:   lower rate at higher temperature

To be specific, the "apparent" activation energy of a reaction is the quantity E_a calculated with eqn 1.3 or 2.2 above from the temperature variation of the apparent rate coefficient in an empirical power-law rate equation that fits observation reasonably well. It is obtained without any assumptions as to a mechanism.

Barring participation of anomalous non-Arrhenius steps, the rate coefficients of all individual steps increase with temperature. A decrease of the reaction rate with increasing temperature then can only be produced by reverse steps whose coefficients increase more sharply than do those of the forward ones:

A decrease in rate with increase in temperature is generally possible only if the reaction involves at least one reverse step.

The overall reaction, however, need not be reversible.

Consider as the simplest example a reaction with pathway

$$A \;\longleftrightarrow\; X \;\longrightarrow\; P \tag{12.1}$$

where X remains at trace level. The rate equation, obtained with the Bodenstein approximation for X is

$$r_P = \frac{k_{AX} k_{XP}}{k_{XP} + k_{XA}} C_A$$

so that the apparent rate coefficient is given by

$$k_{app} = \frac{k_{AX} k_{XP}}{k_{XP} + k_{XA}} \qquad (12.2)$$

The rate decreases with increasing temperature if the denominator increases by a larger factor than does the numerator.

To establish a more restrictive criterion in terms of activation energies of steps, we need a rule for activation energies of combinations of rate coefficients. As can be deduced from eqn 2.2 above, the activation energy $(E_a)_{1*2}$ of a *product* $k_1 k_2$ of rate coefficients k_1 and k_2 with activation energies $(E_a)_1$ and $(E_a)_2$ is

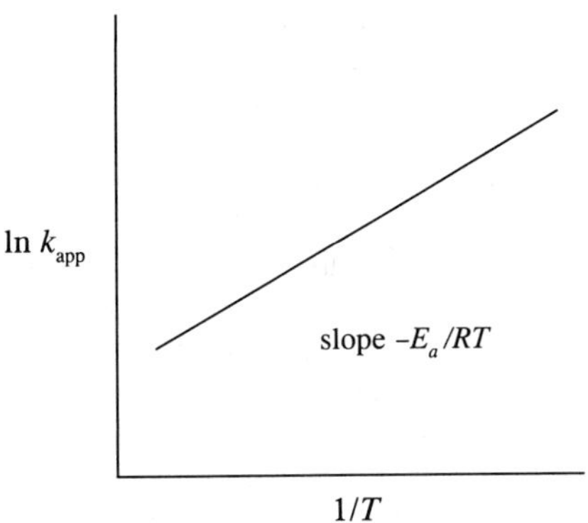

Figure 12.1. Arrhenius plot for reaction with negative apparent activation energy (schematic).

$$(E_a)_{1*2} = -R \frac{d \ln(k_1 k_2)}{d(1/T)} = -R \frac{d}{d(1/T)}(\ln k_1 + \ln k_2) = (E_a)_1 + (E_a)_2 \qquad (12.3)$$

Similarly, for a *ratio* k_1/k_2 of rate coefficients:

$$(E_a)_{1/2} = -R \frac{d \ln(k_1/k_2)}{d(1/T)} = -R \frac{d}{d(1/T)}(\ln k_1 - \ln k_2) = (E_a)_1 - (E_a)_2 \qquad (12.4)$$

In words:

> The activation energy of a product of rate coefficients equals the sum of the activation energies of the coefficients.
>
> The activation energy of a ratio of rate coefficients equals the difference of the activation energies of the coefficients.

(granted no non-Arrhenius steps participate).

To return to the reaction with pathway 12.1 and apparent rate coefficient given by eqn 12.2: For the rate to decrease with increasing temperature, at least one of the two denominator terms in eqn 12.2 must increase by a larger factor than does the numerator, that is, have a higher activation energy than the latter. This can only be the second term, the reverse coefficient k_{XA}, because the activation energy $(E_a)_{XP}$ of the first term is necessarily lower than that of the numerator, $(E_a)_{AX} + (E_a)_{XP}$. Accordingly:

- *The rate of a reaction with pathway* A $\longleftrightarrow$ X $\longrightarrow$ P *may decrease with increasing temperature if the activation energy of its reverse step is higher than the sum of those of its two forward steps.*

This is a necessary condition (in the absence of non-Arrhenius steps), but not a sufficient one: Even if it is met, the rate may increase with increasing temperature because the activation energy of the first denominator term in eqn 12.2 is lower than that of the numerator and may exert the stronger influence.

The condition above can be generalized for simple pathways

$$A \longleftrightarrow \ \ldots\ \longleftrightarrow\ X_j\ \longleftrightarrow\ X_{j+1}\ \longleftrightarrow\ \ldots\ \longleftrightarrow\ P$$

(arbitrary number of steps, co-reactants and co-products not shown; see Section 6.3 for simple pathways and their rate equations):

- *The forward rate of a simple reaction may decrease with increasing temperature if the pathway contains at least one intermediate* X_j *whose enthalpy* $(-\Delta H°)_{0j}$ *of exothermic formation from the reactants is larger than the activation energy* $(E_a)_{j,j+1}$ *of the subsequent step* $X_j \longrightarrow X_{j+1}$.

This ensures that at least one denominator term, the j'th, in the rate equation increases with temperature by a larger factor than does the numerator, and is a necessary condition, but not a sufficient one (again granted absence of non-Arrhenius steps).

Derivation. Apart from concentrations, the ratio of the numerator and the j'th denominator term is the ratio of the products of the first $j+1$ forward coefficients and the first j reverse coefficients. The product of the ratios of forward and reverse coefficients of the first j steps equals the equilibrium constant of formation of X_j from the reactants or a power of it. According to the van't Hoff equation

$$\mathrm{d}\ln K / \mathrm{d}(1/T) \ = \ -\Delta H° / R$$

and the Arrhenius equation 2.2, this constant decreases more steeply with increasing temperature than $k_{j,j+1}$ increases if $(-\Delta H°)_{0j}$ is larger than $(E_a)_{j,j+1}$.

For more complex cases, specific conditions are not readily derived beyond the obvious requirement that the denominator in the equation for the apparent rate coefficient must increase with temperature by a larger factor than does the numerator.

Negative apparent activation energies, if within a limited temperature range, are most often found in heterogeneous catalysis, where adsorption must precede the reaction on the catalyst surface. Adsorption equilibria are less favorable at higher temperature, and this effect may outweigh that of higher surface reaction rate coefficients if the (negative) adsorption enthalpy is larger than the activation energy of the surface reaction [12].

12.1.2. Change in rate control with change in temperature

As the steps of a multistep reaction have different activation energies, a change in temperature may affect their balance and cause a shift in rate control, leading to a rate behavior not in accordance with the Arrhenius equation.

From a point of view of reaction mechanisms, two situations can be distinguished:

- *shift of rate control to a different step within a pathway,* or
- *shift of rate control to an alternative, parallel pathway.*

In both instances, the algebraic form of the rate equation and, with it, the reaction orders are apt to change also. However, this is not necessarily always so.

Shift of rate control within a pathway. As with electric resistances in sequence or traffic on a single road, rate control within one and the same pathway rests with the *slowest* step, the "bottleneck" (see Section 4.1). If the step that is slowest at low temperature has a high activation energy, an increase in temperature may let it become faster than another with lower activation energy, which then assumes rate control.

More often than not, the two steps have different co-reactants, or one may have a co-reactant and the other not. The algebraic form of the rate equation then changes with the change in rate control, and the experimental determination of the rate equations at low and high temperatures makes it apparent that rate control shifts with temperature. An Arrhenius plot covering both temperature regions cannot be drawn because the rate equations differ; the apparent rate coefficient may even have different dimensions in the two regions. However, since a change in control requires the step that is rate-controlling at low temperature to have the higher activation energy, it can be said that the low-temperature Arrhenius plot must have the steeper slope.

It is possible, of course, that the algebraic form of the rate equation does *not* change with the change in rate control. This is true if neither step has a co-reactant or if both have the same co-reactant or co-reactants. If so, the resulting Arrhenius plot shows two distinct temperature regimes with constant, but different slopes, connected by a curved transition region where rate control switches from one step to the other, as shown in Figure 12.2, left diagram, next page. As pointed out above, the slope in the low-temperature region (right in an Arrhenius plot versus reciprocal temperature) is the steeper one [13].

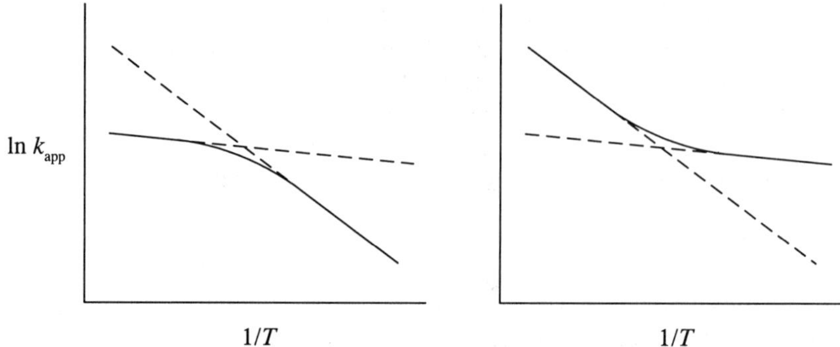

Figure 12.2. Arrhenius plots for reactions with shift in rate control but no change in form of rate equation (schematic). Left: shift to different step within same pathway; right: shift to alternative, parallel pathway. Dashed lines are linear extrapolations into respective other temperature regions (from Helfferich and Savage [14]).

The diagrams in Figure 12.2 also show linear extrapolations of the Arrhenius straight lines into the respective other temperature regions. These represent the behavior that would be observed if rate control would not change with temperature. It can be seen that, in the left diagram, the solid lines are below the dashed ones in both temperature regions; that is, in both regions the observed apparent rate coefficient (solid line) is smaller than the extrapolated one (dashed line). Accordingly, in both regions the *slower* mechanism is realized.

A behavior as in Figure 12.2, left diagram, is a certain indication of a shift in rate control, even though no change in the form of the rate equation is observed.

To summarize:

Rate control in a pathway may switch to a different step upon temperature variation if the step controlling at low temperature has a higher activation energy than at least one other. The apparent activation energy of the pathway then is *higher* in the low-temperature region.

Behavior of this general type is also occasionally found in systems in which a reactant must be supplied from another phase by mass transfer, e.g., in heterogeneous catalysis or homogeneous hydrogenation, air oxidation, etc. Usually, the activation energy of the reaction is fairly high, that of mass transfer only a few kilojoule per mole. Rate control may then shift from reaction at low temperature to mass transfer or a combination of mass transfer and reaction at high temperature.

Shift of rate control to alternative pathway. Often, two or more possible pathways from the same reactant or reactants to the same product or products exist side-by-side. As with electric resistances in parallel or traffic along parallel roads, the rate then is dominated by flow along the *fastest* pathway. A dirt track parallel to a freeway contributes little to traffic flow! The rate-controlling step along the fastest pathway may have a low apparent activation energy, so that a temperature increase may cause another pathway to become faster and assume control. Usually, such a shift also produces a change in the algebraic form of the rate equation. As with shift of control within a pathway, a determination of the rate equations in the two temperature regions then makes the shift in control obvious, and an Arrhenius plot covering both temperature regions cannot be drawn. However, since a shift in control requires the pathway that is rate-controlling at low temperature to have the lower activation energy, the Arrhenius plot for the low-temperature region must have the gentler slope. This is the opposite of what happens with shift in rate control within one and the same pathway (see above).

Again, the algebraic form of the rate equation might not change with the shift in control. If it does not, the Arrhenius plot shows two straight-line portions connected by a curve in the temperature range where the shift occurs. This time, however, the slope is steeper in the high-temperature region, as shown in Figure 12.2, right diagram. The diagram also includes linear extrapolations into the respective other temperature regions. In contrast to the behavior upon shift of control within a pathway (left diagram), the observed apparent rate coefficient (solid line) in both regions now is larger than the extrapolated one (dashed line): In both regions, the *faster* mechanism is realized. Such behavior is strong evidence for a shift of rate control to another pathway despite the fact that no change in the algebraic form of the rate equation is observed [13].

To summarize:

Rate control in a network with parallel pathways may switch from one to another if the pathway controlling at lower temperature has a lower apparent activation energy than the other. The apparent activation energy of the network then is *lower* in the low-temperature region.

12.1.3. Rate maxima and minima

A shift of rate control with temperature may occur from a mechanism with positive to one with negative apparent activation energy or vice versa. If so, the rate has a maximum or minimum in the temperature range of the shift. A rate maximum is sometimes seen, for example, in heterogeneous catalysis [12,15].

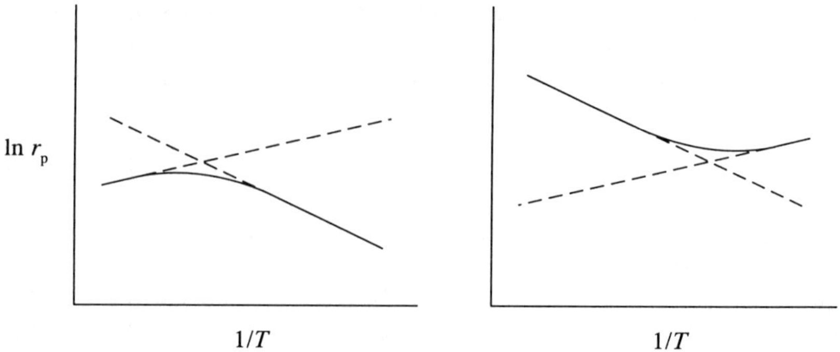

Figure 12.3. Maximum and minimum in temperature dependence of rate (schematic). Logarithmic plots of rate versus reciprocal absolute temperature. Dashed lines are linear extrapolations into respective other temperature regions.

If the rate has a maximum, the controlling mechanism in each of the two temperature regions is the slower one; if the rate has a minimum, it is the faster one (see Figure 12.4). In view of the rules established in the previous section, a corollary is:

A maximum or minimum in the temperature dependence of the rate indicates a shift of rate control: a maximum, a shift to another step in the same pathway; a minimum, a shift to an alternative, parallel pathway.

12.1.4. *Activation energies of phenomenological coefficients*

The phenomenological coefficients in a one-plus rate equation derived for a specific mechanism are composites of individual rate coefficients of steps. Activation energies for them can be established from their temperature dependence with the Arrhenius equation. With regard to their activation energies, two types of phenomenological coefficients can be distinguished: those consisting entirely of products, ratios, or ratios of products of individual rate coefficients; and those which also involve additive terms.

As was shown in Section 12.1.1 and with an addition for ratios of products:

- *the activation energy of a product of individual rate coefficients is given by the sum of the activation energies of the latter;*

- *the activation energy of a ratio of individual rate coefficients is given by the difference of the activation energies of the latter;* and

- *the activation energy of a ratio of products of individual rate coefficients is given by the sum of the activation energies of the coefficients in the numerator minus the sum of the activation energies of those in the denominator.*

(granted the absence of non-Arrhenius steps). In all these three cases, the activation energy of the phenomenological coefficient as the sum, or difference, or difference of sums of constant activation energies is itself constant. In contrast, if the phenomenological coefficient involves *additive* terms, the same cannot be said. Here, a constant activation energy results only if all but one of the additive terms remain insignificant or if all significant terms have the same activation energy. Accordingly

A significant variation of the activation energy of a phenomenological coefficient with temperature indicates that the coefficient involves additive terms of individual rate coefficients of steps.

The reverse is not necessarily true: If the coefficient is composed of additive terms, these may have very similar activation energies or one of them may dominate, and the activation energy of their sum then remains approximately constant.

The curvature of an Arrhenius plot for a phenomenological coefficient can sometimes be used to distinguish between rival models: A model in which a phenomenological coefficient is a product, ratio, or ratio of products of individual rate coefficients of steps can in general not be correct if an Arrhenius plot of that coefficients is distinctly curved.

12.2. Uncommon heat-transfer problems

In most instances in practice, the rate of and heat evolution by an exothermic chemical reaction are smooth functions of temperature and reactant concentrations, with an apparent activation energy rarely above 200 kJ mol^{-1} and apparent overall reaction order not higher than three. Engineers designing reactors tend to take this for granted. Simultaneous or multistep reactions, however, may behave differently. Two quite common situations are those in which the starting material is a mixture of reactants of very different reactivities, or the reactant is a highly reactive isomer, most of which is quickly converted to much less reactive ones as conversion to products progresses. In both cases, the initial very fast conversion of a highly reactive component in the starting material results in a short burst of very high heat release, a fact a reactor designer ignores at his peril.

> A strong burst of heat release at beginning conversion may result if the starting material is a mixture containing a component that is much more reactive than the others, or is a highly reactive isomer that undergoes isomerization to less reactive ones while conversion to products proceeds.

The first of the two situation above is occasionally encountered in petroleum processing, where multicomponent mixtures of reactants are common. The second may arise in reactions with coupled parallel steps [14] (see Section 5.3).

Example 12.1.   Hydroformylation of long-chain 1-olefins with phosphine-substituted cobalt hydrocarbonyl catalysts.   Hydroformylation of long-chain 1-olefins with phosphine-substituted cobalt hydrocarbonyl catalysts provides a striking example of coupled parallel steps and the potential of an uncommon heat-transfer problem. The network is of the type 12.5 below, with the A_i as the olefin isomers and the P_i as the isomeric alcohol products (arrows represent multistep pathways;   see also Example 5.3, Figure 5.9, and network 5.43 in Section 5.3 and network 7.40 in Section 7.4).

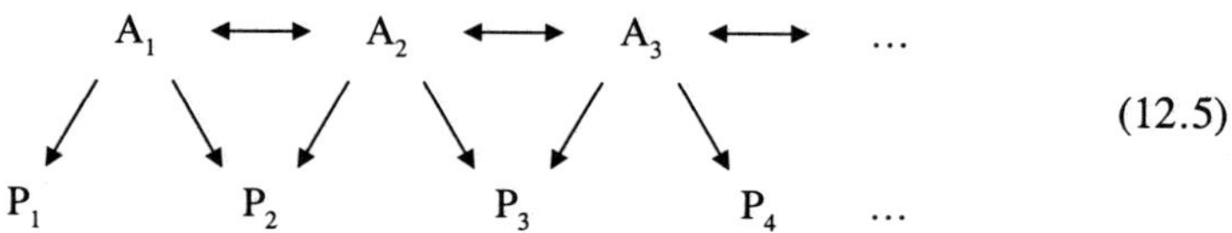

(12.5)

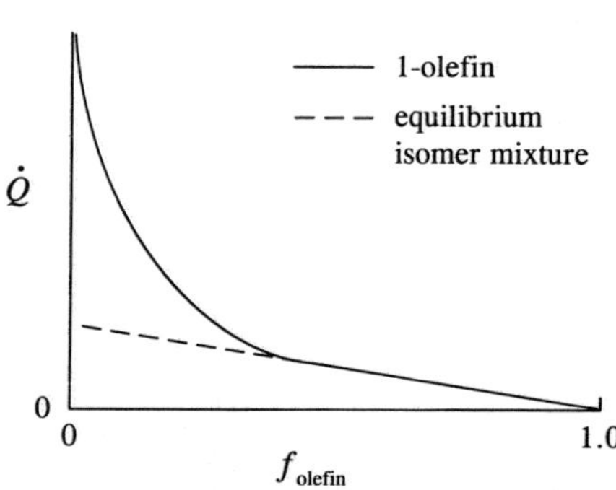

Figure 12.4.   Instantaneous isothermal heat release $\dot{Q}$ as function of olefin conversion for hydroformylation of long-chain olefins with cobalt phosphino-hydrocarbonyl catalysts (schematic).

The 1-olefin, A_1, is about two orders of magnitude more reactive than the other isomers with the double bonds in internal positions, but in thermal equilibrium at typical reaction temperatures (150 to 200°C) it constitutes only a few percent of total olefin. Fast double bond migration lets a steady-state olefin isomer distribution be closely approached before much olefin has been converted. As a result, if 1-olefin is charged to the reactor, the abundance of this highly reactive isomer in the feed stream causes reaction and heat release rates at the inlet of a continuous reactor or at start in a batch reactor to be abnormally high. However, most of the 1-olefin is consumed very quickly by isomerization and reaction to products, making the rate (at constant temperature) drop to a few percent of that at the inlet or start. Figure 12.4 shows such behavior and, for comparison, that of the same olefin with double bonds in isomerization equilibrium.

Interestingly, the initial burst of reaction also aggravates a very serious mass-transfer problem, as will be shown in the next section.

12.3. Uncommon mass-transfer problems

Many homogeneous reactions occur in the liquid phase, but consume reactants that must be supplied by mass transfer from a gas phase (or occasionally from another liquid phase). This is a typical problem of reaction engineering and is treated in some detail in most modern texts of that field [1,3,4,9,16,17]. Customarily, a power law is assumed for the rate of the chemical reaction and is then combined with a standard linear-driving force or Fickian diffusion treatment of mass transfer. A mass-transfer limitation lowers the rate, which in some extreme situations can become entirely mass transfer-controlled. Certain types of multistep reactions, however, can produce a totally different and very interesting behavior that may involve instability.

In a great majority of cases of practical interest, the reaction orders are positive for reactants and negative or zero for products, as is taken for granted in standard texts. If mass transfer from another phase is not fast enough to deliver as much of a reactant as the reaction consumes, the concentration of that reactant drops somewhat and, as a result, the reaction slows down to some extent. Similarly, in a reversible reaction whose product leaves the phase in which the reaction occurs, slow mass transfer of the product reduces the rate. However, if the reaction is product-promoted (positive reaction order with respect to a product or major intermediate) or reactant-inhibited (negative reaction order with respect to a reactant), a mass-transfer limitation may increase rather than decrease the rate. It does so in a product-promoted reaction if mass transfer is slow in removing a product whose reaction order is positive. It does so in a reactant-inhibited reaction if it is slow in supplying a reactant whose order is negative. A lagging of mass transfer increases the concentration difference between the phases that acts as driving force and, thereby, boosts the mass-transfer rate. Also, an increase of conversion rate lowers the concentrations of reactants whose orders are positive (no reaction can have negative orders with respect to all reactants). Both these effects work against a continuing acceleration of the reaction rate. In a product-promoted reaction, lagging removal of the accelerating product merely maintains the autocatalytic effect, which, of course, would not come into play at all if the respective product were removed as soon as it is formed. In a reactant-inhibited reaction, however, the accelerating effect of lagging supply of the respective reactant may win out over the increased mass transfer rate and cause a catastrophic depletion: The phase runs out of reactant or some other process takes over (see also Section 8.9.2). There is a sharp limit of stability beyond which that happens.

In a product-promoted reaction, *fast* mass transfer of the accelerating product into another phase *slows* the rate by depressing autocatalysis.

In a reactant-inhibited reaction, *slow* mass transfer of the inhibiting reactant from another phase *accelerates* the rate and may cause instability.

Example 12.2.   Potential mass transfer-induced instability in olefin hydroformylation [14]. The rate of olefin hydroformylation with cobalt hydrocarbonyl catalysts in a liquid phase obeys in good approximation the Martin equation

$$r_{ald} \;=\; \frac{k_1 C_{ole} C_{cat}}{1 + k_2 p_{CO}/p_{H_2}} \tag{6.12}$$

(see Example 6.2 in Sections 6.3). The rate is of negative order with respect to CO and positive order with respect to H_2, both being reactants that must be supplied by mass transfer from the gas phase. H_2 as the much smaller molecule has a much higher mass-transfer coefficient than does CO. Accordingly, any mass-transfer limitation makes itself felt with respect to CO first. Because of the negative order with respect to CO, a decrease in CO concentration in the liquid increases the rate. As described above, this acceleration may win out over the counteracting effects of decreases in olefin and H_2 concentrations. If so, reactant depletion of the liquid becomes self-accelerating. However, as the liquid becomes depleted of CO, the hydrocarbonyl catalyst becomes unstable and metallic cobalt is plated out. This catalyst loss eventually reins in the accelerating reaction.

One might say, a mass-transfer limitation under normal conditions acts as a gentle brake on the reaction, slowing it down at worst to the rate that mass transfer to the reacting phase can sustain, whereas in hydroformylation with cobalt hydrocarbonyl catalysts, mass transfer imposes an upper limit on the amount of catalyst the system will tolerate. Assume a small amount of catalyst is used; mass transfer then has no trouble supplying as much CO as the reaction consumes (note conversion is first order in catalyst). However, if now the amount of catalyst and thereby the conversion rate are increased, the point may be reached where mass transfer can no longer keep up with CO consumption. The self-accelerating conversion then depletes the liquid of CO to the extent that the catalyst added beyond the limit of mass-transfer stability decomposes.

The example of the mass-transfer effect in olefin hydroformylation is interesting in still another respect. If a phosphine-substituted cobalt hydrocarbonyl is used as catalyst in combination with 1-olefin as reactant, a very strong burst of reaction ensues at the reactor inlet or at start of a batch reaction (see Example 12.1).

This places extreme demands on the supply of CO from the gas phase by mass transfer and can easily lead to instability and high catalyst losses unless appropriate precautions were taken in the design.

A reasonably rigorous mathematical treatment of a reacting system under conditions of mass-transfer instability is difficult and complex and requires a detailed knowledge of fluid dynamics in the reactor, information hard to come by. The best way to proceed in process development and design therefore is to establish approximately where the limit of stability lies, and then to keep the design a safe margin away from it. No plant operates without occasional deviations from design operating conditions, and the risk that it could stray into the region of instability must be minimized.

Summary

Apparent activation energies of reactions may be negative, corresponding to a decrease in rate with increase in temperature, if the pathway contains at least one reverse step with an activation energy that is high compared with those of the forward steps. Also, the rates of a few reaction steps of free radicals decrease with increasing temperature in a non-Arrhenius fashion.

A change in temperature may cause a shift in rate control to another step in the same pathway or to an alternative, parallel pathway. Usually, this entails a change in the algebraic form of the rate equation, and determination of the latter at different temperatures then makes it obvious that such a shift occurs. It is also, possible, however, that the form of the rate equation remains the same. If so, the apparent activation energy changes from one approximately constant value to another over a relatively narrow temperature range, giving an Arrhenius plot with two straight-line portions connected by a curve. Whether the form of the rate equation changes or not, the apparent activation energy is higher (steeper slope of Arrhenius plot) in the low-temperature region if rate control shifts within the pathway, and is higher (steeper slope) in the high-temperature region if the shift is to another pathway. In the first case, the slower of the two possible mechanisms dominates the rate; in the second case, the faster one.

Because of a shift in rate control, the temperature dependence of the rate may show a maximum or, more rarely, a minimum. A rate maximum indicates a shift of rate control to a another step in the same pathway; a minimum, a shift to an alternative, parallel pathway.

Phenomenological coefficients in reduced rate equations are combinations of individual rate coefficients of steps. If a phenomenological coefficient consists entirely of a product, a ratio, or a ratio of products of individual rate coefficients, its activation energy is essentially temperature-independent. In contrast, if the phenomenological coefficient involves additive terms, its activation energy varies with temperature unless the terms have similar activation energies or one of them dominates.

An initial burst of high heat release may occur if the starting material of an exothermic reaction is a mixture of components of very different reactivities, or is a highly reactive isomer that is quickly isomerized to less reactive ones as conversion to products progresses.

In product-promoted reactions in which the accelerating species exits into another phase, or in reactant-inhibited reactions in which the respective reactant enters from another phase, slow mass transfer may boosts rather than depresses the reaction rate. In reactant-inhibited reactions, slow supply of the inhibiting reactant may cause the system to become unstable: There may be a sharp stability limit beyond which catastrophic self-acceleration occurs until the phase has become depleted of the reactant or reactants or some other phenomenon has come into play.

The anomalous heat- and mass-transfer problems are illustrated with examples from olefin hydroformylation with cobalt hydrocarbonyl catalysts.

References

1.   G. Astarita, *Mass transfer with chemical reaction*, Elsevier, Amsterdam, 1967.

2.   J. J. Carberry, *Chemical and catalytic reaction engineering*, McGraw-Hill, New York, 1973, ISBN 0070097909, Sections 3-10 and 4-2 and Chapter 6.

3.   A. R. Cooper and G. V. Jeffreys, *Chemical kinetics and reactor design,* Prentice-Hall, Englewood Cliffs, 1971, ISBN 0131286781, Chapter 9.

4.   P. V. Danckwerts, *Gas-liquid reactions*, McGraw-Hill, New York, 1970.

5.   H. S. Fogler, *Elements of chemical reaction engineering*, Prentice-Hall, Englewood Cliffs, 3rd ed., 1999, ISBN 0135317088, Chapters 8 and 10.

6.   G. F. Froment and K. B. Bischoff, *Chemical reactor analysis and design*, Wiley, 2nd ed., 1990, ISBN 0471510440, Chapters 3, 6, and 7.

7.   C. G. Hill, Jr., *An introduction to chemical engineering kinetics & reactor design*, Wiley, New York, 1977, ISBN 0471396095, Chapter 10.

8.   O. Levenspiel, *Chemical reaction engineering*, Wiley, New York, 3nd ed., 1999, ISBN 047125424X, Chapters 6 and 13.

9.   O. Levenspiel, *The chemical reactor omnibook*, O.S.U. Book Stores, Corvallis, OR, 1989, ISBN 0882461648, Section 41.

10.   G. D. Ulrich, *A guide to chemical engineering reactor design and kinetics*, Ulrich Research and Consulting, Lee, 1993, Chapter 5.

11.   I. R. Sims, *Res. Chem. Kinetics*, **4** (1997) 121.

12.   T. J. Yoon and M. A. Vannice, *J. Catal.*, **82** (1983) 457.

13.   J. A. Christiansen, *Adv. Catal.*, **5** (1953) 311, Section 5.2.

14.   F. G. Helfferich and P. E. Savage, *Reaction kinetics for the practical engineer*, Course #195, AIChE Educational Services, New York, 7th ed., 1999, Chapter 11.

15.   R. Badilla-Ohlbaum, H. J. Neuberg, W. F. Graydon, and M. J. Phillips, *J. Catal.*, **47** (1977) 273,

16.   Carberry (ref. 2), Chapter 6.

17.   Levenspiel (ref. 8), Chapter 13.

Glossary of Symbols

Quantities

A	preexponential factor in Arrhenius equation 1.3	same as rate coefficient
C_i	concentration of species i or intermediate X_i	M
$\mathcal{C}$	denominator in Christiansen rate equation [Section 8.4]	depends on network
D_{jk}	denominator in rate eqn for simple pathway or segment $j \longleftrightarrow k$ [eqn 6.6]	depends on pathway
D_{jj}	sum of terms of row $j+1$ in Christiansen matrix [Section 8.4]	depends on catalytic cycle
e	$= 2.71828$ (basis of natural logarithms)	$-$
E_a	molar activation energy	kJ mol^{-1}
f	effectiveness factor of free radicals for initiation [p. 312]	$-$
f_i	fractional conversion of reactant i [eqn 1.4, 1.5]	$-$
F_i	molar flow rate of species i	mol t^{-1}
g	geometric progression factor [eqn 10.82]	$-$
$(\Delta G_f^{\circ})_i$	standard Gibbs free energy of formation of species i	kJ mol^{-1}
ΔH°	standard enthalpy change of reaction	kJ mol^{-1}
H_o	Hammett acidity function [Section 8.1]	$-$
k	rate coefficient	depends on rate eqn
$k_a, k_b, \ldots$	phenomenological coefficient in rate eqn	depends on rate eqn
$k_{AA}, k_{AB}, k_{BA}, k_{BB}$	propagation rate coefficients in copolymerization [p. 341]	M$^{-1}t^{-1}$
k_{app}	apparent rate coefficient	depends on rate eqn
$k_{c\ldots}$	rate coefficient of coupling of ...	M$^{-1}t^{-1}$
$k_{ch\ldots}$	rate coefficient of chain transfer to ...	M$^{-1}t^{-1}$
k_d	rate coefficient of disproportionation	M$^{-1}t^{-1}$
k_{ij}	rate coefficient of step $i \longrightarrow j$	depends on molecularity
k_{iso}	characteristic rate coefficient of isomerization [eqn 5.39]	t^{-1}
k_p	rate coefficient of propagation	M$^{-1}t^{-1}$
k_{trm}	rate coefficient of termination	M$^{-1}t^{-1}$
$k_{t\ldots}$	rate coefficient of termination by ... in copolymerization	M$^{-1}t^{-1}$
k	number of forward steps in simple pathway or catalytic cycle	$-$
K_{cat}	dissociation constant of complex catalyst	M
K_d	equilibrium constant of dissociation	M
K_{ij}	equilibrium constant of reaction $i \longleftrightarrow j$	depends on stoichiometry
$K^{\ddagger}$	equilibrium constant of formation of activated complex [Section 2.2]	depends on stoichiometry
$\mathcal{L}_{ij}$	loop coefficient of reduced loop segment $i \longrightarrow j$ [eqn 6.15]	t^{-1}
m, n	reaction orders	$-$
MW	molecular weight	$-$
$\overline{MW}$	number-average molecular weight (monomer included in mole count)	$-$
n_i	stoichiometric number of species i	$-$
$\overline{n}_f$	average effective functionality of monomer [p. 308]	$-$
$\overline{N}$	number-average degree of polymerization (monomer included in mole count)	$-$

lapc	low-abundance propagating center in ionic polymerization
L	ligand
macs	most abundant catalyst-containing species
masi	most abundant surface intermediate in heterogeneous catalysis
M	collision partner (or reaction vessel wall)
M	monomer in polymerization
M_i	monomer i in copolymerization
M_{A*}, M_{B*}	reactive end groups in chain-growth copolymerization
$M^{\ddagger}$	activated complex
Me	metal
2-MGN	2-methylglutaronirtrile
ole	olefin
par	paraffin
poi	catalyst poison
P, Q, ...	reaction products
P^-	anionic propagation center in anionic polymerization
P_i	polymer with i monomer units
P_i^*	catalyst adduct with i monomer units in coordination polymerization
Ph	organic phosphine
4-PN	4-pentenenitrile
R·	free radical
S	solvent or other molecule in chain reaction
Tr	transfer agent in chain reaction
X, X_i	trace-level intermediate, chain carriers in chain reactions
X_L	lumped species [Section 8.5.2]
Y	chain carrier in chain reaction
Σcat	total catalyst material (including amounts bound as intermediates) [Section 8.3]
Σ°	total catalyst material in cycle (excluding external pathways)
$\Sigma P\cdot$	total free-radical population in chain-growth polymerization
ΣP^-	total population of propagating centers in anionic polymerization
$\Sigma P*$	total population of catalyst adducts in coordination polymerization

Author Index